METHODS OF DIFFERENTIAL GEOMETRY IN ANALYTICAL MECHANICS

NORTH-HOLLAND MATHEMATICS STUDIES 158

(Continuation of the Notas de Matemática)

NORTH-HOLLAND – AMSTERDAM • NEW YORK • OXFORD • TOKYO

METHODS OF DIFFERENTIAL GEOMETRY IN ANALYTICAL MECHANICS

Manuel de LEÓN

CECIME
Consejo Superior de Investigaciones Cientificas
Madrid, Spain

Paulo R. RODRIGUES

Departamento de Geometria
Instituto de Matemática
Universidade Federal Fluminense
Niterói, Brazil

1989

NORTH-HOLLAND – AMSTERDAM • NEW YORK • OXFORD • TOKYO

ELSEVIER SCIENCE PUBLISHERS B.V.
Sara Burgerhartstraat 25
P.O. Box 211, 1000 AE Amsterdam, The Netherlands

Distributors for the U.S.A. and Canada:
ELSEVIER SCIENCE PUBLISHING COMPANY, INC.
655 Avenue of the Americas
New York, N.Y. 10010, U.S.A.

ISBN: 0 444 88017 8

Printed in the Netherlands

To our parents

Contents

Preface

The purpose of this book is to make a contribution to the modern development of Lagrangian and Hamiltonian formalisms of Classical Mechanics in terms of differential–geometric methods on differentiable manifolds. The text is addressed to mathematicians, mathematical physicists concerned with differential geometry and its applications, and graduate students.

Chapter 1 is a review of some topics in Differential Geometry. It is included in the text to state its main properties and to help the reader in subsequent chapters.

Chapters 2 and 3 are devoted to the study of several geometric structures which are closely related to Lagrangian mechanics. Almost tangent structures and tangent bundles are examined in Chapter 2. The theory of vertical, complete and horizontal lifts of tensor fields and connections to tangent bundles are also included.

In Chapter 4 we study the differential calculus on the tangent bundle of a manifold given by its canonical almost tangent structure. Connections in tangent bundles, in the sense of Grifone, are examined and other approaches to connections are briefly considered.

In Chapter 5 we study symplectic structures and cotangent bundles. In fact, the canonical symplectic structure of the cotangent bundle of a manifold is the (local) model for symplectic structures (Darboux theorem). Lifts of tensor fields and connections to cotangent bundles are also included.

In Chapter 6 we examine Hamiltonian systems. As there are many specialized books where this topic is extensively dealt with we decided to reduce the material to some essential results. This chapter may be considered as an introduction to the subject.

Chapter 7 is devoted to Lagrangian systems on manifolds. We apply the main results of our previous chapters to Lagrangian systems. It is usual to find in the literature regular Lagrangian systems obtained by pulling back to the tangent bundle the canonical symplectic form of the cotangent bundle of a given manifold, using for this the fiber derivative of the Lagrangian function. In this vein we do not need to use the tangent bundle geometry. Nevertheless there is an alternative approach for Lagrangian systems which consists of using the structures directly underlying the tangent bundle manifold. This gives an independent approach, i.e., an independent formulation of the Hamiltonian theory. This point of view is that of J. Klein which was adopted in the French book of C. Godbillon (1969). More recently some points which use this kind of geometric formulation have also been presented in the book of G. Marmo et al. (1985). We think that this viewpoint gives a more powerful and elegant exposition of the subject. In fact we may say that almost tangent geometry has a similar role in Lagrangian theories to the role of symplectic geometry in Hamiltonian theories.

Chapter 8 is concerned with presymplectic structures. As the reader will see in Chapter 7 the almost tangent formulation of classical lagrangian systems does not require regularity conditions on the Lagrangian functions. Thus, in general, if we wish the Euler–Lagrange equations to define a vector field describing the dynamics (as it occurs in the regular case) we are lead into constrained Lagrangians. Presymplectic forms also appear in the Hamiltonian formalism, originated, for example, by degenerate Lagrangians, and lead to the so–called Dirac–Bergmann constraint theory. In this chapter we describe the geometric tools for such situations which have been inspired by many authors.

We conclude the book with two Appendices. One is concerned with Particle Mechanics in local coordinates and is addressed to students who are not very familiar with the classical approach. The other is devoted to a brief summary on the theory of Jet–bundles, an important topic in modern differential geometry.

We would like to express our gratitude to the Conselho Nacional de Desenvolvimento Cientifico e Tecnologico, CNPq (Brazil) Proc. 31.1115/79, the Fundaçao de Amparo a Pesquisa do Rio de Janeiro (FAPERJ), Proc. E–29/170.662/88 and the Consejo Superior de Investigaciones Científicas, CSIC (Spain) for their financial support during the preparation of the manuscript.

We thank Pilar Criado for her very careful typing of the text on a microcomputer using TeX. Our thanks are also due to Luis A. Cordero and Alfred Gray who helped us to use this typesetting system and to John Butterfield for his valuable suggestions. To the Editor of Notas de Matematica, Leopoldo Nachbin and to the Mathematics Acquisitions Editor of Elsevier Science Publishers B.V./Physical Sciences and Engineering Division, Drs. Arjen Sevenster, our thanks for including this volume in their series.

Chapter 1

Differential Geometry

1.1 Some main results in Calculus on R^n

In this section we review briefly some facts about partial derivatives from advanced calculus.

Let $f : U \subset R^n \longrightarrow R$ be a function defined on an open subset U of R^n. Then

$$f(x) = f(x^1, \ldots, x^n), x = (x^1, \ldots, x^n) \in U.$$

At each point $x_0 \in U$, we define the **partial derivative** $(\partial f/\partial x^i)_{x_0}$ of f with respect to x^i as the following limit (if it exists):

$$\left(\frac{\partial f}{\partial x^i}\right)_{x_0} = \lim_{h \longrightarrow 0} \frac{f(x^1, \ldots, x^i + h, \ldots, x^n) - f(x^1, \ldots, x^i, \ldots, x^n)}{h}.$$

If $\partial f/\partial x^i$ is defined at each point of U, then $\partial f/\partial x^i$ is a new function on U. When the n functions $\partial f/\partial x^1, \ldots, \partial f/\partial x^n$ are continous on U, we say that f is **differentiable of class** C^1.

Now, we define inductively the notion of differentiability of class C^k : f is of class C^k on U if its first derivatives $\partial f/\partial x^i, 1 \leq i \leq n$, are of class C^{k-1}. If f is of class C^k for every k, then f is said to be C^∞ (or simply differentiable).

Then we have the partial derivatives of order k defined on U by

$$\frac{\partial^k f}{\partial x^{i_1} \ldots \partial x^{i_k}} = \frac{\partial}{\partial x^{i_1}} \left(\frac{\partial^{k-1} f}{\partial x^{i_2} \ldots \partial x^{i_k}} \right)$$

We can easily prove that the value of the derivatives of order k is independent of the order of differentiation, that is, if $(j_1, \ldots, j_k)$ is a permutation of $(i_1, \ldots, i_k)$,, then

$$\frac{\partial^k f}{\partial x^{j_1} \ldots \partial x^{j_k}} = \frac{\partial^k f}{\partial x^{i_1} \ldots \partial x^{i_k}}.$$

Next, let $F : U \subset R^n \longrightarrow R^m$ be a mapping (or map). If $\pi^a : R^m \longrightarrow R, 1 \leq a \leq m$ denotes the canonical projection $\pi^a(x^1, \ldots, x^m) = x^a$, then we have m functions $F^a : U \subset R^n \longrightarrow R$ given by $F^a = \pi^a \circ F$. We say that F is **differentiable of class** C^1, C^k or C^∞ if each F^a is C^1, C^k or C^∞, respectively. We may sometimes call a C^∞ map F **smooth** or **differentiable.**

If F is differentiable on U, we have the $m \times n$ **Jacobian matrix**

$$J(F) = \begin{bmatrix} \frac{\partial F^1}{\partial x^1} & \cdots & \frac{\partial F^1}{\partial x^n} \\ \vdots & & \vdots \\ \frac{\partial F^m}{\partial x^1} & \cdots & \frac{\partial F^m}{\partial x^n} \end{bmatrix}$$

at each point of U.

Let $F : U \subset R^n \longrightarrow R^m$ and $G : V \subset R^m \longrightarrow R^p$ so that $H = G \circ F$ is defined on U.

Theorem 1.1.1 *(1) H is differentiable; (2)* $J(H) = J(G)J(F)$, *that is,*

$$\frac{\partial H^\alpha}{\partial x^i} = \frac{\partial G^\alpha}{\partial x^a}\frac{\partial F^a}{\partial x^i}, 1 \leq i \leq n, 1 \leq a \leq m, 1 \leq \alpha \leq p.$$

Let $F : U \subset R^n \longrightarrow V \subset R^n$ be a mapping. We say that F is a **diffeomorphism** if: (1) F is a homeomorphism and (2) F and F^{-1} are differentiable. Obviously, if F is a diffeomorphism, then F^{-1} is a diffeomorphism too.

Theorem 1.1.2 (Inverse Function Theorem) *Let U be an open subset of R^n and $F : U \longrightarrow R^n$ a differentiable mapping. If $J(F)$ at $x_0 \in U$ is non-singular, then there exists an open neighborhood V of $x_0, V \subset U$, such that $F(V)$ is open and $F : V \longrightarrow F(V)$ is a diffeomorphism. (See Boothby [9] for a proof).*

Let $F : U \subset R^n \longrightarrow R^m$ be a differentiable mapping. The **rank of** F **at** $x_0 \in U$ is defined as the rank of the Jacobian matrix $J(F)$ at x_0. Obviously, rank (F) at $x_0 \leq inf(m,n)$. Then, if the rank of F at x_0 is k, we deduce that the rank of F is greater or equal to k on some open neighborhood V of x_0. In particular, if $F : U \subset R^n \longrightarrow F(U) \subset R^n$ is a diffeomorphism, then F has constant rank n.

Theorem 1.1.3 (Rank Theorem) *Let $U_0 \subset R^n, V_0 \subset R^m$ be open sets, $F : U_0 \longrightarrow V_0$ a differentiable mapping and suppose the rank of F to be equal to k on U_0. If $x_0 \in U_0$ and $y_0 = F(x_0) \in V_0$, then there exist open sets $U \subset U_0$ and $V \subset V_0$ with $x_0 \in U$ and $y_0 \in V$, and there exist diffeomorphisms $G : U \longrightarrow G(U) \subset R^n, H : V \longrightarrow H(V) \subset R^m$ such that $(H \circ F \circ G^{-1})(G(U)) \subset V$ and*

$$(H \circ F \circ G^{-1})(x^1, \ldots, x^n) = (x^1, \ldots, x^k, 0, \ldots, 0).$$

(see Boothby [9] for a proof).

Remark 1.1.4 We can easily check that Theorems 1.1.2 and 1.1.3 are equivalent.

1.2 Differentiable manifolds

Definition 1.2.1 *A* **topological manifold** *M of dimension m is a Hausdorff space with a countable basis of open sets such that for each point of M there is a neighborhood homeomorphic to an open set of R^m. Each pair (U, ϕ) where $\phi : U \longrightarrow \phi(U) \subset R^m$ is called a* **coordinate neighborhood**. *If $x \in U$, then $\phi(x) = (x^1(x), \ldots, x^m(x)) \in R^m; x^i(x), 1 \leq i \leq m$, is called the* **ith coordinate** *of x, and the functions $x^1, \ldots, x^m$, are called the* **coordinate functions** *corresponding to (U, ϕ) (or* **local coordinate system***).*

Now, let $(U, \phi), (V, \psi)$ be two coordinate neighborhoods of M. Then

$$\psi \circ \phi^{-1} : \phi(U \cap V) \longrightarrow \psi(U \cap V)$$

is a homeomorphism with inverse $\phi \circ \psi^{-1}$. In local coordinates, if $(x^i), (y^i)$ are the local coordinates corresponding to $(U, \phi), (V, \psi)$, respectively, then we have

$$(x^1, \ldots, x^m) \longrightarrow (y^1(x^i), \ldots, y^m(x^i)).$$

Definition 1.2.2 *$(U,\phi),(V,\psi)$ are said to be C^∞–***compatible** *if $\psi\circ\phi^{-1}$ and $\phi\circ\psi^{-1}$ are C^∞ mappings.*

Definition 1.2.3 *A* **differentiable** *or* (C^∞) **structure** *on a topological manifold M is a family $U = \{(U_\alpha,\phi_\alpha\}$ of coordinate neighborhoods such that:*
(1) the U_α cover M;
(2) for any α,β (U_α,ϕ_α) and (U_β,ϕ_β) are C^∞–compatible;
(3) any coordinate neighborhood (V,ψ) C^∞–compatible with every $(U_\alpha,\phi_\alpha)\in U$ belong to U.
*A C^∞ (***differentiable***)* **manifold** *is a topological manifold endowed with a C^∞–structure.*

Remark 1.2.4 Suppose that M is a topological manifold. If $U=\{(U_\alpha,\phi_\alpha)\}$ is a family of C^∞–compatible coordinate neigborhoods which cover M, we define a set $\bar{U}$ by $\bar{U}=\{(U,\phi)/\ (U,\phi)$ is a coordinate neighborhood C^∞–compatible with any $(U_\alpha,\phi_\alpha)\in U\}$. Obviously, $U\subset\bar{U}$ and $\bar{U}$ is the unique C^∞ structure on M which contains U; U is called a C^∞ **atlas** and $\bar{U}$ a **maximal** C^∞ **atlas.**

Remark 1.2.5 Let (U,ϕ) be a coordinate neighborhood on a C^∞ manifold M. If $V\subset U$ is an open set of M, then $(V,\phi_{/V})$ is a **new** coordinate neighborhood (its coordinate functions are the restriction of the coordinate functions corresponding to (U,ϕ)). If $x\in U$, then we may choose $V\subset U$ such that $x\in V$ and $\phi(V)$ is an open ball $B(\phi(X),\epsilon)$ with radius ϵ or a cube $C(\phi(x),\epsilon)$ of side $2\epsilon,\epsilon>0$, in R^m. Moreover, we may compose ϕ with a translation ψ such that $\psi(\phi(x))=0\in R^m$.

Examples
(1) **The Euclidean space R^m.**
In fact, the canonical Cartesian coordinates define a C^∞ structure on R^m with a single coordinate neighborhood.
(2) Furthermore, let V be an m–dimensional vector space over R. If $\{e_i\}$ is a basis of V, then V may be identified with R^m. By means of the identification

$$v=x^1e_1+\ldots+x^me_m\longrightarrow(x^1,\ldots,x^m)\in R^m,$$

V becomes a C^∞ manifold of dimension m and this C^∞ structure is independent of the choice of the basis $\{e_i\}$.
(3) Let $gl(m,R)$ be the set of $m\times m$ matrices $A=(a_i^j)$ over R. Then $gl(m,R)$ is a vector space over R of dimension m^2. With the identification

$$(a_i^j) \longrightarrow (a_1^1, \ldots, a_1^m, \ldots a_m^1, \ldots a_m^m) \in R^{m^2},$$

then $gl(m, R)$ becomes a C^∞ manifold of dimension m^2.

(4) **Open submanifolds.**

Let U be an open set of a differentiable manifold M of dimension m. Then U is a C^∞ manifold of the same dimension. To see this, it is sufficient to restrict the coordinate neighborhoods of M to U. The manifold U is called an **open submanifold** of M.

(5) **The general linear group $Gl(m, R)$.**

A particular case of (4) is the following. Let $Gl(m, R)$ be the group of all non–singular $m \times m$ matrices over R. Then $Gl(m, R)$ is an open set of $gl(m, R)$. In fact, let

$$\det : gl(m, R) \longrightarrow R$$

be the determinant map. Then

$$Gl(m, R) = gl(m, R) - (\det)^{-1}(0).$$

Thus, $Gl(m, R)$ is an open submanifold of $gl(m, R)$.

(6) **The sphere S^n.**

The sphere S^n is the set

$$S^n = \{x = (x^1, \ldots, x^{n+1}) \in R^{n+1} / \sum_{i=1}^{n+1} (x^i)^2 = 1\}$$

Let $N = (0, \ldots, 0, 1)$ and $S = (0, \ldots, 0, -1)$. Then the standard C^∞ structure on S^n is obtained by taking the C^∞ atlas

$$U = \{(S^n - N, \varphi_N), (S^n - S, \varphi_S)\},$$

where φ_N and φ_S are stereographic projections from N and S, respectively.

(7) **Product manifolds.**

Let M, N be two C^∞ manifolds of dimension m, n respectively. We consider the product space $M \times N$. If $(U, \phi), (V, \psi)$ are coordinate neighborhoods of M, N, respectively, we may define a coordinate neighborhood $(U \times V, \phi \times \psi)$ on $M \times N$ by

$$(\phi \times \psi)(x, y) = (\phi(x), \psi(y)), x \in U, y \in V.$$

Then $M \times N$ becomes a C^∞ manifold of dimension $m+n$. A particular case is the m–torus $T^m = S^1 \times \ldots \times S^1$, the m–fold product of circles S^1.

Remark 1.2.6 In the sequel, we will say simply manifold for C^∞ manifold; we may also sometimes say differentiable manifold.

1.3 Differentiable mappings. Rank Theorem

Definition 1.3.1 *Let $F : M \longrightarrow N$ be a mapping. F is said to be (C^∞)* **differentiable** *if for every $x \in M$ there exist coordinate neighborhoods (U, ϕ) of x and (V, ψ) of $y = F(x)$ with $F(U) \subset V$ such that*

$$\bar{F} = \psi \circ F \circ \phi^{-1} : \phi(U) \longrightarrow \psi(V)$$

is a differentiable mapping.

This means that $F_{/U} : U \longrightarrow V$ may be written in local coordinates $x^1, \ldots, x^m$ and $y^1, \ldots, y^n$ as follows:

$$\bar{F} : (x^1, \ldots, x^m) \longrightarrow (y^1(x^1, \ldots, x^m), \ldots, y^n(x^1, \ldots, x^m)),$$

where each $y^a = y^a(x^1, \ldots, x^m), 1 \leq a \leq n$, is C^∞ on $\phi(U)$.

Remark 1.3.2 Obviously, every differentiable mapping is continous.

Remark 1.3.3 Let $f : M \longrightarrow R$ be a function on M. Then f is differentiable if there exists, for each $x \in M$ a coordinate neighborhood (U, ϕ) of x such that $f \circ \phi^{-1} : \phi(U) \longrightarrow R$ is differentiable. Here, we consider the canonical differentiable structure on R. We denote by $C^\infty(M)$ the set of all differentiable functions on M. Obviously, all the definitions rest valid for mappings and functions defined on open sets of M.

Definition 1.3.4 *A differentiable mapping $F : M \longrightarrow N$ is a* **diffeomorphism** *if it is a homeomorphism and F^{-1} is differentiable. In such a case, M and N are said to be* **diffeomorphic**. *A diffeomorphism $F : M \longrightarrow M$ is said to be a* **transformation** *of M.*

Let $F : M \longrightarrow N$ be a differentiable mapping and let $x \in M$. If (U, ϕ) and (V, ψ) are coordinate neighborhoods of x and $F(x)$, respectively, with $F(U) \subset V$, then F is locally expressed by

$$\bar{F}(x^i) = (y^a(x^i)), 1 \leq i \leq m, 1 \leq a \leq n.$$

Definition 1.3.5 *The* **rank** of F **at** x *is defined to be rank of $\bar{F}$ at $\phi(x)$.*

Hence the rank of F at x is the rank at $\phi(x)$ of the Jacobian matrix

$$(\partial y^a/\partial x^i).$$

One can easily prove that this definition is independent of the choice of coordinates.

The most important case for us will be that in which the rank is constant. In fact, from the Rank Theorem in Section 1.1, we have the following.

Theorem 1.3.6 (Rank Theorem).-*Let $F : M \longrightarrow N$ be as above and suppose that F has constant rank k at every point of M. If $x \in M$ then there exist coordinate neighborhoods (U, ϕ) and (V, ψ) as above such that $\phi(x) = 0 \in R^m, \psi(F(x)) = 0 \in R^n$ and $\bar{F}$ is given by*

$$\bar{F}(x^1, \ldots, x^m) = (x^1, \ldots, x^k, 0, \ldots, 0).$$

Furthermore, we may suppose that (U, ϕ) and (V, ψ) are cubic neighborhoods centered at x and $F(x)$, respectively.

Corollary 1.3.7 *A necessary condition for F to be a diffeomorphism is that $\dim M = \dim N = \operatorname{rank} F$.*

1.4 Partitions of unity

Partitions of unity will be very useful in the sequel, for instance, in order to construct Riemannian metrics on an arbitrary manifold.

First, let us recall some definitions and results.

Definition 1.4.1 *A covering $\{U_\alpha\}$ of a topological space M is said to be* **locally finite** *if each $x \in M$ has a neighborhood U which intersects only a finite number of sets U_α. If $\{U_\alpha\}$ and $\{V_\beta\}$ are covering of M such that if $V_\beta \subset U_\alpha$ for some α, then $\{V_\beta\}$ is called a* **refinement** *of $\{U_\alpha\}$.*

Definition 1.4.2 *A topological space M is called* **paracompact** *if every open covering has a locally finite refinement.*

Now, let M be a manifold of dimension m. Then M is locally compact (in fact, M is locally Euclidean; so, M **has all the local properties of** R^m). A standard result of general topology shows that every locally compact Hausdorff space with a countable basis of open sets is paracompact (see Willard [127], for instance). Then we have.

Proposition 1.4.3 *Every manifold is a paracompact space.*

Definition 1.4.4 *Let $f \in C^\infty(M)$. The* **support** *of f is the closure of the set on which f does not vanishes, that is,*

$$supp\ (f) = cl\{x \in M / f(x) \neq 0\}.$$

We say that f has **compact support** *if supp (f) is compact in M.*

Definition 1.4.5 *A* **partition of unity** *on a manifold M is a set $\{(U_i, f_i)\}$, where*
(1) $\{U_i\}$ is a locally finite open covering of M;
(2) $f_i \in C^\infty(M), f_i \geq 0, f_i$ has compact support, and supp $(f_i) \subset U_i$ for all i;
(3) for each $x \in M, \sum_i f_i(x) = 1$.
(Note that by virtue of (1) the sum is a well-defined function on M).
A partition of unity is said to be **subordinate** *to an atlas $\{U_\alpha\}$ of M if $\{U_i\}$ is a refinement of $\{U_\alpha\}$.*

Lemma 1.4.6 *Let $U_i = B(0,1)$, $U_2 = B(0,2)$ in R^m. Then there is a C^∞ function $g : R^m \longrightarrow R$ such that g is 1 on U_1 and 0 outside U_2. We call g a* **bump function.**

Proof: Let $\theta : R \longrightarrow R$ be given by

$$\theta(t) = \begin{cases} \exp((-1)/(1-t^2)), & |\, t \,| < 1 \\ 0, & |\, t \,| \geq 1 \end{cases}$$

Now, we put

$$\theta_1(s) = (\int_{-\infty}^{s} \theta(t)dt)/(\int_{-\infty}^{+\infty} \theta(t)dt).$$

Then θ_1 is a C^∞ function such that $\theta_1(s) = 0$, if $s < -1$, and $\theta_1(s) = 1$ if $s > 1$. Let

$$\theta_2(s) = \phi_1(-s-2).$$

Thus, θ_2 is a C^∞ function such that $\theta_2(s) = 1$ if $s > 1$, and $\theta_2(s) = 0$ if $s > 2$. Finally, let

$$g(x) = \theta_2(\|\, x \,\|), x \in R^m.$$

Then g is the required function. □

Lemma 1.4.7 *Let U_1, U_2 be open sets of an m-dimensional manifold M such that $cl(U_1) \subset U_2$. Then there exists $g \in C^\infty(M)$ such that g is 1 on U_1 and is 0 outside U_2.*

Proof: The proof is a direct consequence of the Lemma 1.4.6. □

Proposition 1.4.8 *If $\{V_\alpha\}$ is an atlas of an m-dimensional manifold M, there is a partition of unity subordinate to $\{V_\alpha\}$.*

Proof: Let $\{W_\lambda\}$ be an open covering. Since M is paracompact, then there is a locally finite refinement consisting of coordinate neighborhoods $\{(U_i, \phi_i)\}$ such that $\phi_i(U_i)$ is the open ball centered at 0 and of radius 3 in R^m, and such that $(\phi_i)^{-1}(B(0,1))$ cover M. Now, let $\{V_\alpha\}$ be an atlas of M and let $\{(V_i, \phi_i)\}$ be a locally finite refinement with these properties. From the Lemma 1.4.7, there is a function $g_i \in C^\infty(M)$ such that supp $(g_i) \subset V_i$ and $g_i \geq 0$. We now put

$$f_i(x) = (g_i(x))/(\Sigma_j g_j(x)), \quad x \in M.$$

Then $\{f_i\}$ are the required functions. □

1.5 Immersions and submanifolds

In this section we shall consider some special kinds of differentiable mappings with constant rank.

Definition 1.5.1 *Let $F : N \longrightarrow M$ be a differentiable mapping with $n =$ dim $N \leq m =$ dim M. F is said to be an* **immersion** *if rank $F = n$ at every point of N. If an immersion F is injective, then N (or its image $F(N)$), endowed with the topology and differentiable structure which makes $F : N \longrightarrow F(N)$ a diffeomorphism, is called an* **(immersed) submanifold** *of M.*

From the theorem of rank, we deduce that, if $F : N \longrightarrow M$ is an immersion, then, for each $x \in N$, there exist cubical coordinate neighborhoods $(U, \phi), (V, \psi)$ centered at x and $F(x)$, respectively, such that F is locally given by

$$\bar{F} : (x^1, \ldots, x^n) \longrightarrow (x^1, \ldots, x^n, 0, \ldots, 0).$$

Hence F is locally injective.

Remark 1.5.2 We note that an immersion need not be injective. For instance, the mapping

$$F : R \longrightarrow R^2$$

given by

$$F(t) = (\cos 2\pi t, \sin 2\pi t)$$

is a immersion, but $F(t + 2\pi) = F(t)$.

Definition 1.5.3 *An* **embedding** *is an injective immersion $F : N \longrightarrow M$ which is a homeomorphism of N onto its image $F(N)$, with its topology as a subspace of M. Then N (or $F(N)$) is said to be an* (**embedded**) **submanifold** *of M.*

Remark 1.5.4 We note that an injective immersion need not be an embedding. For instance, let $F : R \longrightarrow R^2$ be given by

$$F(t) = (2\cos(t - \frac{1}{2}\pi), \sin 2(t - \frac{1}{2}\pi)).$$

The image of F is a figure eight denoted by E; the image point making a complet circuit starting at $(0,0) \in R^2$ as t goes from 0 to 2π. $E = F(R)$ is compact considered as subspace of R^2, but R is the real line. Then E and R are not homeomorphic.

Let M be a differentiable manifold of dimension m.

Definition 1.5.5 *A subset N of M is said to have the* n–**submanifold property** *if, for each $x \in N$, there exists a coordinate neighborhood (U, ϕ) with local coordinates $(x^1, \ldots, x^m)$ such that*

$$\phi(U \cap N) = \{x \in \phi(U) / x^{n+1} = \ldots = x^m = 0\}.$$

Now, we consider the subspace topology on N. We put $U' = U \cap N$ and

$$\phi'(x) = (x^1, \ldots, x^n) \in R^n, x \in U'.$$

Let $(U, \phi), (V, \psi)$ be coordinate neighborhoods as above. Then $\psi' \circ (\phi')^{-1} : \phi'(U') \longrightarrow \psi'(V')$ is a C^∞ mapping. Then N is a n–dimensional manifold and the natural inclusion $i : N \longrightarrow M$ is an embedding. Thus, N is an embedded submanifold of M. It is not hard to prove the converse, that is, if $F : N \longrightarrow M$ is an embedding, then $F(N)$ has the n–submanifold property. We leave the proof to the reader as an exercise.

To end this section, we shall describe a useful method of finding examples of manifolds.

Proposition 1.5.6 *Let $F : N \longrightarrow M$ be a differentiable mapping, with dim $N = n$, dim $M = m, n \geq m$. Suppose that F has constant rank k on N and let $y \in F(N)$. Then $F^{-1}(y)$ is a closed embedded submanifold of N.*

Proof: First, $F^{-1}(y)$ is closed since F is continous. Furthermore, let $x \in F^{-1}(y)$. By the theorem of rank, there exist coordinate neighborhoods $(U, \phi), (V, \psi)$ of x and y, respectively, such that F is locally given by

$$(x^1, \ldots, x^n) \longrightarrow (x^1, \ldots, x^n, 0, \ldots, 0).$$

Hence we have

$$U \cap F^{-1}(y) = \{x \in U / x^1 = \ldots = x^k = 0\}.$$

Therefore, $F^{-1}(y)$ has the $(n-k)$–submanifold property. □

Corollary 1.5.7 *Let $F : N \longrightarrow M$ be as above. If rank $F = m$ at every point of $F^{-1}(y)$, then $F^{-1}(y)$ is a closed embedded submanifold of N.*

Proof: In fact, if rank $F = m$ at every point of $F^{-1}(y)$, then F has rank m on an neighborhood of $F^{-1}(y)$. Then we apply the Proposition 1.5.6. □

Example.- Let $F : R^n \longrightarrow R$ be the mapping defined by

$$F(x^i, \ldots, x^n) = \sum_{i=1}^{n} (x^i)^2.$$

Then F has rank 1 on $R^n - \{0\}$. But $S^{n-1} \subset R^n - \{0\}$. Thus, S^{n-1} is a closed embedded submanifold of R^n. It is not hard to prove that this structure of manifold on S^{n-1} coincides with the one given in Section 1.2.

1.6 Submersions and quotient manifolds

Definition 1.6.1 *Let $F : M \longrightarrow N$ be a differentiable mapping with $m =$ dim $M \geq n =$ dim N. F is said to be a* **submersion** *if rank $F = n$ at every point of M.*

If $F : M \longrightarrow N$ is a submersion, then F is locally given by

$$\bar{F} : (x^1, \ldots, x^m) \longrightarrow (x^1, \ldots, x^n).$$

This fact is a direct consequence from the Rank Theorem. Hence, a submersion is locally surjective.

Definition 1.6.2 *Let $F : M \longrightarrow N$ be a submersion. Let $y \in N$. Then $F^{-1}(y)$ is called a* **fibre** *of the submersion F.*

From Proposition 1.5.6, we deduce that, if $y \in F(M)$, then the fibre $F^{-1}(y)$ is a closed embedded submanifold of dimension $m - n$ of M.

Now, let M be a topological space and $\sim$ an equivalence relation on M. We denote by $M/\sim$ the quotient space of M relative to $\sim$. Let $\pi : M \longrightarrow M/\sim$ be the canonical projection. It is easy to prove (see Willard [127]) the following.

Proposition 1.6.3 *(1) If $\pi : M \longrightarrow M/\sim$ is an open mapping and M has a countable basis of open sets, then $M/\sim$ has a countable basis also. (2) Put $R = \{(x,y)/x \sim y\}$. Then $M/\sim$ is Hausdorff if and only if R is a closed subset of $M \times M$.*

Next, let M be a differentiable manifold of dimension m. Proposition 1.6.3 is useful in determining those equivalence relations $\sim$ on M whose quotient space is again a manifold.

If $M/\sim$ is a manifold such that $\pi : M \longrightarrow M/\sim$ is a submersion, then $M/\sim$ is said to be a **quotient manifold** of M. It is not hard to prove that, if such a manifold structure exists, then it is unique.

Example (Real projective space RP^n).- Let $M = R^{n+1} - \{0\}$. We define $x \sim y$ if there is $t \in R - \{0\}$ such that $y = tx$, that is,

$$y^i = tx^i, \quad 1 \leq i \leq n+1.$$

Hence the equivalence class $[x]$ of x is the line trough the origin and x. We denote $M/\sim$ by RP^n, the **real projective space** of dimension n. Let $\pi : M \longrightarrow RP^n$ be the canonical projection. For each $t \in R - \{0\}$, we define $\phi_t : M \longrightarrow M$ by

$$\phi_t(x) = tx.$$

Then ϕ_t is a diffeomorphism with $\phi_t^{-1} = \phi_{1/t}$. If $U \subset M$ is an open set, then

$$\pi^{-1}(\pi(U)) = \bigcup_{t \in R-\{0\}} \phi_t(U).$$

Then $\pi(U)$ is an open set of RP^n. Hence π is open.

Now, we define a real continous function f on $M \times M$ by

$$f(x,y) = \sum_{i \neq j} (x^i y^j - x^j y^i)^2.$$

Then $R = f^{-1}(0)$ is a closed set of $M \times M$. Hence RP^n is Hausdorff.

We define $n+1$ coordinate neighborhoods (U_i, ϕ_i) as follows:

$$U_i = \pi(U_i'),$$

where $U_i' = \{x \in M / x^i \neq 0\}$, and

$$\phi_i : U_i \longrightarrow R^n$$

by

$$\phi_i(x) = (x^1/x^i, \ldots, x^{i-1}/x^i, x^{i+1}/x^i, \ldots, x^{n+1}/x^i).$$

It is easy to prove that $\phi_j \circ (\phi_i)^{-1}$ is C^∞. Then RP^n is a quotient manifold of M of dimension n.

1.7 Tangent spaces. Vector fields

In this section we introduce the tangent space of an m–dimensional manifold M.

Definition 1.7.1 *Let $x \in M$. A* **curve** *at x is a differentiable map $\sigma : I \longrightarrow M$ from an open interval $I \subset R$ into M with $0 \in I$ and $\sigma(0) = x$. Let σ and τ be curves at x. We say that σ and τ are* **tangent at** *x if there exists a coordinate neighborhood (U, ϕ) of x with local coordinates (x^i) such that*

$$(dx^i \circ \sigma / dt)_{/t=0} = (dx^i \circ \tau / dt)_{/t=0}, \quad 1 \leq i \leq m.$$

One can easily check that the definition is independent of the choice of local coordinates. The equivalence class $[\sigma]$ of σ is called the **tangent vector of** σ **at** x; sometimes, it is denoted by $\dot{\sigma}(0)$. If $\sigma : I \longrightarrow M$ is a curve in M (where $I = (-\epsilon, \epsilon), \epsilon > 0$ is an open interval in R) then the **tangent vector of** σ **at** t is defined by $d\sigma/dt = \dot{\sigma}(t) = \dot{\tau}(0), \tau(s) = \sigma(s+t)$.

Definition 1.7.2 *The* **tangent space** T_xM *of* M *at* x *is the set of equivalence classes of curves at* x.

Now, we define a map

$$\phi' : T_xM \longrightarrow R^m$$

by

$$\phi'([\sigma]) = ((dx^i \circ \sigma/dt)_{/t=0}, \ldots, (dx^m \circ \sigma/dt)_{/t=0}).$$

It is clear that ϕ' is injective. Furthermore, it is surjective. In fact, let $v = (v^i, \ldots, v^m) \in R^m$ and consider the curve σ given by

$$\sigma(t) = \phi^{-1}(\phi(x) + tv).$$

Hence $\phi'([\sigma]) = v$. Then we consider a vector space structure on T_xM such that ϕ' is a linear isomorphism. One can easily prove that this vector structure is independent of the choice of (U, ϕ). Thus T_xM is a vector space of dimension m.

Let σ^i be the curve at x defined by

$$\sigma^i(t) = \phi^{-1}(\phi(x) + te_i),$$

where $(e_1, \ldots, e_m)$ is the canonical basis of R^m. Then the tangent vectors $(\dot{\sigma}^i(0), \ldots, \dot{\sigma}^m(0))$ form a basis for T_xM. In the sequel, we shall use the notation

$$(\partial/\partial x^i)_x = \dot{\sigma}^i(0), 1 \leq i \leq m.$$

Obviously,

$$\phi'((\partial/\partial x^i)_x) = e_i.$$

Now, let $F : M \longrightarrow N$ be a differentiable mapping. Then we define a linear mapping

$$dF(x) : T_xM \longrightarrow T_{F(x)}N$$

by

$$dF(x)([\sigma]) = [F \circ \sigma].$$

Definition 1.7.3 *The linear mapping $dF(x)$ (also denoted by $F_\star$) will be called the* **differential of F at x.**

Now, let $x \in M$ and $(U, \phi), (V, \psi)$ coordinate neighborhoods of x and $F(x)$ with local coordinates $(x^1, \ldots, x^m)$ and $(y^1, \ldots, y^n)$, respectively. A direct computation shows that

$$dF(x)((\partial/\partial x^i)_x) = \sum_{a=1}^{n} (\partial y^a/\partial x^i)(\partial/y^a)_{F(x)}.$$

Hence $dF(x)$ is represented by the Jacobian matrix of F. We deduce that

$$\text{rank } F \textit{ at } x = \text{rank } dF(x).$$

Consequently, we have

Proposition 1.7.4 *(1) F is an immersion if $dF(x)$ is an injective linear mapping, for each $x \in M$. (2) F is a submersion if $dF(x)$ is a surjective linear mapping, for each $x \in M$. (3) If F is a diffeomorphism, then $dF(x)$ is a linear isomorphism, for each $x \in M$. Conversely, if $dF(x)$ is a linear isomorphism, then F is a local diffeomorphism on a neighborhood of x.*

The following result generalizes the well-known chain rule.

Theorem 1.7.5 *Let $F : M \longrightarrow N$ and $G : N \longrightarrow P$ be differentiable mappings. Then $G \circ F$ is differentiable and*

$$d(G \circ F)(x) = dG(F(x)) \circ dF(x), \text{for each } x \in M.$$

Proof: Obviously, if F and G are C^∞, then $G \circ F$ is C^∞ also. Now, let σ be a curve at $x \in M$; then

$$d(G \circ F)(x)([\sigma]) = [(G \circ F) \circ \sigma].$$

On the other hand, we have

$$dG(F(x))(dF(x)([\sigma]) = dG(F(x))([F \circ \sigma]) = [G \circ (F \circ \sigma)].$$

This ends the proof. □

Definition 1.7.6 *A* **vector field** *X on a manifold M is a function assigning to each point $x \in M$ a tangent vector $X(x)$ of M at x. (in Section 1.9 we shall precise the word function employed here).*

Let (U, ϕ) be a coordinate neighborhood with local coordinates $(x^i), 1 \leq i \leq m = dim\ M$. For each $x \in U$, we have

$$X(x) = \sum_{i=1}^{m} X^i(x)(\partial/\partial x^i)_x.$$

Then we have

$$X = \sum_{i=1}^{m} X^i \ \partial/\partial x^i$$

on U.

Definition 1.7.7 *A vector field X is said to be* $\mathbf{C}^\infty$ *if the functions X^i are C^∞ for each coordinate neighborhood (U, ϕ).*

Let X be a vector field on a manifold M. A curve $\sigma : I \longrightarrow M$ in M is called an **integral curve** of X if, for every $t \in I$, the tangent vector $X(\sigma(t))$ is the tangent vector to the curve σ at $\sigma(t)$.

Proposition 1.7.8 *For any point $x_0 \in M$, there is a unique integral curve of X, defined on $(-\epsilon, \epsilon)$ for some $\epsilon > 0$, such that $x_0 = \sigma(0)$.*

Proof: In fact, let (U, ϕ) be a coordinate neighborhood of x_0 with local coordinates (x^i) and let

$$X = \sum_{i=1}^{m} X^i \ \partial/\partial x^i$$

on U. Then an integral curve of X is a solution of the following system of ordinary differential equation

$$dx^i/dt = X^i(x^i(t), \ldots, x^m(t)), \ \ 1 \leq i \leq m.$$

Now, the result follows from the fundamental theorem for systems of ordinary differential equations. □

Remark 1.7.9 From Proposition 1.7.8 a vector field is also called a **first order differential equation**.

Definition 1.7.10 *A 1–parameter* **group of transformations** *of M is a mapping*

$$\Phi : R \times M \longrightarrow M,$$

such that
(1) for each $t \in R, \Phi_t : x \longrightarrow \Phi(t,x)$ is a transformation of M;
(2) for all $s, t \in R$ and $x \in M$, $\Phi_{s+t}(x) = \Phi_t(\Phi_s(x))$.

Each 1–parameter group of transformations $\Phi = (\Phi_t)$ induces a vector field X as follows. Let $x \in M$. Then $X(x)$ is the tangent vector to the curve $t \longrightarrow \Phi_t(x)$ (called the **orbit** of x) at $x = \Phi_0(x)$. Hence the orbit $\Phi_t(x)$ is an integral curve of X starting at x. X is called the **infinitesimal generator** of Φ_t.

A **local 1–parameter group of local transformations** can be defined in the same way. Actually, $\Phi_t(x)$ is defined only for t in a neighborhood of 0 and x in an open set of M. More precisely, a local 1–parameter group of local transformations defined on $(-\epsilon, \epsilon) \times U$, where $\epsilon > 0$ and U is an open set of M, is a mapping

$$\Phi : (-\epsilon, \epsilon) \times U \longrightarrow M$$

such that
(1) for each $t \in (-\epsilon, \epsilon), \Phi_t : x \longrightarrow \Phi(t,x)$ is a diffeomorphism of U onto the open set $\Phi_t(U)$;
(2) If $s, t, s+t \in (-\epsilon, \epsilon)$ and if $x, \Phi_s(x) \in U$, then

$$\Phi_{s+t}(x) = \Phi_t(\Phi_s(x)).$$

As above, Φ_t induces a vector field X defined on U. Now, we prove the converse.

Proposition 1.7.11 *Let X be a vector field on M. Then, for each $x_0 \in M$, there exists a neigborhood U of x_0, a positive real number ϵ and a local 1–parameter group of local transformations $\Phi : (-\epsilon, \epsilon) \times U \longrightarrow M$ which induces X on U.*

Proof: Let (V, ϕ) a coordinate neighborhood of x_0 with local coordinates (x^i) such that $x^i(x_0) = 0, 1 \leq i \leq m$. Consider the following system of ordinary differential equations:

$$df^i/dt = X^i(f^1(t), \dots, f^m(t)), \tag{1.1}$$

where $X = X^i\, \partial/\partial x^i$ on V. From the fundamental theorem for systems of ordinary differential equation, there is a unique set of functions $(f^1(t,x), \dots, f^m(t,x))$ defined for x with $x^i \in (-\delta, \delta)$ and for $t \in (-\lambda, \lambda)$ such that (f^i) is a solution of (1.1) for each fixed x satisfying the initial condition

$$f^i(0,x) = x^i, 1 \leq i \leq m.$$

We set

$$\Phi_t(x) = (f^1(t,x), \dots, f^m(t,x))$$

and

$$W = \{x/x^i \in (-\delta, \delta)\}.$$

Now, if $s, t, s+t \in (-\lambda, \lambda)$ and x, $\Phi_t(x) \in W$, then the functions $g^i(t) = f^i(s+t, x)$ form a solution of (1.1) with initial conditions

$$g^i(0) = f^i(s,x).$$

From the uniqueness of the solution, we deduce that $g^i(t) = f^i(t, \Phi_s(x))$. This proves that $\Phi_t(\Phi_s(x)) = \Phi_{s+t}(x)$. Now, since $\Phi_0 : W \longrightarrow W$ is the identity, there exist $\mu > 0$ and $\nu > 0$ such that, if $U = \{x/x^i \in (-\mu, \mu))$ then $\Phi_t(U) \subset W$, if $t \in (-\nu, \nu)$. Consequently, we have

$$\Phi_{-t}(\Phi_t(x)) = \Phi_t(\Phi_{-t}(x)) = \Phi_0(x) = x,$$

for every $x \in U, t \in (-\nu, \nu)$. Hence Φ_t is a local 1–parameter group of local transformations defined on $(-\nu, \nu) \times U$ which induces X on U.□

Definition 1.7.12 *A vector field X on M is called* **complete** *if X generates a global 1–parameter group of transformations on M.*

Proposition 1.7.13 *On a compact manifold M, every vector field is complete.*

We leave the proof to the reader as an exercise.

To end this section, we interpret a vector field as an operator on functions.

Let $v \in T_xM$ be a tangent vector to M at $x \in M$. If f is a differentiable function defined on a neighborhood U of x, then we can define a real number vf by

$$vf = (d(f \circ \sigma)/dt)_{/t=0},$$

where $\sigma \in v$. We have the following properties:
(1) $v(f+g) = vf + vg$,
(2) $v(\alpha f) = \alpha(vf), \alpha \in R$,
(3) **(Leibniz rule)** $v(fg) = f(x)(vg) + (vf)g(x)$.

Now, let $C^\infty(x)$ be the set of differentiable functions defined on a neighborhood of x. Two functions f and g of $C^\infty(x)$ are related if they agree on some neighborhood of x, that is, if f and g define the same **germ** at x. The quotient set is denoted by $\bar{C}^\infty(x)$. Hence $\bar{C}^\infty(x)$ is a real algebra. By a **derivation** on $\bar{C}^\infty(x)$ we mean a linear operator

$$D : \bar{C}^\infty(x) \longrightarrow \bar{C}^\infty(x)$$

such that $D(fg) = (Df)g + f(Dg)$, $f, g \in \bar{C}^\infty(x)$. Then each tangent vector v of M at x is a derivation on $\bar{C}^\infty(x)$. We prove the converse. First, if D is a derivation on $\bar{C}^\infty(x)$, then $D\alpha = 0$, for each constant function α. Now let $f \in C^\infty(x)$. The Taylor expansion of f, with respect to a local coordinate system (y^i) at x, is

$$f(y) = f(x) + (\partial f/\partial y^i)(y^i - x^i) + \omega_{ij}(y^i - x^i)(y^j - x^j),$$

where $x^i = y^i(x), x^i = y^i(x)$ and $\omega_{ij} \longrightarrow (\partial^2 f/\partial y^i \partial y^j)$ when $y \longrightarrow x$. Then we have

$$Df = (\partial f/\partial y^i)(Dy^i).$$

Hence, we deduce that

$$D = X^i(\partial/\partial y^i)_x,$$

where $X^i = Dy^i, 1 \leq i \leq m$.

Now, let X be a vector field on M. If $f \in C^\infty(M)$, then we can define a new C^∞ function Xf by

$$Xf(x) = X(x)f.$$

(Obviously, if f is constant, then $Xf = 0$). Then we have $X(fg) = (Xf)g + f(Xg)$. Thus, X acts as a derivation on the algebra $C^\infty(M)$. Denote by $\chi(M)$ the set of all vector fields on M. Obviously, $\chi(M)$ is a vector space over R and a $C^\infty(M)$–module. Now, let $X, Y \in \chi(M)$. Then we can define a new vector field $[X, Y]$ as follows:

$$[X,Y](x)(f) = X(x)(Yf) - Y(x)(Xf), \quad x \in M, f \in C^\infty(x).$$

Then $[X, Y]$ is a vector field on M, which is called the **Lie bracket** of X and Y. A simple computation shows that
(1) $[X,Y] = -[Y,X]$;
(2) $[fX, gY] = f(Xg)Y - g(Yf)X + (fg)[X,Y]$;
(3) (**Jacobi identity**) $[[X,Y],Z] + [[Y,Z],X] + [[Z,X],Y] = 0$.

Remark 1.7.14 This properties show that $(\chi(M), [\ ,\ ])$ is a Lie algebra (see Section 1.19).

In terms of local coordinates, we have

$$[X,Y] = (X^i(\partial Y^j/\partial x^i) - Y^i(\partial X^j/\partial x^i))\ \partial/\partial x^j,$$

where $X = X^i\ \partial/\partial x^i, Y = Y^i, Y = Y^i\ \partial/\partial x^i$.

1.8 Fibred manifolds. Vector bundles

Definition 1.8.1 *A* **bundle** *is a triple* (E, p, M)*, where* $p : E \longrightarrow M$ *is a surjective submersion. The manifold E is called the* **total space,** *the manifold M is called the* **base space,** *p is called the* **projection** *of the bundle. For each* $x \in M$*, the submanifold* $p^{-1}(x) = E_x$ *is called the* **fibre** *over x. We also say that E is a* **fibred manifold** *over M.*

Example.- Let M, F be manifolds. Then $(M \times F, p, M)$ is a bundle, where $p : M \times F \longrightarrow M$ is the canonical projection on the first factor. This bundle is called a **trivial bundle**.

Definition 1.8.2 *Let (E, p, M) be a bundle. A mapping* $s : M \longrightarrow E$ *such that* $p \circ s = id$ *is called a (global)* **section** *of E. If s is defined on an (open) subset U of M, then* $s : U \longrightarrow E$ *is called a local section of E over U.*

Note that there always exist local sections since p is a surjective submersion.

Definition 1.8.3 *Let (E, p, M) and (E',p',M') be two bundles. A* **bundle morphism** $(H,h) : (E,p,M) \longrightarrow (E',p',M')$ *is a pair of differentiable maps* $H : E \longrightarrow E'$ *and* $h : M \longrightarrow M'$ *such that* $p' \circ H = h \circ p$. *(Roughly speaking, a bundle morphism is a fibre preserving map).*

From Definition 1.8.3 one easily deduces that H maps the fibre of E over x into the fiber of E' over $h(x)$.

Definition 1.8.4 *A bundle morphism* $(H,h) : (E,p,M) \longrightarrow (E',p',M')$ *is an* **isomorphism** *if there exists a bundle morphism* $(H',h') : (E',p',M') \longrightarrow (E,p,M)$ *such that* $H' \circ H = id_E$ *and* $h' \circ h = id_M$. *Then* (E,p,M) *and* (E',p',M') *are said to be* **isomorphic**.

Now, we consider bundles (or fibred manifolds) with an additional vector space structure on each fibre.

Definition 1.8.5 *Let M be a differentiable m–dimensional manifold. A real* **vector bundle** *E of rank n over M is a bundle (E,p,M) such that:*
(1) For each $x \in M$, E_x has the structure of a real vector space of dimension n;
(2) **Local triviality***) For each $x \in M$ there exists a neighborhood U of x and a diffeomorphism*

$$H : U \times R^n \longrightarrow p^{-1}(U)$$

such that, for each $y \in U$, the correspondence $v \longrightarrow H(y,v)$ defines an isomorphism between the vector space R^n and the vector space E_y.

Examples.- (1) Let M be a differentiable manifold. Then $M \times R^n$ is a (trivial) vector bundle of rank n over M.
(2) The **tangent** and **cotangent bundles** of M (see Section 1.9).

Definition 1.8.6 *Let* $(E,p,M),(E',p',M")$ *be vector bundles. Then a* **vector bundle homomorphism** *is a bundle morphism (H,h) such that the restriction* $H_x : E_x \longrightarrow E'_{h(x)}$ *is linear for each* $x \in M$. *(H,h) is called a* **vector bundle isomorphism** *if there exists a vector bundle homomorphism* $(H',h') : (E',p',M') \longrightarrow (E,p,M)$ *such that* $H' \circ H = id_E$ *and* $h' \circ h = id_M$. *In such a case, E and E' are said to be* **isomorphic**. *If M'=M, a M–***vector bundle homomorphism** *(or* **vector bundle homomorphism over** *M) is defined by a vector bundle homomorphism of the form* (H, id_M).

If $(H, h) : (E, p, M) \longrightarrow (E', p', M')$ is a vector bundle isomorphism, then: (1) H and h are diffeomorphisms; (2) the restriction $H_x : E_x \longrightarrow E'_{h(x)}$ is a linear isomorphism, for each $x \in M$. The converse is true for vector bundles with the same base.

Proposition 1.8.7 *Let $H : E \longrightarrow E'$ be a vector bundle homomorphism over M. If for each $x \in M$, $H_x : E_x \longrightarrow E'_x$ is a linear isomorphism then H is a vector bundle isomorphism.*

We leave the proof to the reader as an exercise.

Definition 1.8.8 *A vector bundle $p : E \longrightarrow M$ of rank n is called* **trivial** *if it is isomorphic to $M \times R^n \longrightarrow M$.*

Remark 1.8.9 Hence the local triviality property means that $p^{-1}(U)$ is a trivial vector bundle over U isomorphic to $U \times R^n$.

Next, we will describe a number of basic constructions involving vector bundles (see Godbillon [63], Milnor and Stasheff [96]).

(1) **Restricting a bundle to a subset of the base space**.

Let (E, p, M) be a vector bundle over M and $N \subset M$ a submanifold of M. We set $\bar{E} = p^{-1}(N)$ and denote by $\bar{p} : \bar{E} \longrightarrow N$ the restriction of p to $\bar{E}$. Then one obtains a new vector bundle $(\bar{E}, \bar{p}, N)$ called the **restriction** of E to N. Each fiber $\bar{E}_x, x \in N$, is equal to the corresponding fiber E_x.

(2) **Induced bundles**.

Let N be an arbitrary manifold and (E, p, M) a vector bundle. For any map $f : N \longrightarrow M$ we can construct the **induced bundle** $f^*(E) = (\bar{E}, \bar{p}, N)$ over N as follows. The total space $\bar{E} \subset N \times E$ consists of all pairs (x, e) such that $f(x) = p(e)$. The projection $\bar{p} : \bar{E} \longrightarrow N$ is defined by $\bar{p}(x, e) = x$. Then one obtains a commutative diagram

$$\begin{array}{ccc} \bar{E} & \xrightarrow{\bar{f}} & E \\ {\scriptstyle \bar{p}}\downarrow & & \downarrow{\scriptstyle p} \\ N & \xrightarrow{f} & M \end{array}$$

where $\bar{f}(x, e) = e$. The vector space structure in $\bar{E}_x$ is defined by

$$\alpha(x, e) + \beta(x, e') = (x, \alpha e + \beta e'), \quad \alpha, \beta \in R.$$

Thus $\bar{f}$ is a vector bundle homomorphism over f.

(3) **Cartesian products.**

Given two vector bundles (E_1, p_1, M_1) and (E_2, p_2, M_2) the **Cartesian product** is the vector bundle $(E_1 \times E_2, p_1 \times p_2, M_1 \times M_2)$. Obviously, each fiber $(E_1 \times E_2)_{(x_1,x_2)}$ is identified in a natural way with $(E_1)_{x_1} \times (E_2)_{x_2}, x_1 \in M_1, x_2 \in M_2$.

(4) **Whitney sums.**

Let $(E_1, p_1, M), (E_2, p_2, M)$ be two vector bundles over M. Let $\Delta : M \longrightarrow M \times M$ be the diagonal mapping defined by $\Delta(x) = (x, x)$. The vector bundle $\Delta^*(E_1 \times E_2)$ over M is denoted by $E_1 \oplus E_2$, and called the **Whitney sum** of E_1 and E_2. Each fiber $(E_1 \oplus E_2)_x$ is canonically identified with the direct sum $(E_1)_{x_1} \oplus (E_2)_{x_2}$.

(5) In general, the algebraic opeations on vector spaces can be extended in a natural way to vector bundles. Details of the corresponding constructions are left to the reader (see Godbillon [63]).

Definition 1.8.10 *Let (E, p, M) and (E', p', M) be two vector bundles over M and $H : E' \longrightarrow E$ a vector bundle homomorphism (over M) such that the restriction $H_x : E'_x \longrightarrow E_x$ of H to any fiber E'_x is injective. We say that (E', p', M) is a* **vector subbundle** *of (E, p, M) (Obviously, we may identity E' with $H(E')$).*

Definition 1.8.11 *Let $(E, p, M), (E', p', M)$ be two vector bundles over M and $H : E \longrightarrow E'$ a vector bundle homomorphism over M. Then*

$$Ker\ H = \bigcup_{x \in M} ker\ H_x$$

is a vector subbundle of E which will be called the **kernel** *of H and*

$$Im\ H = \bigcup_{x \in M} Im\ H_x$$

is a vector subbundle of E' which will be called the **image** *of H.*

Moreover, if E' is a vector subbundle of E, then we can define a new vector bundle E'' over M by setting

$$E'' = \bigcup_{x \in M} (E_x / E'_x);$$

E'' is called the **quotient vector bundle** of E by E'.

Let now $(E,p,M),(E',p',M)$ and (E'',p'',M) be vector bundles and $H : E \longrightarrow E', G : E' \longrightarrow E''$ vector bundle homomorphisms over M. The sequence

$$E \xrightarrow{H} E' \xrightarrow{G} E''$$

is said to be **exact** if for each $x \in M$ the sequence of vector spaces

$$E_x \xrightarrow{H_x} E'_x \xrightarrow{G_x} E''_x$$

is exact. In such a case, we writte

$$0 \longrightarrow E \xrightarrow{H} E' \xrightarrow{G} E'' \longrightarrow 0.$$

For instance, if E is a vector subbundle of E' and E'' is the quotient vector bundle of E' by E, then the sequence

$$0 \longrightarrow E \xrightarrow{i} E' \xrightarrow{p} E'' \longrightarrow 0$$

is exact, where i is the canonical inclusion and p the canonical projection.

1.9 Tangent and cotangent bundles

Let M be an m–dimensional manifold. We set

$$TM = \bigcup_{x \in M} T_xM.$$

Let $\tau_M : TM \longrightarrow M$ be the canonical projection defined by

$$\tau_M(v) = x, \quad v \in T_xM.$$

Hence $(\tau_M)^{-1}(U) = TU$, for each open set U of M. Let (U,ϕ) be a coordinate neighborhood on M with local coordinates $(x^1,\ldots,x^m)$. Then we can define a mapping

$$\Phi : U \times R^m \longrightarrow TU$$

given by

$$\Phi(x,a) = a^i(\partial/\partial x^i)_x,$$

where $a = (a^1, \ldots, a^m) \in R^m, \phi(x) = (x^1, \ldots, x^m)$.$\Phi$ is a bijective mapping, since, if $v \in T_xM, x \in U$, then

$$v = v^i(\partial/\partial x^i)_x.$$

Consequently, $\Phi(x^i, \ldots, x^m, v^1, \ldots, v^m) = v$. Hence Φ defines a bijective mapping

$$\Phi' : \phi(U) \times R^m \longrightarrow TU$$

given by $\Phi'(x^1, \ldots, x^m, a^1, \ldots, a^m) = \Phi(x, a)$. Now, it is clear that there is a unique topology on TM such that for each coordinate neighborhood (U, ϕ) of M, the set TU is an open set of TM and $\Phi : U \times R^m \longrightarrow TU$, defined as above, is a homeomorphism. Thus we have local coordinates (x^i, v^i) on TU called the **induced coordinates** in TM. Next, we prove that, in fact, TM has the structure of manifold of dimension $2m$.

Let (U, ϕ), (V, ψ) be two coordinate neighborhoods on M such that $U \cap V \neq \emptyset$; then $TU \cap TV \neq \emptyset$. Let $v \in T_xM, x \in U \cap V$. Then, if $(x^i), (y^i)$ are local coordinates corresponding to $(U, \phi), (V, \psi)$, respectively, we have

$$v = v^i(\partial/\partial x^i)_x = w^i(\partial/\partial y^i)_x,$$

where $w^i = v^j(\partial y^i/\partial x^j)_x$. Hence

$$(\psi')^{-1} \circ \Phi' : \phi(U) \times R^m \longrightarrow \psi(V) \times R^m$$

is given by

$$((\psi')^{-1} \circ \Phi')(x^i, a^i) = (y^i, a^j(\partial y^i/\partial x^j)).$$

Hence the neighborhoods $(TU, (\Phi')^{-1})$ **determine a C^∞–structure on TM of dimension $2m$ relative to which τ_M is a submersion.** In fact, τ_M is locally given by

$$\tau_M(x^i, v^i) = (x^i).$$

Moreover, (TM, τ_M, M) is a vector bundle of rank m, which will be called the **tangent bundle** of M.

Actually, it is clear that a vector field X on M defines a section of TM, and conversely. It is easy to prove that a vector field $X : M \longrightarrow TM$ is C^∞ if and only if X is C^∞ as a mapping from M into TM.

Let $F : M \longrightarrow N$ be a differentiable mapping. Then we can define a map $TF : TM \longrightarrow TN$ as follows:

$$TF(v) = dF(x)(v),$$

for $v \in T_xM, x \in M$. Thus TF is a vector bundle homomorphism such that the following diagram

$$\begin{array}{ccc} TM & \xrightarrow{TF} & TN \\ {\scriptstyle \tau_M}\downarrow & & \downarrow{\scriptstyle \tau_N} \\ M & \xrightarrow{F} & N \end{array}$$

is commutative. Sometimes, we shall employ the notation $TF(v)$ for $dF(x)(v)$ if there are no danger of confusion.

Now, let x be a point of M. We set

$$T_x^*M = (T_xM)^*,$$

i.e., T_x^*M is the dual vector space of T_xM; T_x^*M is called the **cotangent vector space** of M at x and an element $\alpha \in T_x^*M$ is called a **tangent covector** (or **1–form**) of M at x.

Let $f \in C^\infty(M)$. Then the differential $df(x)$ of f at $x \in M$ is a linear mapping

$$df(x) : T_xM \longrightarrow T_{f(x)}R.$$

Since $T_{f(x)}R$ may be canonically identified with R, we may consider $df(x)$ as a tangent covector at x. Let $v \in T_xM$ and σ a curve in M such that $\sigma(0) = x$ and $\dot\sigma(0) = v$. Then we have

$$df(x)(v) = df(x)[\sigma] = [f \circ \sigma].$$

On the other hand, we have

$$v(f) = (d(f \circ \sigma)/dt)_{t=0}.$$

But $[f \circ \sigma]$ is the tangent vector of R at $f(x)$ defined by the curve $f \circ \sigma$. If t denotes the coordinate of R, we obtain

$$[f \circ \sigma] = (d(f \circ \sigma)/dt)_{t=0}(\partial/\partial t)_{f(x)}$$

since $a\,(\partial/\partial t)_{f(x)} \in T_{f(x)}R$ is identified with $a \in R$. Hence we deduce that

$$df(x)(v) = v(f).$$

Now, let (U, x^i) be a local system of coordinates at x and consider the 1–forms $(dx^i)(x)$ at $x, 1 \leq i \leq m$. Then we have

$$(dx^i)(x)(\partial/\partial x^j)_x = (\partial/\partial x^j)_x(x^i) = \delta^i_j.$$

Therefore $\{(dx^1)(x), \ldots, (dx^m)(x)\}$ is a basis for $T^*_x M$. In fact, this is the dual basis of $\{(\partial/\partial x^1)_x, \ldots, (\partial/\partial x^m)_x\}$. Next, we compute the components of $df(x)$ with respect to the dual basis $\{(dx^1)(x), \ldots, (dx^m)(x)\}$. We have

$$df(x)((\partial/\partial x^i)_x) = (\partial/\partial x^i)_x f = (\partial f/\partial x^i)_x.$$

Hence we obtain

$$df(x) = \sum_{i=1}^{m} (\partial f/\partial x^i)_x (dx^i)(x).$$

Definition 1.9.1 *A 1–form α on M is a function assigning to each point $x \in M$ a tangent covector $\alpha(x) \in T^*_x M$.*

Let (U, x^i) be a local system of coordinates in $M, \leq i \leq m$. For each $x \in U$ we have

$$\alpha(x) = \sum_{i=1}^{m} \alpha_i(x)(dx^i)(x).$$

Then we have

$$\alpha = \sum_{i=1}^{m} \alpha_i dx^i$$

on U, where $\alpha_i : U \longrightarrow R$. We say that α is C^∞ if the functions α_i are C^∞ for each coordinate neighborhood U.

Now we set

$$T^*M = \bigcup_{x \in M} T^*_x M$$

and define a mapping

$$\pi_M : T^*M \longrightarrow M$$

by

$$\pi_M(\alpha) = x, \quad \alpha \in T_x^*M.$$

We can introduce local coordinates in $\pi_M^{-1}(U) = T^*U$ as follows. If $\alpha \in T_x^*M, x \in U$, we have

$$\alpha = \sum_{i=1}^{m} \alpha_i(x)(dx^i)(x), \quad x = (x^i).$$

Then the coordinates of α are $(x^i, \alpha_i(x))$. These coordinates are usually denoted by (x^i, p_i) and called the **induced coordinates** in T^*M. Thus T^*M is a $2m$–dimensional manifold. Moreover, we can prove that (T^*M, π_M, M) is a vector bundle over M of rank m, called the **cotangent bundle** of M. Obviously, the canonical projection π_M is locally given by

$$\pi_M(x^i, p_i) = (x^i).$$

We note that a 1–form on M may be interpreted as a section of T^*M over M.

1.10 Tensor fields. The tensorial algebra. Riemannian metrics

In this section we study the basic material on tensor fields on manifolds which is used in the succeding sections.

Algebraic preliminaries

Let A and B be finite–dimensional vector spaces over R (or C). We denote by $M(A, B)$ the vector space generated by the Cartesian product $A \times B$, i.e., the free vector space generated by the pairs $(a, b), a \in A, b \in B$. Let $N(A, B)$ be the vector subspace of $M(A, B)$ generated by elements of the form

$$(\alpha a, b) - \alpha(a, b), \quad (a, \alpha b) - \alpha(a, b),$$

$$(a_1 + a_2, b) - (a_1, b) - (a_2, b),$$
$$(a, b_1 + b_2) - (a, b_1) - (a, b_2),$$
$$\alpha \in R \text{ (or } C), \; a, a_1, a_2 \in A, \; b, b_1, b_2 \in B.$$

We set $A \otimes B = M(A, B)/N(A, B)$ and define $\rho : A \times B \longrightarrow A \otimes B$ by $\rho(a, b) = a \otimes b$; $A \otimes B$ is called the **tensor product** of A and B.

Let C be a vector space and $q : A \times B \longrightarrow C$ a bilinear mapping. We say that the couple (C, q) has the **universal factorization property** of $A \times B$ if for every space D and every bilinear mapping $r : A \times B \longrightarrow D$ there exists a unique linear mapping $\phi : C \longrightarrow D$ such that the following diagram

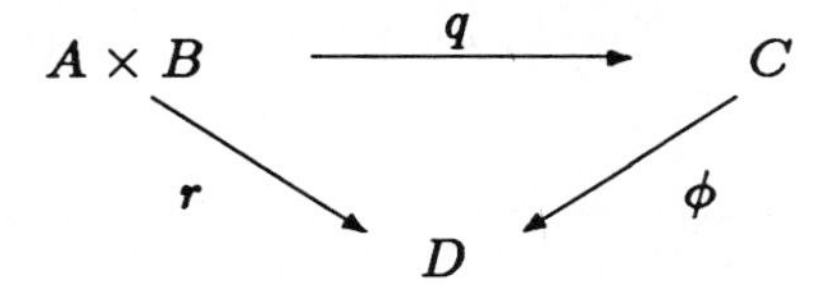

is commutative.

Proposition 1.10.1 *The couple $(A \otimes B, \rho)$ has the universal factorization property of $A \times B$ and the vector space $A \otimes B$ is unique up to an isomorphism.*

Proof: It is clear that $\rho : A \times B \longrightarrow A \otimes B$ is bilinear. Now, let $r : A \times B \longrightarrow D$ be a bilinear mapping. Then r can be uniquely extended to an homomorphism $r' : M(A, B) \longrightarrow D$ such that $r'(N(A, B)) = 0$. Thus r' induces a mapping $\phi : A \otimes B \longrightarrow D$ such that

$$\phi \circ \rho = r.$$

The proof of the uniqueness is left to the reader as an exercise. □

Proposition 1.10.2 *We have canonical isomorphisms:*
(1) $A \otimes B \cong B \otimes A$, $a \otimes b \longrightarrow b \otimes a$;
(2) $(A \otimes B) \otimes C \cong A \otimes (B \otimes C)$, $(a \otimes b) \otimes c \longrightarrow a \otimes (b \otimes c)$;
(3) $R \otimes A \cong A \otimes R \cong A$, $\alpha \otimes a \longrightarrow a \otimes \alpha \longrightarrow a$.

Proof: (1) Let $\varphi : A \times B \longrightarrow B \otimes A$ be the bilinear mapping defined by $\varphi(a, b) = b \otimes a$. From Proposition 1.10.1 we deduce that there exists a unique linear mapping $\psi : A \otimes B \longrightarrow B \otimes A$ such that $\psi(a \otimes b) = b \otimes a$. Similarly, there exists a unique linear mapping $\psi' : B \otimes A \longrightarrow A \otimes B$ such that $\psi'(b \otimes a) = a \otimes b$. Hence $\psi' \circ \psi = id_{A \otimes B}, \psi \circ \psi' = id_{B \otimes A}$. So ψ is the desired isomorphism. The proofs of (2) and (3) are similar and hence omitted. □

From (2) of Proposition 1.10.2, we can writte $A \otimes B \otimes C$ and $a \otimes b \otimes c$. Furthermore, if $A_1, \ldots, A_s$ are vector spaces we may inductively define its tensor product $A_1 \otimes \ldots \otimes A_s$.

Since $M(A,B)/N(A,B)$ is the quotient vector space, we have for any element $(a_1, a_2) \otimes (b_1 + b_2) \in A \otimes B$ the distribution law

$$(a_1 + a_2) \otimes (b_1 + b_2) = a_1 \otimes b_1 + a_1 \otimes b_2 + a_2 \otimes b_1 + a_2 \otimes b_2.$$

Hence, let $\{a_1, \ldots, a_m\}$ be a basis for A and $\{b_1, \ldots, b_n\}$ a basis for B. Then

$$\{a_i \otimes b_j; 1 \leq i \leq m, 1 \leq j \leq n\}$$

is a basis for $A \otimes B$. In fact, they generate $A \otimes B$ and are linearly independent. Hence

$$dim\ (A \otimes B) = (dim\ A)(dim\ B) = mn.$$

Now, let $A^\star$ be the dual vector space of A. For $a \in A$ and $a^\star \in A^\star$ we denote by $< a, a^\star >$ the value of $a^\star : A \longrightarrow R$ on a, i.e.,

$$< a, a^\star >= a^\star(a).$$

Proposition 1.10.3 *Let $Hom(A^\star, B)$ be the vector space of linear mappings of $A^\star$ into B. Then there exists a canonical isomorphism $\psi : A \otimes B \longrightarrow Hom(A^\star, B)$ such that*

$$(\psi(a \otimes b))(a^\star) =< a, a^\star > b.$$

Proof: Consider the bilinear mapping

$$\varphi : A \times B \longrightarrow Hom(A^\star, B)$$

defined by

$$(\varphi(a,b))(a^\star) =< a, a^\star > b.$$

From Proposition 1.10.1, there exists a unique homomorphism $\psi : A \otimes B \longrightarrow Hom(A^\star, B)$ such that

$$(\psi(a \otimes b))(a^\star) = < a, a^\star > b.$$

Finally we prove that ψ is an isomorphism. In fact, let $\{a_1, \ldots, a_m\}$ be a basis for $A, \{a_1^\star, \ldots, a_m^\star\}$ the dual basis for $A^\star$ and $\{b_1, \ldots, b_n\}$ a basis for B. It is sufficient to prove that $\{\psi(a_i \otimes b_j)\}$ are linearly independent. If $\sum \lambda_{ij}\psi(a_i \otimes b_j) = 0$, then

$$0 = (\sum \lambda_{ij}\psi(a_i \otimes b_j))(a_k^\star) = \sum \lambda_{kj} b_j.$$

Hence $\lambda_{kj} = 0$. Since $dim\ (A \otimes B) = dim\ Hom(A^\star, B), \psi$ is an isomorphism. □

The following result can be proved in a similar way.

Proposition 1.10.4 *There exists a unique isomorphism* $\psi : A^\star \otimes B^\star \longrightarrow (A \otimes B)^\star$ *such that*

$$(\psi(a^\star \otimes b^\star))(a \otimes b) = < a, a^\star >< b, b^\star > .$$

Now, let V be an m–dimensional vector space. For a positive integer r, we call

$$T^r V = V \otimes \ldots \otimes V (r - \text{times})$$

the **contravariant tensor space of degree** r. An element of $T^r V$ is called a **contravariant tensor of degree** r. $T^1 V$ is V itself, and $T^0 V$ is defined to be R. Similarly, $T_s V = V^\star \otimes \ldots \otimes V^\star$ (s times) is called the **covariant tensor space of degree** s. An element of $T_s V$ is called a **covariant tensor of degree** s. Then $T_1 V = V^\star$ and $T_0 V$ is defined to be R.

Let $\{e_1, \ldots, e_m\}$ be a basis for V and $\{e^1, \ldots, e^m\}$ the dual basis for $V^\star$. Then

$$\{e_{i_1} \otimes \ldots \otimes e_{i_r}; 1 \leq i_1, \ldots, i_r \leq m\}$$

$$(\text{resp. } \{e^{j_1} \otimes \ldots \otimes e^{i_s}; 1 \leq j_1, \ldots, j_s \leq m\})$$

is a basis for $T^r V$ (resp. $T_s V$). Then, if $K \in T^r V$ (resp. $L \in T_s V$) we have

$$K = K^{i_1 \ldots i_r} e_{i_1} \otimes \ldots \otimes e_{i_r}$$

$$(\text{resp. } L = L_{j_1 \ldots j_s} e^{j_1} \otimes \ldots \otimes e^{j_s}),$$

where $K^{i_1 \ldots i_r}$(resp. $L_{j_1 \ldots j_s}$) are the **components** of K (*resp.* L).

We define the **(mixed) tensor space of type** (r,s), or **tensor space of contravariant degree r and covariant degree** s as the tensor product

$$T_s^r V = T^r V \otimes T_s V = V \otimes \ldots \otimes V \otimes V^* \otimes \ldots \otimes V^*$$

(V r–times and V^* s–times). In particular, we have

$$T_0^r V = T^r V,\ T_s^0 V = T_s V,\ T_0^0 V = T^0 V = T_0 V = R.$$

It is obvious that the set

$$\{e_{i_1} \otimes \ldots \otimes e_{i_r} \otimes e^{j_1} \otimes \ldots \otimes e^{j_s}; 1 \leq i_1, \ldots, i_r, j_1, \ldots, j_s \leq m\}$$

is a basis for $T_s^r V$. Then $dim\, T_s^r V = m^{r+s}$.

An element $K \in T_s^r V$ is called a **tensor of type (r,s)** or **tensor of contravariant degree r and covariant degree s**. We have

$$K = K_{j_1 \ldots j_s}^{i_1 \ldots i_r} e_{i_1} \otimes \ldots \otimes e_{i_r} \otimes e^{j_1} \otimes \ldots \otimes e^{j_s}, \tag{1.2}$$

where $K_{j_1 \ldots j_s}^{i_1 \ldots i_r}$ are the **components** of K. For a change of basis $\bar{e}_i = A_i^j e_j$, we easily obtain

$$\bar{K}_{j_1 \ldots j_s}^{i_1 \ldots i_r} = A_{k_1}^{i_1} \ldots A_{k_r}^{i_r} B_{j_1}^{l_1} \ldots B_{j_s}^{l_s} K_{l_1 \ldots l_s}^{k_1 \ldots k_r}, \tag{1.3}$$

where B is the inverse matrix of A so that

$$A_j^i B_k^j = \delta_k^i.$$

We set

$$TV = \bigoplus_{r,s} T_s^r V.$$

Then an element of TV is of the form $K = \sum_{r,s} K_s^r$, where $K_s^r \in T_s^r V$ are zero except for a finite number of them. If we define the product $K \otimes L \in T_{s+q}^{r+p} V$ of two tensors $K \in T_s^r V$ and $L \in T_q^p V$ as follows:

$$(K \otimes L)^{i_1 \ldots i_{r+p}}_{j_1 \ldots j_{s+q}} = K^{i_1 \ldots i_r}_{j_1 \ldots j_s} L^{i_{r+1} \ldots i_{r+p}}_{j_{s+1} \ldots j_{s+q}}, \tag{1.4}$$

a simple computation from (1.3) shows that (1.4) is independent on the choice of the basis $\{e_i\}$. Then TV becomes an associate algebra over R which is called the **tensor algebra** on V.

In TV we introduce the operation called **contraction**. Let $K \in T^r_s V$ given by (1.2) and (i, j) a pair of integers such that $1 \leq i \leq r, 1 \leq j \leq s$. We define the **contraction operator** C^i_j as follows: $C^i_j K$ is the tensor of type $(r-1, s-1)$ whose components are given by

$$(C^i_j K)^{i_1 \ldots i_{r-1}}_{j_1 \ldots j_{s-1}} = \sum_k K^{i_1 \ldots k \ldots i_{r-1}}_{j_1 \ldots k \ldots j_{s-1}}, \tag{1.5}$$

where the superscript k appears at the i–th position and the subscript k appears at the j–th position. (As above, (1.5) does not depends on the choice of the basis).

Next we shall interpret tensors as multilinear mappings.

Proposition 1.10.5 *$T_s V$ is canonically isomorphic to the vector space of all s–linear mappings of $V \times \ldots \times V$ into R.*

Proof: By generalizing Proposition 1.10.4, we see that $T_s V = V^* \otimes \ldots \otimes V^*$ is the dual vector space of $T^s V = V \otimes \ldots \otimes V$, the isomorphism given by

$$(a^*_1 \otimes \ldots \otimes a^*_s)(b_1 \otimes \ldots \otimes b_s) = < b_1, a^*_1 > \ldots < b_s, a^*_s > .$$

Now, from the universal factorization property, it follows that $(T^s V)^*$ is isomorphic to the space of s–linear mappings of $V \times \ldots \times V$ into R.□

If $K = K_{j_1 \ldots j_s} e^{j_1} \otimes \ldots \otimes e^{j_s} \in T_s V$, then K corresponds to an s–linear mapping of $V \times \ldots \times V$ into R such that

$$K(e_{j_1}, \ldots, e_{j_s}) = K_{j_1 \ldots j_s}.$$

Proposition 1.10.6 *$T^1_s V$ is canonically isomorphic to the vector space of s–linear mappings of $V \times \ldots \times V$ into V.*

Proof: We have $T^1_s V = V \otimes T_s V$. From Proposition 1.10.2, $V \otimes T_s V \cong T_s V \otimes V$. But $T_s V \otimes V \cong Hom((T_s V)^*, V) \cong Hom(T^s V, V)$ by Proposition 1.10.3. By the universal factorization property, $Hom(T^s V, V)$ can be identified with the space of s–linear mappings of $V \times \ldots \times V$ into V.□

If $K = K^i_{j_1 \dots j_s} e_i \otimes e^{j_1} \otimes \dots \otimes e^{j_s} \in T^1_s V$, then K corresponds to an s–linear mapping of $V \times \dots \times V$ into V such that

$$K(e_{j_1}, \dots, e_{j_s}) = K^i_{j_1 \dots j_s} e_i.$$

Now, let $K \in T_s V$ (or $T^1_s V$). We say that K is **symmetric** if for each pair $1 \leq i, j \leq s$, we have

$$K(v_1, \dots, v_i, \dots, v_j, \dots, v_s) = K(v_1, \dots, v_j, \dots, v_i, \dots, v_s).$$

Similarly, if interchanging the i–th and j–th variables, changes the sign:

$$K(v_1, \dots, v_i, \dots, v_j, \dots, v_s) = -K(v_1, \dots, v_j, \dots, v_i, \dots, v_s),$$

then we say that K is **skew–symmetric**. We can easily prove that K is symmetric (resp. skew–symmetric) if

$$K(v_1, \dots, v_s) = K(v_{\sigma(1)}, \dots, v_{\sigma(s)})$$

$$(\text{resp. } K(v_1, \dots, v_s) = \epsilon_\sigma K(v_{\sigma(1)}, \dots, v_{\sigma(s)})),$$

where σ is a permutation of $(1, \dots, s)$ and ϵ_σ denotes the sign of σ.

Tensor fields on manifolds

Definition 1.10.7 *A* **tensor field** K **of type** (r, s) *on a manifold* M *is an assignement of a tensor* $K(x) \in T^r_s(T_x M)$ *to each point* x *of* M.

Let (U, x^i) be a local coordinate system on M. Then a tensor field K of type (r, s) on M may be expressed on U by

$$K = K^{i_1 \dots i_r}_{j_1 \dots j_s} \partial / \partial x^{i_1} \otimes \dots \otimes \partial / \partial x^{i_r} \otimes dx^{j_1} \otimes \dots \otimes dx^{j_s},$$

where $K^{i_1 \dots i_r}_{j_1 \dots j_s}$ are functions on U which will be called the **components** of K with respect to (U, x^i). We say that K is C^∞ if its components are functions of class C^∞ with respect to any local coordinate system. The change of components is given by (1.3), where $A^i_j = (\partial \bar{x}^i / \partial x^j)$ is the Jacobian matrix between two local coordinate systems. From now on, we shall mean by a tensor field that of class C^∞ unless otherwise stated.

From Propositions 1.10.5 and 1.10.6, we can interpret a tensor field K of type $(0, s)$ (resp. $(1, s)$) as a s–linear mapping

$$K : \chi(M) \times \ldots \times \chi(M) \longrightarrow C^\infty(M)$$

$$(\text{resp. } K : \chi(M) \times \ldots \times \chi(M) \longrightarrow \chi(M))$$

defined by

$$(K(X_1, \ldots, X_s))(x) = K(x)(X_1(x), \ldots, X_s(x)).$$

We denote by $\mathcal{T}_s^r(M)$ the vector space of all tensor fields of type (r, s) on M. We note that $\mathcal{T}_s^r(M)$ is a $C^\infty(M)$–module.

Given two tensor fields A and B on M we may construct a new tensor field $[A, B]$ given by

$$[A, B](X, Y) = [AX, BY] + [BX, AY] + AB[X, Y]$$

$$+BA[X, Y] - A[X, BY] - A[BX, Y] - B[X, AY] - B[AX, Y],$$

$$X, Y \in \chi(M).$$

Then $[A, B]$ is a tensor field of type (1,2) satisfying $[A, B] = -[B, A]$. We call $[A, B]$ the **Nijenhuis torsion** of A and B (see Nijenhuis [101]).

Remark 1.10.8 A tensor field K of type $(1, p)$ on a manifold M, $p \geq 1$, is sometimes called a vector p–form on M.

Riemannian metrics

Definition 1.10.9 *A* **Riemannian metric** *on M is a covariant tensor field g of degree 2 which satisfies:*
(1) $g(X, X) \geq 0$ and $g(X, X) = 0$ if and only if $X = 0$, and
(2) g is symmetric, i.e., $g(X, Y) = g(Y, X)$, for all vector fields X, Y on M.

If g is a symmetric covariant tensor field of degree 2 which satisfies

$$(1)' \; g(X, Y) = 0 \text{ for all } Y \text{ implies } X = 0,$$

then g is called a **pseudo–Riemannian metric on** M.

In other words, g assigns an inner product g_x in each tangent space $T_xM, x \in M$. If (U, x^i) is a local coordinate system then the components of g are given by

$$g_{ij} = g(\partial/\partial x^i, \partial/\partial x^j),$$

or equivalently,

$$g = g_{ij} dx^i \otimes dx^j.$$

We shall give an application which illustrates the utility of the partitions of unity.

Proposition 1.10.10 *On any manifold there exists a Riemannian metric.*

Proof: Let M be an m–dimensional manifold and $\{(U_\alpha, \varphi_\alpha)\}$ an atlas on M. There exists a partition of unity $\{(U_i, f_i)\}$ subordinate to $\{U_\alpha\}$. Since each U_i is contained in some U_α, we set $\varphi_i = \varphi_\alpha / U_i$. Then we define a covariant tensor field g_i of degree 2 on M by

$$g_i(x)(X, Y) = \begin{cases} f_i(x) & < d\varphi_i(x)X, \ d\varphi_i(x)Y >, \ \text{if } x \in U_i \\ 0, & \text{if } x \notin U_i \end{cases}$$

for all $X, Y \in T_xM$, where $< \, , >$ in the standard inner product on R^m. Hence $g = \sum_i g_i$ is a Riemannian metric on M. □

1.11 Differential forms. The exterior algebra

Algebraic preliminaries

Let V be an m–dimensional vector space. We denote by $\wedge^p V$ (resp. $S^p V$) the subspace of $T_p V$ which consists of all skew–symmetric (resp. symmetric) covariant tensors on V. Obviously, $\wedge^0 V = S^0 V = R$, $\wedge^1 V = S^1 V = V^*$.

We now define two linear transformations on $T_p V$:

$$\textbf{alternating mapping } A : T_p V \longrightarrow T_p V$$
$$\textbf{symmetrizing mapping } S : T_p V \longrightarrow T_p V$$

as follows:

$$(AK)(v_1,\ldots,v_p) = \frac{1}{p!}\sum_\sigma \epsilon_\sigma\ K(v_{\sigma(1)},\ldots,v_{\sigma(p)}),$$

$$(SK)(v_1,\ldots,v_p) = \frac{1}{p!}\sum_\sigma K(v_{\sigma(1)},\ldots,v_{\sigma(p)}),$$

where the summation is taken over all permutations σ of $(1,2,\ldots,p)$. One can easily check that AK (resp. SK) is skew–symmetric (resp. symmetric) and that K is skew–symmetric (resp. symmetric) if and only if $AK = K$ (resp. $SK = K$).

If $\omega \in \wedge^p V$ and $\tau \in \wedge^q V$, we define the **exterior product** $\omega \wedge \tau \in \wedge^{p+q} V$ by

$$\omega \wedge \tau = \frac{(p+q)!}{p!q!} A(\omega \otimes \tau).$$

Similarly, one can define the **symmetric product** of $\omega \in S^p V$ and $\tau \in S^q V$ as given by

$$\omega \odot \tau = ((p+q)!/p!q!)S(\omega \otimes \tau)).$$

The proofs of the following propositions are left to the reader as an exercise.

Proposition 1.11.1 *We have*

$$(\omega \wedge \tau)(v_1,\ldots,v_{p+q}) = \Sigma'\epsilon_\sigma\ \omega(v_{\sigma(1)},\ldots,v_{\sigma(p)})\ \tau(v_{\sigma(p+1)},\ldots,v_{\sigma(p+q)}),$$

where $\sum'$ *denotes the sum over all* **shuffles**, *i.e., permutations* σ *of* $(1,\ldots,p+q)$ *such that* $\sigma(1) < \ldots < \sigma(p)$ *and* $\sigma(p+1) < \ldots < \sigma(p+q)$.

Proposition 1.11.2 *The exterior product is bilinear and associative, i.e.,*

$$(\lambda_1\omega_1 + \lambda_2\omega_2) \wedge \tau = \lambda_1(\omega_1 \wedge \tau) + \lambda_2(\omega_2 \wedge \tau),$$

$$\omega \wedge (\lambda_1\tau_1 + \lambda_2\tau_2) = \lambda_1(\omega \wedge \tau_1) + \lambda_2(\omega \wedge \tau_2),$$

$$\omega \wedge (\tau \wedge \eta) = (\omega \wedge \tau) \wedge \eta.$$

From Proposition 1.11.2, we can writte $\omega \wedge \tau \wedge \eta$, or, more generally, $\omega_1 \wedge \ldots \wedge \omega_s$.

Let

$$\wedge V = \bigoplus_{p=0} \wedge^p V = R \oplus \wedge^1 V \oplus \wedge^2 V \oplus \ldots$$

Then $\wedge V$ becomes an associate algebra over R, which will be called the **exterior** or **Grassman algebra** over R.

Proposition 1.11.3 *If $\omega \in \wedge^p V$ and $\tau \in \wedge^q V$, then*

$$\omega \wedge \tau = (-1)^{pq}\, \tau \wedge \omega.$$

Proof: This is equivalent to prove that

$$A(\omega \otimes \tau) = (-1)^{pq} A(\tau \otimes \omega).$$

To prove this we note that

$$A(\omega \otimes \tau)(v_1, \ldots, v_{p+q}) = \frac{1}{(p+q)!} \sum_\sigma \epsilon_\sigma \, (\omega \otimes \tau)(v_{\sigma(1)}, \ldots, v_{\sigma(p+q)})$$

$$= \frac{1}{(p+q)!} \sum_\sigma \omega(v_{\sigma(1)}, \ldots, v_{\sigma(p)})\, \tau(v_{\sigma(p+1)}, \ldots, v_{\sigma(p+q)})$$

$$= \frac{1}{(p+q)!} \sum_\sigma \tau(v_{\sigma(p+1)}, \ldots, v_{\sigma(p+q)})\, \omega(v_{\sigma(1)}, \ldots, v_{\sigma(p)}),$$

since $(\omega \otimes \tau)(w_1, \ldots, w_{p+q}) = \omega(w_1, \ldots, w_p)\tau(w_{p+1}, \ldots, w_{p+q})$. Let α be the permutation given by

$$(1, \ldots, p+q) \longrightarrow (p+1, \ldots, p+q, 1, \ldots, p).$$

Then we have

$$A(\omega \otimes \tau)(v_1, \ldots, v_{p+q})$$

$$= \frac{1}{(p+q)!} \sum_\sigma \epsilon_\sigma (\epsilon_\alpha)^2\, \tau(v_{\sigma\alpha(1)}, \ldots, v_{\sigma\alpha(p)})\, \omega(v_{\sigma\alpha(p+1)}, \ldots, v_{\sigma\alpha(p+q)})$$

$$= \epsilon \left(\frac{1}{(p+q)!} \sum_{\sigma\alpha} \epsilon_{\sigma\alpha}\, \tau(v_{\sigma\alpha(1)}, \ldots, v_{\sigma\alpha(p)})\, \omega(v_{\sigma\alpha(p+1)}, \ldots, v_{\sigma\alpha(p+q)}) \right)$$

$$= \epsilon_\alpha\, A(\tau \otimes \omega)(v_1, \ldots, v_{p+q})$$

$$= (-1)^{p+q} A(\tau \otimes \omega)(v_1, \ldots, v_{p+q}),$$

since $\epsilon_\alpha = (-1)^{pq}$ and $\epsilon_{\sigma\alpha} = \epsilon_\sigma \epsilon_\alpha$.□

Next we shall determine a basis for $\wedge V$.

Proposition 1.11.4 *If $p > m$, then $\wedge^p V = 0$. For $0 \leq p \leq m$, dim $\wedge^p V = \binom{m}{p}$. Let $\{e_1, \ldots, e_m\}$ be a basis for V and $\{e^1, \ldots, e^m\}$ the dual basis for V^*. Then the set*

$$\{e^{i_1} \wedge \ldots \wedge e^{i_p} / 1 \leq i_1 < i_2 < \ldots < i_p \leq m\}$$

is a basis for $\wedge^p V$.

Proof: If $p > m$, then $\varphi(e_{i_1}, \ldots, e_{i_p}) = 0$ for any set of basic elements; thus $\wedge^p V = 0$. Suppose that $0 \leq p \leq m$. Since A maps $T_p V$ onto $\wedge^p V$, the image of the basis $\{e^{i_1} \otimes \ldots \otimes e^{i_p}\}$ for $T_p V$ spans $\wedge^p V$. We have

$$A(e^{i_1} \otimes \ldots \otimes e^{i_p}) = \frac{1}{k!} e^{i_1} \wedge \ldots \wedge e^{i_p}.$$

Permuting the order of $i_1, \ldots, i_p$ leaves the right side unchanging, except for a possible change of sign according to Proposition 1.11.3. Then the set $\{e^{i_1} \wedge \ldots \wedge e^{i_p} / 1 \leq i_1 < i_2 < \ldots < i_p \leq m\}$ spans $\wedge^p V$. On the other hand, they are independent. In fact, suppose that some linear combination of them is zero, namely

$$\sum_{i_1 < \ldots < i_p} \lambda_{i_1 \ldots i_p} e^{i_1} \wedge \ldots \wedge e^{i_p} = 0.$$

Hence, if $j_1 < \ldots < j_p$, we have

$$0 = \Big(\sum_{i_1 < \ldots < i_p} \lambda_{i_1 \ldots i_p} e^{i_1} \wedge \ldots \wedge e^{i_p} \Big)(e_{j_1}, \ldots, e_{j_p})$$

$$= \lambda_{j_1 \ldots j_p}$$

Therefore, each coefficient $\lambda_{i_1 \ldots i_p}$ must be zero. This ends the proof. □

From Proposition 1.11.4, we have

$$\wedge V = \bigoplus_{p=0}^{m} \wedge^p V.$$

Then $\wedge V$ is a finite dimensional vector space of dimension $\sum_{p=0}^{m} \binom{m}{p} = 2^m$.

We next examine what happens when we map one vector space to another.

Let $F : V \longrightarrow W$ be a linear mapping. Then it induces a linear mapping

$$F^\star : T_p^\star W \longrightarrow T_p^\star V$$

given by

$$(F^\star \varphi)(v_1, \ldots, v_p) = \varphi(Fv_1, \ldots, Fv_p), \varphi \in T_p V, v_1, \ldots v_p \in V.$$

If $p = 0$, we set $F^\star \lambda = \lambda$, for every $\lambda \in R$. Obviously, if φ is skew–symmetric (resp. symmetric) then $F^\star \varphi$ is skew–symmetric (resp. symmetric). Thus we have two canonical induced homomorphisms

$$F^\star : \wedge^p W \longrightarrow \wedge^p V, \; F^\star : S^p W \longrightarrow S^p V.$$

Proposition 1.11.5 *We have*

$$F^\star(\omega \wedge \tau) = (F^\star \omega) \wedge (F^\star \tau),$$

for every $\omega \in \wedge^p W, \tau \in \wedge^q W$.

Proof: If $\varphi \in \wedge^s W$ and σ is a permutation of $(1, \ldots, s)$, we denote by φ_σ the covariant tensor of degree s given by

$$\varphi_\sigma(w_1, \ldots, w_s) = \varphi(w_{\sigma(1)}, \ldots, w_{\sigma(s)}).$$

With this notation, we have

$$A\varphi = \frac{1}{s!}\sum_{\sigma} \epsilon_{\sigma}\varphi_{\sigma}.$$

Now, let $\omega \in \wedge^p W, \tau \in \wedge^q W$. Then we obtain

$$F^\star(\omega \wedge \tau) = F^\star\left(\frac{(p+q)!}{p!q!}A(\omega \otimes \tau)\right)$$

$$= \frac{1}{p!q!}F^\star(\sum_{\sigma} \epsilon_{\sigma}(\omega \otimes \tau)_{\sigma}) = \frac{1}{p!q!}\sum_{\sigma} \epsilon_{\sigma}F^\star((\omega \otimes \tau)_{\sigma})$$

$$= \frac{1}{p!q!}\sum_{\sigma} \epsilon_{\sigma}F^\star(\omega_{\sigma} \otimes \tau_{\sigma}) \text{ (since } (\omega \otimes \tau)_{\sigma} = \omega_{\sigma} \otimes \tau_{\sigma})$$

$$= \frac{1}{p!q!}\sum_{\sigma} \epsilon_{\sigma}(F^\star\omega_{\sigma}) \otimes (F^\star\tau_{\sigma}) \text{ (since } F^\star(\omega \otimes \tau) = (F^\star\omega \otimes F^\star\tau))$$

$$= \frac{1}{p!q!}\sum_{\sigma} \epsilon_{\sigma}(F^\star\omega)_{\sigma} \otimes (F^\star\tau)_{\sigma} \text{ (since } F^\star(\varphi_{\sigma}) = (F^\star\varphi)_{\sigma})$$

$$= \frac{(p+q)!}{p!q!}A((F^\star\omega) \otimes (F^\star\tau))$$

$$= (F^\star\omega) \wedge (F^\star\tau). \square$$

Corollary 1.11.6 *$F^\star : \wedge W \longrightarrow \wedge V$ is an algebra homomorphism.*

Differential forms

We introduce the following terminology.

Definition 1.11.7 *A skew-symmetric covariant tensor field of degree p on a manifold M is called a* **differential form of degree** *p (or sometimes simply p–***form***).*

The set $\wedge^p M$ of all such forms is a subspace of $\mathcal{T}_p^0(M)$ (in fact, a $C^\infty(M)$–submodule). If $\omega \in \wedge^p M$ and $\tau \in \wedge^q M$, we define the **exterior product** $\omega \wedge \tau \in \wedge^{p+q} M$ by

$$(\omega \wedge \tau)(x) = \omega(x) \wedge \tau(x), \quad x \in M.$$

(It is obvious that $\omega \wedge \tau$ is differentiable).

Hence, if we set $\wedge M = \bigoplus_{p=0}^m \wedge^p M$, where dim $M = m$, we deduce that $\wedge M$ is an associative algebra such that

$$\omega \wedge \tau = (-1)^{pq} \tau \wedge \omega, \; \omega \in \wedge^p M, \; \tau \in \wedge^q M;$$

$\wedge M$ is called the **algebra of differential forms** or **exterior algebra** on M. We also have

$$(f\omega) \wedge \tau = f(\omega \wedge \tau) = \omega \wedge (f\tau), \text{ for every } f \in C^\infty(M).$$

If (U, x^i) is a local coordinate system, then $\{dx^{i_1} \wedge \ldots \wedge dx^{i_p} / 1 \leq i_1 < i_2 < \ldots < i_p \leq m\}$ is a basis of $\wedge^p U$. Therefore, a p–form ω on M can be locally expressed by

$$\omega = \sum_{i_1 < \ldots < i_p} \omega_{i_1 \ldots i_p} dx^{i_1} \wedge \ldots \wedge dx^{i_p}.$$

Now, let $F : M \longrightarrow N$ be a C^∞ mapping. Then F induces an algebra homomorphism

$$F^* : \wedge N \longrightarrow \wedge M$$

given by

$$(F^*\omega)(x) = (dF(x))^*(\omega(F(x))), \quad x \in M, \; \omega \in \wedge^p N.$$

Definition 1.11.8 *Let ω be a p–form on M. The* **support** *of ω is the closure of the set on which ω does not vanishes, i.e.,*

$$supp\,(\omega) = cl\{x \in M / \omega(x) \neq 0\}.$$

We say that ω has **compact support** *is $supp\,(\omega)$ is compact in M.*

Derivations of $\wedge M$

Let $\wedge M$ be the algebra of differential forms on M.

Definition 1.11.9 *A* **derivation** *(resp.* **skew–derivation***) of $\wedge M$ is a linear mapping $D : \wedge M \longrightarrow \wedge M$ which satisfies.*

$$D(\omega \wedge \tau) = D\omega \wedge \tau + \omega \wedge D\tau \text{ for } \omega, \tau \in \wedge M$$

$$(\text{resp. } D(\omega \wedge \tau) = D\omega \wedge \tau + (-1)^p \omega \wedge D\tau \text{ for } \omega \in \wedge^p M,\ \tau \in \wedge M).$$

A derivation or a skew–derivation D of $\wedge M$ is said to be **of degree** k *if it maps $\wedge^p M$ into $\wedge^{p+k} M$ for every k.*

Proposition 1.11.10 *(1) If D and D' are derivations of degree k and k' respectively, then the bracket product defined by $[D, D'] = DD' - D'D$ is a derivation of degree $k + k'$.*
(2) If D is a derivation of degree k and D' is a skew–derivation of degree k', then the bracket product defined by $[D, D'] = DD' - D'D$ is a skew–derivation of degree $k + k'$.
(3) If D and D' are skew–derivations of degree k and k' respectively, then the bracket product defined by $[D, D'] = DD' + D'D$ is a derivation of degree $k + k'$.

Proposition 1.11.11 *A derivation or a skew–derivation D is completely determined by its action on $C^\infty(M)$ and $\wedge^1 M$.*

Proof: We prove the result for derivations of $\wedge M$. The proof for skew–derivations of $\wedge M$ is similar and it is left to the reader as an exercise.

First, we notice that a derivation D can be localized, i.e., if $\omega \in \wedge M$ vanishes on an open set U, then $D\omega$ vanishes on U. In fact, for each $x \in U$, let f be a function such that $f(x) = 0$ and $f = 1$ outside U. Hence $\omega = f\omega$ and we have

$$D\omega = D(f\omega) = (Df)\omega + f(D\omega)$$

Since $f(x) = 0$ and $\omega(x) = 0$, we deduce that $(D\omega)(x) = 0$. Then, if ω and τ coincide on U, then $D\omega/_U = D\tau/_U$.

Now let D_1 and D_2 be two derivations of $\wedge M$ and set $D = D_1 - D_2$. Suppose that D is zero on $C^\infty(M)$ and $\wedge^1 M$. Let be $\omega \in \wedge^p M$ and x an arbitrary point of M. Choose a coordinate neighborhood V of x with local coordinates $x^1, \ldots, x^m$. Hence

$$\omega = \sum_{i_1<\ldots<i_p} \omega_{i_1\ldots i_p} dx^{i_1} \wedge \ldots \wedge dx^{i_p} \tag{1.6}$$

on V. We may extend $\omega_{i_1\ldots i_p}$ and dx^i to M and assume that (1.6) holds in a smaller neighborhood U of x. Therefore we have

$$\begin{aligned} D\omega &= \sum_{i_1<\ldots<i_p} D(\omega_{i_1\ldots i_p} dx^{i_1} \wedge \ldots \wedge dx^{i_p}) \\ &= \sum_{i_1<\ldots<i_p} (D\omega_{i_1\ldots i_p}) dx^{i_1} \wedge \ldots \wedge dx^{i_p} \\ &+ \sum_{i_1<\ldots<i_p} \omega_{i_1\ldots i_p} (Ddx^{i_1}) \wedge dx^{i_2} \wedge \ldots \wedge dx^{i_p} \\ &+ \ldots + \sum_{i_1<\ldots<i_p} \omega_{i_1\ldots i_p} dx^{i_1} \wedge \ldots \wedge (Ddx^{i_p}) = 0. \square \end{aligned}$$

Remark 1.11.12 There is another approach to the theory of derivations due to Frölicher and Nijenhuis [56]. They define a **derivation** of $\wedge M$ of degree r as a linear mapping $D : \wedge M \longrightarrow \wedge M$ such that:
(1) degree $(D\omega) =$ degree $\omega + r$
(2) $D(\omega \wedge \tau) = D\omega \wedge \tau + (-1)^{pr} \omega \wedge D\tau$, where $\omega \in \wedge^p M$.

The commutator of two derivation D and D' of degree r and r' is defined by

$$[D, D'] = DD' - (-1)^{rr'} D'D$$

and it is a new derivation of degree $r + r'$.

A derivation D is said to be of type i_* (resp. d_*) if it acts trivially on $\wedge^o M$ (resp. $[D, d] = 0$). Frölicher and Nijenhuis prove that a derivation D of $\wedge M$ can be decomposed in a unique way as a sum of a derivation of type i_* and a derivation of type d_*.

1.12 Exterior differentiation

Let $f \in C^\infty(M)$ be a differentiable function on M. Define the **exterior derivative** df of f by

$$df : x \in M \longrightarrow df(x) \in T_x^* M.$$

Then df is a 1–form on M locally given by

$$df = (\partial f / \partial x^i) dx^i.$$

Now we shall extend the operator d to an arbitrary p–form.

Theorem 1.12.1 *There exists a unique linear operator $d : \wedge M \longrightarrow \wedge M$ such that*
(1) $d : \wedge^p M \longrightarrow \wedge^{p+1} M$,
(2) $d^2 = 0$,
(3) df is the differential of f, for each $f \in C^\infty(M)$,
(4) if $\omega \in \wedge^p M$, then

$$d(\omega \wedge \tau) = d\omega \wedge \tau + (-1)^p \omega \wedge d\tau.$$

Proof: EXISTENCE. Let $x \in M$. Consider a local coordinate system (x^i) on a neighborhood U of x. If $\omega \in \wedge^p M$, then

$$\omega / U = \omega_{i_1 \dots i_p} dx^{i_1} \wedge \dots \wedge dx^{i_p},$$

where the summation is over $1 \leq i_1 < i_2 < \dots < i_p \leq m$. We define $d\omega$ at x by setting

$$d\omega(x) = (d\omega_{i_1 \dots i_p})(x) \wedge dx^{i_1}(x) \wedge \dots \wedge dx^{i_p}(x).$$

Next we show that the definition of $d\omega(x)$ is independent of the choice of coordinates. First we prove the following properties:
(a) $d\omega(x) \in \wedge^{p+1}(T_x^* M)$.
(b) $d\omega(x)$ depends only on the germ of ω at x.
(c) If $\omega, \tau \in \wedge^p M$ and $a, b \in R$, then

$$d(a\omega + b\tau)(x) = a(d\omega)(x) + b(d\tau)(x).$$

(d) If $\omega \in \wedge^p M, \tau \in \wedge^q M$, then

$$d(\omega \wedge \tau)(x) = (d\omega)(x) \wedge \tau(x) + (-1)^p \omega(x) \wedge (d\tau)(x).$$

From the linearity of d, to check (d) it is enough to consider forms

$$\omega = a\, dx^{i_1} \wedge \ldots \wedge dx^{i_p} \text{ and } \tau = b\, dx^{j_1} \wedge \ldots \wedge dx^{j_q}.$$

Then we have

$$d(\omega \wedge \tau)(x) = d((ab)(x)((dx^{i_1} \wedge \ldots \wedge dx^{i_p}) \wedge (dx^{j_1} \wedge \ldots \wedge dx^{j_p}))(x)$$

$$= [(da)(x)b(x) + a(x)(db)(x)]((dx^{i_1} \wedge \ldots \wedge dx^{i_p}) \wedge (dx^{j_1} \wedge \ldots \wedge dx^{j_q}))(x)$$

$$= ((da)(x) \wedge (dx^{i_1} \wedge \ldots \wedge dx^{i_p})(x)) \wedge (b(x)(dx^{j_1} \wedge \ldots \wedge dx^{j_q})(x))$$

$$+(-1)^p((a(x)(dx^{i_1} \wedge \ldots \wedge dx^{i_p})(x)) \wedge ((db(x) \wedge (dx^{j_1} \wedge \ldots \wedge dx^{j_q})(x)),$$

which completes the proof. The $(-1)^p$ is due to the fact that

$$db(x) \wedge (dx^{i_1})(x) \wedge \ldots \wedge (dx^{i_p})(x)$$

$$= (-1)^p (dx^{i_1})(x) \wedge \ldots \wedge (dx^{i_p})(x) \wedge db(x).$$

(e) If $f \in C^\infty(M)$, then

$$d(df)(x) = d(\partial f/\partial x^i)(x) \wedge (dx^i)(x)$$

$$= (\partial^2 f/\partial x^j \partial x^i)(x)(dx^j)(x) \wedge (dx^i)(x) = 0,$$

since $(\partial^2 f/\partial x^j \partial x^i)(x) = (\partial^2 f/\partial x^i \partial x^j)(x)$ and $(dx^j)(x) \wedge (dx^i)(x) = -(dx^i)(x) \wedge (dx^j)(x)$.

Now, we prove the independence of the choice of coordinates. In fact, let (x'^i) be another local coordinate system on a neighborhood U' of x. Then, since d' satisfy (a)–(e), we have

$$(d'\omega)(x) = d'(\omega_{i_1 \ldots i_p} dx^{i_1} \wedge \ldots \wedge dx^{i_p})(x)$$

$$= (d'\omega_{i_1\dots i_p})(x) \wedge (dx^{i_1})(x) \wedge \dots \wedge (dx^{i_p})(x)$$

$$+\sum_{r=1}^{p}(-1)^r \omega_{i_1\dots i_p}(x)(dx^{i_1})(x) \wedge \dots \wedge d'(dx^{i_r})(x) \wedge \dots \wedge (dx^{i_p})(x)$$

$$= (d\omega_{i_1\dots i_p})(x) \wedge (dx^{i_1})(x) \wedge \dots \wedge (dx^{i_p})(x)$$

$$= d\omega(x).$$

Now, if $\omega \in \wedge^p M$, we define $d\omega \in \wedge^{p+1} M$ by

$$d\omega : x \in M \longrightarrow d\omega(x) \in \wedge^{p+1}(T_x^* M).$$

It follows that d satisfy (1)–(4).

UNIQUENESS. Let d' be another linear operator on $\wedge M$ satisfying (1)–(4). First, we prove that if $\omega \in \wedge^p M$ vanishes on a neighborhood U of x, then $d'\omega(x) = 0$. To see this we choose a function f on M which is 0 on a neighborhood of x and 1 on a neighborhood of $M - U$. Then $f\omega = \omega$ and

$$d'\omega(x) = d'(f\omega)(x) = (d'f)(x) \wedge \omega(x) + f(x)(d'(\omega)(x) = 0.$$

Now let (x^i) be a local coordinate system on a coordinate neighborhood U. We define a new linear operator d' on $\wedge U$ as follows. If $x \in M$, let f be a function on M which is 1 on a neighborhood $V \subset U$ of x and has support in U. If $\omega \in \wedge^p U$, then $f\omega$ is a p–form on M which is 0 outside of U and it agrees with ω on V. We define $d'\omega$ by

$$d'\omega(x) = d'(f\omega)(x).$$

From the above remark $d'\omega$ is independent of the choice of f. It is easy to prove that d' verifies (a)–(e). Consequently we have

$$d'\omega(x) = d\omega(x),$$

for each $\omega \in \wedge^p M$ and $x \in M$. So $d' = d$.□

Remark 1.12.2 It is clear from the above proof that

$$(d\omega)/U = d(\omega/U),$$

for all open set U in M.

From Theorem 1.12.1 we deduce that the exterior differentiation d is a skew–derivation of degree 1 of $\wedge M$ and a derivation of degree 1 and type d_* in the sense of Frölicher and Nijenhuis.

The following result will be very useful in the sequel.

Proposition 1.12.3 *If ω is a p–form on M, then*

$$(d\omega)(X_0, X_1, \ldots, X_p) = \sum_{i=0}^{p}(-1)^i X_i(\omega(X_0, \ldots, \check{X}_i, \ldots, X_p))$$

$$+ \sum_{0 \leq i < j \leq p} (-1)^{i+j}\omega([X_i, X_j], X_0, \ldots, \check{X}_i, \ldots, \check{X}_j, \ldots, X_p),$$

where the symbol ˇ *means that the term is omitted. Particularly, if ω is a 1–form, then*

$$(d\omega)(X, Y) = X(\omega(Y)) - Y(\omega(X)) - \omega([X, Y]),$$

$X, Y \in \chi(M)$.

We leave the proof as an exercise.

Now, let $F : M \longrightarrow N$ be a differentiable map and $F^\star : \wedge N \longrightarrow \wedge M$ the induced algebra homomorphism. Then we have.

Proposition 1.12.4 *$F^\star$ and d commute, that is $F^\star d = dF^\star$.*

Proof: For forms f of degree 0, that is, functions, the results holds:

$$F^\star(df) = d(F^\star f).$$

In fact, if $v \in T_xM, x \in M$, we have

$$F^\star(df)(v) = df(dF(x)(v)) = dF(x)(v)(f) = v(f \circ F) = v(F^\star(f))$$

$$= d(F^\star(f))(x)(v) = d(F^\star(f))(v).$$

Next, we shall prove the result for an arbitrary p–form. Suppose that the proposition is true for all forms of degree less than p. It is enough to prove it for p–forms locally given by

$$\omega = a\, dx^{i_1} \wedge \ldots \wedge dx^{i_p}.$$

Then we have

$$d(F^\star\omega) = d(F^\star((a\, dx^{i_1}) \wedge (dx^{i_2} \wedge \ldots \wedge dx^{i_p})))$$

$$= d((F^\star(a\, dx^{i_1}) \wedge F^\star(dx^{i_2} \wedge \ldots \wedge dx^{i_p}))$$

$$= d(F^\star(a\, dx^{i_1})) \wedge F^\star(dx^{i_2} \wedge \ldots \wedge dx^{i_p})$$

$$-(F^\star(a\, dx^{i_1}) \wedge d(F^\star(dx^{i_2} \wedge \ldots \wedge dx^{i_p}))$$

$$= F^\star(d(a\, dx^{i_1})) \wedge F^\star(dx^{i_2} \wedge \ldots \wedge dx^{i_p})$$

$$-F^\star(a\, dx^{i_1}) \wedge F^\star(d(dx^{i_2} \wedge \ldots \wedge dx^{i_p}))$$

$$= F^\star(d(a\, dx^{i_1}) \wedge \ldots \wedge dx^{i_p})) = F^\star(d\omega),$$

since $ddx^{i_r} = 0$, $1 \leq r \leq m$.□

1.13 Interior product

For each vector field X on a manifold M, we define the **interior product** $i_X\omega$ of a p–form ω by X as follows:
(1) $i_X\omega = 0$, when $p = 0$;
(2) $i_X\omega = \omega(X)$, when $p = 1$; and
(3) $(i_X\omega)(Y_1, \ldots, Y_{p-1}) = \omega(X, Y_1, \ldots, Y_{p-1})$, $Y_1, \ldots, Y_{p-1} \in \chi(M)$. Then $i_X\omega \in \wedge^{p-1}M$.

In a similar way we can define the interior product $i_X G$ of a tensor field G of type $(0, p)$, $p \geq 0$.

The proof of the following proposition is left to the reader as an exercise.

Proposition 1.13.1 *We have*
(1) $(i_X)^2 = 0$; *and*
(2) $i_X(\omega \wedge \tau) = (i_X\omega) \wedge \tau + (-1)^p \omega \wedge (i_X\tau)$, $\omega \in \wedge^p M$.

From Proposition 1.13.1 we deduce that i_X is a skew–derivation of degree -1 of $\wedge M$ and a derivation of degree -1 and type i_* in the sense of Frölicher and Nijenhuis.

Now, let F be a tensor field of type (1,1) on M. We define the **interior product** $i_F\omega$ of a p–form ω by F as follows:
(1) $i_F\omega = 0$, when $p = 0$;
(2) $(i_F\omega)(Y_1, \ldots, Y_p) = \sum_{i=1}^p \omega(Y_1, \ldots, FY_i, \ldots, Y_p)$, for all $Y_1, \ldots, Y_p \in \chi(M)$.

Then $i_F\omega \in \wedge^p M$. We can easily check that i_F is a derivation of degree 0 of $\wedge M$ and a derivation of degree 0 and type i_* in the sense of Frölicher and Nijenhuis.

1.14 The Lie derivative

Let X be a vector field on M and ϕ_t a local 1–parameter group of local transformations generated by X. We shall define the **Lie derivative** $L_X\omega$ **of a** p**–form** ω **with respect to** X as follows. For the sake of simplicity, we suppose that ϕ_t is a global 1– parameter group of transformations of M. Hence, for each $t \in R$, $\phi_t^* : \wedge M \longrightarrow \wedge M$ is an automorphism of the exterior algebra $\wedge M$. Then we set

$$(L_X\omega)(x) = \lim_{t \longrightarrow 0} (1/t)[\omega(x) - (\phi_{-t}^*\omega)(x)]$$

and $L_X\omega$ is a p–form on M.

Proposition 1.14.1 *Let X be a vector field on M. Then*
(1) $L_X f = Xf$, *for every function f;*
(2) if $\omega \in \wedge^p M, \tau \in \wedge^q M$, *then*

$$L_X(\omega \wedge \tau) = L_X\omega \wedge \tau + \omega \wedge L_X\tau;$$

(3) L_X *commutes with d, that is,*

$$L_X d = dL_X.$$

Proof: (1) in fact,

$$(L_X f)(x) = \lim_{t \longrightarrow 0} (1/t)[f(x) - f(\phi_{-t}(x))]$$

$$= - \lim_{t \longrightarrow 0} (1/t)[f(\phi_{-t}(x) - f(x)].$$

But $\phi_{-t} = \phi_t^{-1}$ is a local 1–parameter group of local transformations generated by $-X$. Hence

$$L_X f = -((-X)f) = Xf.$$

(2) Let $\omega \in \wedge^p M, \tau \in \wedge^q M$. Then we have

$$(L_X(\omega \wedge \tau))(x) = \lim_{t \longrightarrow 0} (1/t)[(\omega \wedge \tau)(x) - (\phi^\star_{-t}(\omega \wedge \tau))(x)]$$

$$= \lim_{t \longrightarrow 0} (1/t)[\omega(x) \wedge \tau(x) - (\phi^\star_{-t}\omega)(x) \wedge (\phi^\star_{-t}\tau)(x)]$$

$$= \lim_{t \longrightarrow 0} (1/t)[\omega(x) \wedge \tau(x) - (\phi^\star_{-t}\omega)(x) \wedge \tau(x) + (\phi^\star_{-t}\omega) \wedge \tau(x)$$

$$-(\phi^\star_{-t}\omega)(x) \wedge (\phi^\star_{-t}\tau)(x)]$$

$$= (\lim_{t \longrightarrow 0} (1/t)[\omega(x) - (\phi^\star_{-t}\omega)(x)]) \wedge \tau(x)$$

$$+(\lim_{t \longrightarrow 0} (1/t)((\phi^\star_{-t}\omega)(x))) \wedge \lim_{t \longrightarrow 0} (1/t)[\tau(x) - (\phi^\star_{-t}\tau)(x)]$$

$$= L_X\omega \wedge \tau + \omega \wedge L_X\tau.$$

(3) From (2) it is enough to check (3) when applied to functions. Let $f \in C^\infty(M)$. If $v \in T_xM, x \in M$, we have

$$d(L_X f)(x)(v) = v(L_X f) = v(Xf) = v(\lim_{t \longrightarrow 0} (1/t)[f - f \circ \phi_{-t}]),$$

where $f \circ \phi_{-t}$ can be considered as a differentiable function on $R \times M$. On the other hand, we have

$$(L_X(df))(x)(v) = (\lim_{t \longrightarrow 0}(1/t)[(df)(x) - (\phi^{\star}_{-t}(df))(x)])(v)$$

$$= \lim_{t \longrightarrow 0}(1/t)[(df(x)(v) - df(\phi_{-t}(x))((d\phi_{-t}(x)(v))]$$

$$= \lim_{t \longrightarrow 0}(1/t)[vf - v(f \circ \phi_{-t})].$$

If we extend v to a vector field Y on M, then we obtain two vector fields d/dt and Y on $R \times M$ such that $[d/dt, Y] = 0$. This fact implies (3).□

From Proposition 1.14.1, we deduce that L_X is a derivation of degree 0 of $\wedge M$ and a derivation of degree 0 and type d_* in the sense of Frölicher and Nijenhuis.

As relation among d, L_X, and i_X, we have the following (sometimes called **H. Cartan formula**):

Proposition 1.14.2 $L_X = i_X d + d i_X$.

Proof: First, observe that both L_X and $i_X d + d i_X$ are derivations of $\wedge M$ and commute with d because $d^2 = 0$. Hence we need only to check the identity when both sides act on functions. But, for every function f on M, we have

$$(i_X d + d i_X) f = i_X(df) = df(X) = Xf = L_X f.$$

This ends the proof.□

Proposition 1.14.3 *We have*

$$(L_X\omega)(X_1, \ldots, X_p) = X(\omega(X_1, \ldots, X_p))$$

$$- \sum_{i=0}^{p} \omega(X_1, \ldots, [X, X_i], \ldots, X_p),$$

$\omega \in \wedge^p M,\ X, X_1, \ldots, X_p \in \chi(M)$.

Proof: The result follows directly from Propositions 1.12.3 and 1.14.2. □

Proposition 1.14.4 *We have*

$$[L_X, i_Y] = i_{[X,Y]}$$

The proof is left to the reader as an exercise.

Remark 1.14.5 In a similar way we may define the Lie derivative $L_X F$ of a tensor field F of type (r,s) with respect to a vector field X on M. For instance, if F is a tensor field of type (1,1) then $L_X F$ is the tensor field of same type given by

$$(L_X F)(Y) = [X, FY] - F[X,Y].$$

If G is a tensor field of type (0,2), then $L_X G$ is a tensor field of same type given by

$$(L_X G)(Y, Z) = X\, G(Y, Z) - G([X, Y], Z) - G(Y, [X, Z]).$$

1.15 Distributions. Frobenius theorem

The concept of vector fields on a manifold can be used to give an intrinsical treatment of certain first–order linear partial differential equations. First, we introduce some definitions.

Definition 1.15.1 *Let M be a manifold of dimension m. A k–dimensional* **distribution** *D on M is a choice of a k-dimensional subspace $D(x)$ of $T_x M$ for each x in M. D is said to be C^∞ if for each $x \in M$ there is a neighborhood U of x and there are k linearly independent C^∞ vector fields $X_1, \ldots, X_k$ on U which form a basis of $D(y)$ for every $y \in U$. Then we say that $X_1, \ldots, X_k$ is a* **local basis** *of D. A vector field X on M is said to* **belong to** *D if $X(x) \in D(x)$ for each $x \in M$. D is said to be* **involutive** *if $[X, Y] \in D$ for every vector fields X, Y belonging to D.*

Definition 1.15.2 *Let $\phi : N \longrightarrow M$ an immersed submanifold of M. Then N is an* **integral manifold** *of D if*

$$d\phi(y)(T_y N) = D(\phi(y)) \; for \; each \; y \in N.$$

D is said to be **completely integrable** *if there exists an integral manifold of D through each point of M.*

Lemma 1.15.3 *Let D be a completely integrable k–dimensional distribution on a manifold M of dimension m. Then for each $x \in M$ there exists a local coordinate system (x^i) on a neighborhood U of x such that $x^i(x) = 0, 1 \leq i \leq m$ and the submanifold x^i = constant, $k+1 \leq i \leq m$, are integral manifolds of D.*

Proof: Since each immersed submanifold is locally an embedded submanifold the result follows directly from the definitions. □

Theorem 1.15.4 (Frobenius Theorem) *A distribution D on a manifold M is completely integrable if and only if it is involutive.*

Proof: We shall prove that a completely integrable distribution is involutive. In fact, let X, Y be vector fields belonging to D. We must prove that $[X,Y](x) \in D(x)$ for each $x \in M$. Let $\phi : N \longrightarrow M$ be an integral manifold of D trough x, and suppose that $\phi(y) = x$. Since $d\phi(y') : T_{y'} N \longrightarrow D(\phi(y'))$ is a linear isomorphism for each $y' \in N$, then there exist vector fields X', Y' on N such that X', X (resp. Y', Y) are ϕ–related. Therefore $[X', Y'], [X, Y]$ are ϕ–related (see Exercise 1.). Consequently, we have

$$d\phi(y)[X',Y'](y) = [X,Y](x) \in D(x).$$

Hence $[X,Y] \in D$.

In order to prove the converse we proceed by induction on the dimension of D.

Suppose that dim $D = k$ and dim $M = m$. When $k = 1$ then D is a field of line elements and a local basis is given by a nonvanishing vector field X which belongs to D at any point. Therefore there exists a local coordinate system (x^i) such that

$$X = \partial/\partial x^1$$

(See Exercise 1.22.9). Hence D is completely integrable. Furthermore, an integral curve of X is an integral manifold of D.

Now assume that the theorem holds for involutive distributions of dimensions $1, 2, \ldots, k-1$; we prove it for an involutive distribution of dimension k. Since D is smooth, there exists a local basis $X_1, \ldots, X_k$ of D on a neighborhood V' of x. Furthermore, we can find a local coordinate system (y^i) on a neighborhood $V \subset V'$ of x such that $y^i(x) = 0, 1 \leq i \leq m$ and

$$X_1 = \partial/\partial y^1.$$

Since D is involutive we have

$$[X_i, X_j] = \sum_{l=1}^{k} C_{ij}^l X_l, \; 1 \leq i, j \leq k.$$

Define a new local basis of D on V by

$$Y_1 = X_1,$$

$$Y_i = X_i - X_i(y^1)X_1, \; 2 \leq i \leq k.$$

Since $Y_2, \ldots, Y_k$ are tangent to the manifolds $y^1 =$ constant then it follows that $[Y_i, Y_j]$, $2 \leq i, j \leq k$, must be tangent to the submanifolds $y^1 =$ constant also. Therefore the distribution on V defined by $Y_2, \ldots, Y_k$ is involutive on V and on each submanifold $y^1 =$ constant of V including $S \subset V$ defined by $y^1 = 0$. Let

$$Z_i = Y_i/s, \; 2 \leq i \leq k.$$

Then $Z_2, \ldots, Z_k$ is a local basis of a $C^\infty (k-1)$- dimensional involutive distribution on S. By the induction hypothesis, there exists a coordinate system $z^2, \ldots, z^m$ on some neighborhood of x in S such that

$$z^i(x) = 0, \; 2 \leq i \leq m,$$

and the submanifolds defined by $z^i =$ constant, $k+1 \leq i \leq m$, are precisely the integral manifolds of the distribution defined by $Z_2, \ldots, Z_k$ on this neighborhood.

Now, let $p : V \longrightarrow S$ be the canonical projection in the y coordinate system. Then the functions

$$x^1 = y^1,$$

$$x^i = z^i \circ p, \; 2 \leq i \leq m,$$

are defined on some neighborhood of x in M and are independent at x. Furthermore we have $x^i(x) = 0, 1 \leq i \leq m$. Hence there exists a neighborhood U of x in M with the coordinate functions $x^1, \ldots, x^m$, on U. Now we prove that

$$Y_i(x^{k+j}) = 0 \text{ on } U, \; 1 \leq i \leq k, 1 \leq j \leq m-k.$$

In fact, we have

$$\partial x^j/\partial y^1 = \begin{cases} 1, \ if \ j = 1, \\ 0, \ if \ 2 \leq j \leq m \end{cases}$$

on U. Consequently, we deduce that

$$Y_1 = \partial/\partial x^1 \text{ on } U.$$

Hence $Y_1(x^{k+j}) = 0, \ 1 \leq j \leq m-k$. Now when $2 \leq i \leq k, 1 \leq j \leq m-k$, we have

$$(\partial/\partial x^1)(Y_i(x^{k+j})) = Y_1(Y_i(X^{k+j})) = [Y_1, Y_i](x^{k+j}).$$

But since D is involutive we know that

$$[Y_1, Y_i] = \sum_{l=2}^{k} c_{1i}^l Y_l.$$

Thus we deduce that

$$(\partial/\partial x^1)(Y_i(x^{k+j})) = \sum_{l=2}^{k} c_{1i}^l Y_l(x^{k+j}). \tag{1.7}$$

Let us consider the submanifold N of U defined by x^2 = constant, ..., x^m = constant. On N, (1.7) becomes a system of $k-1$ homogeneous linear differential equations with respect to x^1. The functions $Y_i(x^{k+j})$ are solutions of (1.7) satisfying initial conditions $Y_i(x^{k+j}) = Z_i(z^{k+j}) = 0$, when $x^1 = 0$. By the uniqueness of solutions we deduce that $Y_i(x^{k+j})$ must be identically zero on U. It follows that $Y_2, \ldots, Y_k$ are linear combination of the vector fields $\partial/\partial x^2, \ldots, \partial/\partial x^k$. Then $\partial/\partial x^1, \ldots, \partial/\partial x^k$ is a local basis of D and the submanifolds x^i = constant, $k+1 \leq i \leq m$, are integral manifolds of D.□

A k–dimensional distribution D on a manifold M of dimension m is locally defined by k linearly independent vector fields. Alternatively, we may suppose D is locally defined by $m-k$ linearly independent 1–forms $\omega^{k+1}, \ldots, \omega^m$.

Proposition 1.15.5 *The distribution D is involutive if and only if for each $x \in M$ there exist a neighborhood U and m-k linearly independent 1–forms $\omega^{k+1}, \ldots, \omega^m$ on U which vanish on D and satisfy the condition*

$$d\omega^a = \sum_{b=k+1}^{m} \omega_b^a \wedge \omega^b, \quad k+1 \le a \le m,$$

for suitable 1-forms ω_b^a.

Proof: First, we note that, in a neighborhood U of each point x, a local basis $X_1, \dots, X_k$ of D may be completed to a local field of frames $X_1, \dots, X_k$, $X_{k+1}, \dots, X_m$ on M. If $\{\omega^1, \dots \omega^k, \omega^{k+1}, \dots, \omega^m\}$ is the dual field of coframes, then $\omega^{k+1}, \dots, \omega^m$ vanish on $X_1, \dots, X_k$ and hence on D. We have

$$[X_r, X_s] = \sum_{a=1}^{m} C_{rs}^a X_a, \; 1 \le r, s \le m.$$

We know that D is involutive if and only if $C_{rs}^a = 0$ for $1 \le r, s \le k$ and $k+1 \le a \le m$. Now, since $\omega^i(X_j)$ is constant, we deduce that

$$d\omega^i(X_r, X_s) = -\omega^i[X_r, X_s] = -C_{rs}^i, \; 1 \le i, r, s \le m.$$

On the other hand, $d\omega^i$ may be uniquely expressed by

$$d\omega^i = \frac{1}{2} \sum_{p,q} d_{pq}^i \, \omega^p \wedge \omega^q,$$

where $d_{pq}^i = -d_{qp}^i$. Therefore

$$d\omega^i(X_r, X_s) = \frac{1}{2}(d_{rs}^i - d_{sr}^i) = d_{rs}^i.$$

Hence $d_{rs}^i = -C_{rs}^i$. Thus, if D is involutive, we have

$$d\omega^a = \frac{1}{2} \sum_{1 \le p,q \le m} d_{pq}^a \, \omega^p \wedge \omega^q$$

$$= \sum_{q=k+1}^{m} \left\{ \sum_{p=1}^{k} d_{pq}^a \, \omega^p + \frac{1}{2} \sum_{p=k+1}^{m} d_{pq}^a \, \omega^p \right\} \wedge \omega^q,$$

for $k+1 \le a \le m$. If we set

$$\omega_q^a = \sum_{p=1}^{k} d_{pq}^a \, \omega^p + \frac{1}{2} \sum_{p=k+1}^{m} d_{pq}^a \, \omega^p,$$

we deduce that

$$d\omega^a = \sum_{q=k+1}^{m} \omega_q^a \wedge \omega^q.$$

The converse is proved by a similar procedure. □

We now introduce the concept of an ideal of $\wedge M$.

Definition 1.15.6 *An* **ideal** $\mathcal{Y}$ *of* $\wedge M$ *is a vector subspace which satisfy the following property: If* $\omega \in \mathcal{Y}$ *and* $\tau \in \wedge M$*, then* $\omega \wedge \tau \in \mathcal{Y}$.

Let $\mathcal{Y} = \{\omega \in \wedge^1 M/\omega$ vanish on $D\}$. Then $\mathcal{Y}$ is an ideal of $\wedge M$ and we have

Theorem 1.15.7 *D is involutive if and only if* $d\mathcal{Y} = \{d\omega/\omega \in \mathcal{Y}\} \subset \mathcal{Y}$ *(i.e., $\mathcal{Y}$ is a* **differential ideal***).*

The proof follows directly from Proposition 1.15.5.

To end this section, we introduce the concept of foliation (see Molino [97] for more details).

A **foliation** F (of class C^∞) of dimension k on an m–dimensional manifold M is a decomposition of M into disjoint connected subsets $F = \{L_\alpha/\alpha \in A\}$ called the **leaves of the foliation**, such that each point of M has a coordinate neighborhood (U, x^i) such that for each leaf L_α, the components of $L_\alpha \cap U$ are locally given by the equations

$$x^{k+1} = \text{constant}\ , \ldots, x^m = \text{constant}.$$

These coordinates (x^i) are said to be **distinguished**. Then each leaf L_α is a connected immersed submanifold of dimension k. The topology of L_α is given by the basis formed by the sets $\{x \in V/x^{k+1}(x) = \text{constant}, \ldots, x^m(x) = \text{constant}, V \text{ open in } U\}$. Hence the topology of L_α does not necessarily coincide with the induced topology on L_α from M. Thus, in general, L_α is not an embedding submanifold.

Definition 1.15.8 *A* **maximal integral manifold** *N of a distribution D on a manifold M is a connected integral manifold which is not a proper subset of any other connected integral manifold of D.*

The following theorem shows that, globalizing Frobenius theorem, an integrable distribution determines a foliation.

Theorem 1.15.9 *Let D be a k–dimensional involutive distribution on M and $x \in M$. Then through x there passes a unique maximal connected integral manifold of D and every connected integral of D through x is contained in the maximal one.*

An outline of proof: Let N be the set of all those points y in M for which there is a piecewise C^∞ curve joining y to x, whose C^∞ portions are integral curves of D, i.e., their tangent vectors belong to D. Then we can prove that N is the desired maximal integral manifold (see Sternberg [114], Warner [123]).□

From Theorem 1.15.9, we deduce that the maximal integral manifolds of D determine a foliation on M.

1.16 Orientable manifolds. Integration. Stokes theorem

First, we discuss orientation on vector spaces.

Let V be an m–dimensional vector space. We introduce an equivalence relation in the set of all bases of V as follows. Two bases $\{e_1, \ldots, e_m\}$ and $\{\bar{e}_1, \ldots, \bar{e}_m\}$ are said to have the same orientation if $\det(A_i^j) > 0$, where $\bar{e}_i = A_i^j e_j$. It is easy to check that there are exactly two equivalence classes. An **orientation** on V is a choice of one of these classes and then V is called an **oriented vector space**.

We next show that this concept is related to the choice of a basis of $\wedge^m V$.

Proposition 1.16.1 *Let $\omega \in \wedge^m V$ be an m–form on V and let $\{e_1, \ldots, e_m\}$ be a basis of V. If $\{v_1, \ldots, v_m\}$ is a set of vectors with $v_i = A_i^j e_j$, then*

$$\omega(v_1, \ldots, v_m) = \det(A_i^j)\, \omega(e_1, \ldots, e_m).$$

The proof is left to the reader as an exercise.

Corollary 1.16.2 *A non–zero $\omega \in \wedge^m V$ has the same sign (or opposite sign) on two bases if they have the same (resp. opposite) orientation.*

Proof: It is straightforward from Proposition 1.16.1.□

Hence the choice of an m–form $\omega \neq 0$ (i.e., a **volume form** on V) determines an orientation of V given by all the bases $\{e_i\}$ of V such that

$\omega(e_1, \ldots, e_m) > 0$. In such a case $\{e_i\}$ is called a **positively oriented basis**. Moreover, two such forms ω_1 and ω_2 determine the same orientation if and only if $\omega_2 = \lambda\omega_1$, where λ is a positive real number.

To extend the concept of orientation to a manifold M we must try to orient each tangent space T_xM in such a way that orientation of nearly tangent spaces agree.

Definition 1.16.3 *We say that an m-dimensional manifold M is* **orientable** *if there is an m-form ω on M such that $\omega(x) \neq 0$ for all $x \in M$; ω is called a* **volume form**.

Then any such ω orients each tangent space. If $\omega' = \lambda\omega$, where λ is a positive function on M, then ω' gives the same orientation on M.

Example

R^m with the form $\omega = dx' \wedge \ldots \wedge dx^m$ is an orientable manifold. The form ω determines the **standard orientation of R^m**.

Definition 1.16.4 *Let M_1 and M_2 be orientable m-dimensional manifolds with volume forms ω_1 and ω_2, respectively. We say that a diffeomorphism $F : M_1 \longrightarrow M_2$* **preserves** *(resp.* **reverses***)* **orientations** *if $F^\star\omega_2 = \lambda\omega_1$, where $\lambda > 0$ (resp. $\lambda < 0$) is a function on M_1. If $F^\star\omega_2 = \omega_1$, we say that F* **preserves volume forms**.

Now, we interpret the orientability of a manifold in terms of local coordinates.

Proposition 1.16.5 *Let M be a connected manifold. Then M is orientable if and only if M has an atlas $\{(U_\alpha, \varphi_\alpha)\}$ such that the Jacobian (i.e., the determinant of the Jacobian matrix) of $\varphi_\beta \circ \varphi_\alpha^{-1}$ is positive.*

Proof: In fact, suppose that M is orientable with volume form ω. We choose any atlas $\{(U_\alpha, \varphi_\alpha)\}$ by connected coordinate neighborhoods U_α, with local coordinates $x_\alpha^1, \ldots, x_\alpha^m$ such that ω is locally expressed on U_α by

$$\omega = \lambda_\alpha dx_\alpha^1 \wedge \ldots \wedge dx_\alpha^m, \text{ with } \lambda_\alpha > 0.$$

We may easily choose coordinates such that λ_α is positive on U_α, since replacing $(x_\alpha^1, \ldots, x_\alpha^m)$ by $(-x_\alpha^1, x_\alpha^2, \ldots, x_\alpha^m)$ changes the sign of λ_α.

If $U_\alpha \wedge U_\beta \neq \emptyset$, then we obtain

$$\lambda_\beta = \lambda_\alpha \det(\partial x_\alpha^i / \partial x_\beta^j) \text{ (see Proposition 1.16.1).}$$

Since $\lambda_\alpha > 0$, $\lambda_\beta > 0$, we deduce that

$$\det(\partial x^i_\alpha / \partial x^j_\beta) > 0. \tag{1.8}$$

Conversely, suppose that M has an atlas as above. Let $\{U_a, f_a)\}$ be a subordinate partition of unity with respect to $\{(U_\alpha, \varphi_\alpha)\}$. Since each U_a is contained on some U_α, then $\{(U_a, \varphi_a)\}$, where $\varphi_a = \varphi_\alpha / U_a$ is a new atlas of M satisfying (1.8). Define $\omega \in \wedge^m M$ by

$$\omega = \sum_a f_a dx^1_a \wedge \ldots \wedge dx^m_a,$$

extending each summand to all of M by defining it to be zero outside the support of f_a. Let $x \in M$. Then we may choose a coordinate neighborhood (V, ψ) of x with local coordinates $(x^1, \ldots, x^m)$ such that

$$\det(\partial x^i_a / \partial x^j) > 0$$

for all a. Hence we have

$$\omega(x) = \sum_a f_a(x) dx^1_a \wedge \ldots \wedge dx^m_a(x)$$

$$= \sum_a f_a(x) \det(\partial x^i_a / \partial x^j) dx^1 \wedge \ldots \wedge dx^m(x).$$

Now, each $f_a(x) \geq 0$ and at least one of them is positive at x. Moreover $\det(\partial x^i_a / \partial x^j) > 0$. Hence $\omega(x) \neq 0$.□

Definition 1.16.6 *Let M be an orientable m–dimensional manifold with volume form ω. A coordinate neighborhood (U, x^i) is called* **positively oriented** *if $dx^1 \wedge \ldots \wedge dx^m$ and $\omega/_U$ give the same orientation.*

Clearly, (U, x^i) is positively oriented if and only if $\omega/_U = f dx^1 \wedge \ldots \wedge dx^m$, where $f > 0$ is a function on U.

Now, let M be an orientable m–dimensional Riemannian manifold with volume form ω. We define a **natural volume form** Ω on M. Let $\{X_1, \ldots, X_m\}$ be an orthonormal basis of $T_x M$ such that $\omega(X_1, \ldots, X_m) > 0$. We define an m–form Ω by

$$\Omega(x)(X_1, \ldots, X_m) = 1.$$

If $\{Y_1, \ldots, Y_m\}$ is another orthonormal frame at x such that $\omega(Y_1, \ldots, Y_m) > 0$, then $Y_i = A_i^j X_j$. Hence, from Proposition 1.16.1, we have

$$\Omega(x)(Y_1, \ldots, Y_m) = \det(A_i^j)\Omega(X_1, \ldots, X_m) = 1,$$

since $\det(A_i^j) = 1$ (in fact, $(A_i^j) \in O(m)$ and then $|\det(A_i^j)| = 1$, but since $\{X_1, \ldots, X_m\}$ and $\{Y_1, \ldots, Y_m\}$ have the same orientation, one has $\det(A_i^j) > 0$). Thus, $\Omega(x)$ is independent of the orthonormal frame chosen.

Moreover, if $(x^1, \ldots, x^m)$ is a positively oriented coordinate system, we have

$$(\partial/\partial x^i)_x = B_i^k X_k$$

so that

$$g_{ij} = g(\partial/\partial x^i, \partial/\partial x^j) = \sum_{k=1}^{m} B_i^k B_j^k$$

Hence

$$G = \det(g_{ij}) = (\det B)^2$$

Since $\det B > 0$, we obtain

$$\sqrt{G} = \det B$$

and thus

$$\Omega(\partial/\partial x^1, \ldots, \partial/\partial x^m) = \sqrt{G},$$

i.e.,

$$\Omega = \sqrt{G} dx^1 \wedge \ldots \wedge dx^m, \text{ where } G = \det(g_{ij}).$$

Next, we shall define the integral of an m–form on an m–dimensional manifold.

Let us recall that if $f : R^m \longrightarrow R$ is continuous and has compact support then

$$\int_{R^m} f dx^1 \ldots dx^m$$

is defined as the Riemann integral over any rectangle containing the support of f. Moreover, suppose that $G : R^m \longrightarrow R^m$ is a diffeomorphism given by

$$G(x^1, \ldots, x^m) = (y^1(x^i), \ldots, y^m(x^i)).$$

Suppose that $f' : R^m \longrightarrow R$ is the function defined by $f = f' \circ G$. Then f' has compact support and

$$\int f'(y^1, \ldots, y^m) dy^1 \ldots dy^m$$

$$= \int f(x^1, \ldots, x^m) \mid \det(\partial y^i / \partial x^j) \mid dx^1 \ldots dx^m \qquad (1.9)$$

(1.9) is known as the **change of variables rule**.

Now, suppose that (U, x^i) is a positively oriented coordinate neighborhood of an orientable m–dimensional manifold M with volume form ω.

Definition 1.16.7 *Let α be an m–form on M with compact support. If α is locally given by*

$$\alpha = f dx^1 \wedge \ldots \wedge dx^m$$

on U, we define

$$\int_M \alpha = \int_{R^m} f dx^1 \ldots dx^m.$$

(We notice that f has also compact support).

Remark 1.16.8 If (V, y^i) is another positively oriented coordinate neighborhood such that supp $\alpha \subset U \cap V$, and

$$\alpha = g dy^1 \wedge \ldots \wedge dy^m,$$

then from (1.9) we have

$$\int_{R^m} g dy^1 \ldots dy^m = \int_{R^m} f dx^1 \ldots dx^m$$

Now, suppose that α is an arbitrary m–form on M. Let $\mathcal{A} = \{(U_\alpha, \varphi_\alpha)\}$ be an atlas of positively oriented coordinate neighborhoods and $\{(U_i, h_i)\}$ a partition of unity subordinate to $\mathcal{A}$. Define $\alpha_i = h_i\alpha$. So α_i has compact support contained in some U_α. Then we define

$$\int_M \alpha = \sum_i \int_M \alpha_i \tag{1.10}$$

It is not hard to prove that (1.10) does not depend on the choice of $\mathcal{A}$ or the partition of unity $\{h_i\}$.

Definition 1.16.9 *$\int_M \alpha$ is called the* **integral** *of α on M.*

Proposition 1.16.10 *(1) If $-M$ denotes the same underlying manifold with opposite orientation, then*

$$\int_{-M} \alpha = -\int_M \alpha.$$

(2)

$$\int_M a\alpha + b\beta = a\int_M \alpha + b\int_M \beta,$$

for all $a, b \in R$ and $\alpha, \beta \in \wedge^m M$, where α and β have compact support.
(3) If $F : M_1 \longrightarrow M_2$ is a diffeomorphism and α is a m–form on M_2 with compact support, then

$$\int_{M_1} F^\star\alpha = \pm\int_M \alpha,$$

with sign depending on whether F preserves or reserves orientations.

The proof is left to the reader as an exercise.

We now introduce the concept of manifold with boundary.

Let $H^m = \{x = (x^1, \ldots, x^m) \in R^m / x^m \geq 0\}$ with the relative topology of R^m and denote by ∂H^m the subspace defined by $\partial H^m = \{x \in H^m / x^m = 0\}$; ∂H^m is called the **boundary** of H^m. Obviously, ∂H^m is homeomorphic to R^{m-1} by the map $(x^1, \ldots, x^{m-1}) \longrightarrow (x^1, \ldots, x^{m-1}, 0)$.

We notice that differentiability can be defined for maps to R^n of **arbitrary** subsets of R^m in the obvious way. If $F : A \subset R^m \longrightarrow R^n$ is a map defined on a subset A of R^m, we say that F is C^∞ **on** A if, for each point $x \in A$, there exists a C^∞ map F_x on an open neighborhood U_x of

x such that $F = F_x$ on $A \cap U_x$. Then the notion of diffeomorphism applies at one to open subsets U, V of H^m; namely U, V are **diffeomorphic** if there exists a bijective map $F : U \longrightarrow V$ such that F and F^{-1} are both C^∞. If $U, V \subset R^m - \partial H^m$, then U and V are actually open subsets of R^m and this definition coincides with our previous one. On the other hand, if $U \cap \partial H^m \neq \emptyset$, then $V \cap \partial H^m \neq \emptyset$ and $F(U \cap \partial H^m) \subset V \cap \partial H^m$. Similarly $F^{-1}(V \cap \partial H^m) \subset U \cap \partial H^m$. In other words, F maps boundary points to boundary points and interior points to interior points.

We also notice that $U \cap \partial H^m$ and $V \cap \partial H^m$ are open subsets of ∂H^m, a submanifold of R^m diffeomorphic to R^{m-1}, and F, F^{-1} restricted to $U \cap \partial H^m$ and $V \cap \partial H^m$ are diffeomorphisms. Moreover, F and F^{-1} can be extended to open subsets U', V' of R^m such that $U = U' \cap H^m$ and $V = V' \cap H^m$. These extensions will not be unique nor are the extensions in general inverses. However, the differentials of F and F^{-1} on U and V are independent of the extensions chosen and we may suppose that even on the extended domains the Jacobians are of rank m.

Definition 1.16.11 *A C^∞* **manifold with boundary** *is a Hausdorff space M with a contable basis of open sets and a C^∞ differentiable structure $\mathcal{A} = \{(U_\alpha, \varphi_\alpha)\}$ in the following sense: U_α is an open subset of M and φ_α is a homeomorphism from U_α onto an open subset of H^m such that:*
(1) the U_α cover M;
(2) for any α, β the maps $\varphi_\beta \circ \varphi_\alpha^{-1}$ and $\varphi_\alpha \circ \varphi_\beta^{-1}$ are diffeomorphisms of $\varphi_\alpha(U_\alpha \cap U_\beta)$ and $\varphi_\beta(U_\alpha \cap U_\beta)$;
(3) $\mathcal{A}$ is maximal with respect to properties (1) and (2). (Compare with Definition 1.2.3).

The $(U_\alpha, \varphi_\alpha)$ are **coordinate neighborhoods** on M. From the remarks above we see that if $\varphi(x) \in \partial H^m$ in one coordinate system, then this holds for all coordinate systems. The collection of such points is called the **boundary** of M, denoted ∂M, and $M - \partial M$ is an m-dimensional manifold (in the ordinary sense), which we denote by Int M. If $\partial M = \emptyset$, then M is a manifold in the ordinary sense; we call it a manifold **without boundary** when it is necessary to make the distinction.

Proposition 1.16.12 *If M is a manifold with boundary, then the differentiable structure of M determines a differentiable structure of dimension $m-1$ on ∂M such that the inclusion $i : \partial M \longrightarrow M$ is an embedding.*

Proof: In fact, the differentiable structure $\bar{\mathcal{A}}$ on ∂M is determined by the coordinate neighborhoods $(\bar{U}, \bar{\varphi})$, where $\bar{U} = U \cap \partial M$, $\bar{\varphi} = \varphi/_{U\cap\partial M}$ for any coordinate neighborhood (U, φ) of M which contains points of ∂M.□

Hence, if M is a manifold with boundary, then $Int\, M = M - \partial M$ and ∂M are manifolds of dimension m and $m-1$, respectively.

Next we define orientation on a manifold with boundary. Definition 1.16.3 extends in a natural way to the case of manifolds with boundary. Namely, a manifold with boundary M is orientable if there exists an m–form ω on M such that $\omega(x) \neq 0$ for all $x \in M$. The reader may prove that Proposition 1.16.5 holds for manifolds with boundary. Let us recall that an orientation on M is a choice of orientations of all the tangent spaces in such a way that for all positively oriented coordinate neighborhoods, the maps $\varphi_\beta \circ \varphi^{-1}\alpha : \varphi_\alpha(U_\alpha \cap U_\beta) \longrightarrow \varphi_\beta(U_\alpha \cap U_\beta)$ "preserves the natural orientation" of H^m, i.e., $\varphi_\beta \circ \varphi^{-1}$ has positive Jacobian. Thus, we can define an **induced orientation** on ∂M as follows. At every $x \in \partial M$, $T_x(\partial M)$ is an $(m-1)$–dimensional vector subspace of $T_x M$ so that there are, in a coordinate neighborhood intersecting ∂M, exactly two vectors perpendicular to $x^m = 0$; one points inward, the other outward. We say that a basis $\{e_1, \ldots, e_{m-1}\}$ for $T_x(\partial M)$ is **positively oriented** if $\{-\partial/\partial x^m,\ e_1, \ldots, e_{m-1}\}$ is positively oriented with respect to the orientation on $T_x M$.

Theorem 1.16.13 (Stokes' Theorem) *Let M be a manifold with boundary and $\alpha \in \wedge^{m-1}M$ with compact support. Let $i : \partial M \longrightarrow M$ be the inclusion map and $i^*\alpha \in \wedge^{m-1}(\partial M)$. Then*

$$\int_{\partial M} i^*\alpha = \int_M d\alpha.$$

or for sake of simplicity

$$\int_{\partial M} \alpha = \int_M d\alpha$$

Proof: Since both sides of the equation are linear, we may assume without loss of generality that α is a form with compact support contained in some coordinate neighborhood U with local coordinates $x^1, \ldots, x^m$.

There are two cases: $U \cap \partial M = \emptyset$ and $U \cap \partial M \neq \emptyset$. We set

$$\alpha = \sum_{i=1}^{m} (-1)^i \alpha_i dx^1 \wedge \ldots \wedge d\check{x}^i \wedge \ldots \wedge dx^m,$$

where $d\check{x}^i$ means that this differential is omitted in the expression. Then

$$d\alpha = \sum_{i=1}^{m} (\partial\alpha_i/\partial x^i) dx^1 \wedge \ldots \wedge dx^m.$$

Thus

$$\int_M d\alpha = \sum_{i=1}^{m} \int_{R^m} (\partial\alpha_i/\partial x^i) dx^1 \ldots dx^m.$$

If $U \cap \partial M = \emptyset$, we have

$$\int_{\partial M} \alpha = 0.$$

On the other hand, the integration of the ith term in the sum ocurring in $\int_M d\alpha$ is

$$\int_{R^{m-1}} \left[\int_R \left(\frac{\partial\alpha_i}{\partial x^i} \right) dx^i \right] dx^1 \ldots d\check{x}^i \ldots dx^m \text{ (no sum!)} \tag{1.11}$$

But since α_i has compact support we have

$$\int_R \left(\frac{\partial\alpha_i}{\partial x^i} \right) dx^i = \int_{-\infty}^{+\infty} \left(\frac{\partial\alpha_i}{\partial x^i} \right) dx^i = 0.$$

Thus Stokes' theorem holds for this case.

If $U \cap \partial M \neq \emptyset$, then we again have all the integrals in (1.11) equal to zero, except the one corresponding to $i = m$, which is

$$\int_{R^{m-1}} \left[\int_R \left(\frac{\partial\alpha_m}{\partial x^m} \right) dx^m \right] dx^1 \ldots dx^{m-1}$$

$$= \int_{R^{m-1}} \left[\int_0^{+\infty} \left(\frac{\partial\alpha_m}{\partial x^m} \right) dx^m \right] dx^1 \ldots dx^{m-1}$$

$$= -\int_{R^{m-1}} \alpha_m(x^1, \ldots, x^{m-1}, 0) dx^1 \ldots dx^{m-1},$$

since α_m has compact support. Thus

$$\int_M d\alpha = -\int_{R^{m-1}} \alpha_m(x^1, \ldots, x^{m-1}, 0) dx^1 \ldots dx^{m-1}.$$

On the other hand, the local expression of $i^*\alpha$ in the local coordinates $x^1, \ldots, x^{m-1}$, obtained by restriction of $x^1, \ldots, x^m$ to $U \cap \partial M$ is

$$i^*\alpha = (-1)^{m-1}\alpha_m(x^1, \ldots, x^{m-1}, 0)dx^1 \wedge \ldots \wedge dx^{m-1},$$

since $i^*dx^m = 0$.

Now, the basis $\{\partial/\partial x^1, \ldots, \partial/\partial x^{m-1}\}$ is not positively oriented, since the outward unit normal vector is $-\partial/\partial x^m$ in each point of $U \cap \partial M$. Thus

$$\int_{\partial M} i^*\alpha = (-1)^{2m-1} \int_{R^{m-1}} \alpha_m(x^1 \ldots x^{m-1}, 0)dx^1 \ldots dx^{m-1}$$

because the sign of $\{-\partial/\partial x^m,\ \partial/\partial x^1, \ldots, \partial/\partial x^{m-1}\}$ is $(-1)^m$. This ends the proof. □

Remark 1.16.14 From the proof of Stokes theorem, we deduce that if M is a manifold without boundary, i.e., $\partial M = \emptyset$, then we have

$$\int_M d\alpha = 0.$$

Example (Green's Theorem)

Let M be the closure of a bounded open subset of R^2 bounded by simple closed curves (for example, let M be a circular disk or annulus). Then ∂M is the union of these curves (in the examples mentioned above, ∂M is a circle or a pair of concentric circles). If α is a 1–form on M, then

$$\alpha = f dx + g dy,$$

where (x, y) are the canonical coordinates on R^2. Then

$$d\alpha = [(\partial g/\partial x) - (\partial f/\partial y)]dx \wedge dy.$$

By Stokes' theorem, we have

$$\int_M ((\partial g/\partial x) - (\partial f/\partial y))dx \wedge dy = \int_{\partial M} f dx + g dy. \qquad (1.12)$$

But if we cover ∂M with positively oriented coordinate neighborhoods, we deduce that the right–hand of (1.12) is the usual line integral along a curve C (or curves C_i) oriented so that as we transverse the curve the region is on the left. Thus (1.12) becomes

$$\int\int_M ((\partial g/\partial x) - (\partial f/\partial y))dxdy = \sum_i \int_{C_i} f dx + g dy,$$

which is the Green's theorem.

1.17 de Rham cohomology. Poincaré lemma

Definition 1.17.1 *A p–form α on a manifold M is said to be* **closed** *if $d\alpha = 0$. It is said to be* **exact** *if there is a (p-1)–form β such that $\alpha = d\beta$. Since $d^2 = 0$ then every exact form is closed.*

Let $Z^k(M)$ denote the closed p–forms on M; since $Z^k(M)$ is the kernel of the linear map $d : \wedge^p M \longrightarrow \wedge^{p+1} M$, it is a vector subspace of $\wedge^p M$. Similarly the exact p–forms $B^p(M)$ are the image of $d : \wedge^{p-1} M \longrightarrow \wedge^p M$ and then a vector subspace of $\wedge^p M$. Since $B^p(M) \subset Z^p(M)$, we can form the quotient space

$$H^p(M) = Z^p(M)/B^p(M),$$

which is called the **pth de Rham cohomology group** of M. If $dim\, M = m$, then

$$H^p(M) = 0 \text{ when } p > m.$$

If we set

$$H^\star(M) = \bigoplus_{p=0}^{m} H^p(M),$$

then $H^\star(M)$ is a vector space which becomes an algebra over R with the multiplication being that naturally induced by the exterior product of forms, i.e.,

$$[\omega] \smile [\tau] = [\omega \wedge \tau], [\omega] \in H^p(M),\ [\tau] \in H^q(M).$$

$H^\star(M)$ is called the **de Rham algebra** of M ($\smile$ is called the **cup–product**).

Now, let $F : M \longrightarrow N$ be a C^∞ map. Then the algebra homomorphism $F^\star : \wedge N \longrightarrow \wedge M$ commutes with d and hence maps closed forms to closed forms and exact forms to exact forms. Thus it induces a linear map

$$F^\star : H^p(N) \longrightarrow H^p(M)$$

given by

$$F^\star[\alpha] = [F^\star \alpha],\ [\alpha] \in H^p(N).$$

Moreover, we have an algebra homomorphism

$$F^\star : H^\star(N) \longrightarrow H^\star(M).$$

The reader can easily check the following:

Proposition 1.17.2 *(1) If $F : M \longrightarrow M$ is the identity map, then it induces the identity on the de Rham cohomology, i.e., $F^\star = id$.*
(2) Under composition of maps we have

$$(G \circ F)^\star = F^\star \circ G^\star.$$

Corollary 1.17.3 *A diffeomorphism $F : M \longrightarrow N$ induces isomorphisms on the de Rham cohomology.*

Thus two diffeomorphic manifolds have the same de Rham cohomology groups. In other words, the de Rham cohomology is a **differentiable invariant** of a differentiable manifold M. In fact, the **de Rham theorem** proves that the de Rham cohomology is actually a **topological invariant**, i.e., the de Rham cohomology groups depend only on the underlying topological structure of M and do not depend on the differentiable structure. The reader is referred to Warner [123] for a proof of the de Rham theorem.

Furthermore, if M is compact then the de Rham cohomology groups are vector spaces of finite dimension. The dimension b_p of $H^p(M)$ is called the pth Betti number of M and it is a topological invariant from the de Rham theorem.

Proposition 1.17.4 *Let M be a connected manifold. Then $H^0(M) = R$.*

Proof: $\wedge^0 M$ consists of C^∞–function on M and $Z^0(M)$ of those functions f for which $df = 0$. But $df = 0$ if and only if f is constant (see Exercise 1.22.10). Since $B^0(M) = 0$, then $H^0(M) \simeq Z^0(M) \cong R$.□

Next we shall prove the Poincaré lemma.

Proposition 1.17.5 (Poincaré Lemma) *For each $p \geq h$ there is a linear map $h_p : \wedge^p R^m \longrightarrow \wedge^{p-1} R^m$ such that*

$$h_{p+1} \circ d + d \circ h_p = id.$$

Proof: Let $(x^1, \ldots, x^m)$ be the canonical coordinates in R^m. Consider the vector field

$$X = \sum_{i=1}^{m} x^i (\partial / \partial x^i)$$

on R^m. We define a linear map $A_p : \wedge^p R^m \longrightarrow \wedge^p R^m$ by

$$A_p(f\,dx^{i_1}\wedge\ldots\wedge dx^{i_p})(x)=\left(\int_0^1 t^{p-1}f(tx)dt\right)dx^{i_1}\wedge\ldots\wedge dx^{i_p}(x)$$

for all $x\in R^m$ and then we extend it linearly to all $\wedge^p R^m$.

We have

$$(A_p\circ L_X)(f\,dx^{i_1}\wedge\ldots\wedge dx^{i_p})(x)$$

$$=A_p[(L_Xf)dx^{i_1}\wedge\ldots\wedge dx^{i_p}+f\,L_X(dx^{i_1}\wedge\ldots\wedge dx^{i_p})](x)$$

$$=A_p[(pf+\sum_i x^i(\partial f/\partial x^i))dx^{i_1}\wedge\ldots\wedge dx^{i_p}](x)$$

(since $L_X dx^i = dx^i$)

$$=\left[\int_0^1 t^{p-1}(pf(tx)+\sum_i x^i(tx)(\partial f/\partial x^i)_{tx})dt\right]dx^{i_1}\wedge\ldots\wedge dx^{i_p}(x)$$

$$=\left[\int_0^1 \frac{d}{dt}(t^pf(tx)dt\right]dx^{i_1}\wedge\ldots\wedge dx^{i_p}(x)$$

$$=f(x)dx^{i_1}\wedge\ldots\wedge dx^{i_p}(x).$$

Hence

$$A_p\circ L_X=id_{\wedge^pR^m} \tag{1.13}$$

Moreover, A commutes with d, i.e.,

$$A_p\circ d=d\circ A_{p-1}. \tag{1.14}$$

In fact,

$$((A_p\circ d)(f\,dx^{i_1}\wedge\ldots\wedge dx^{i_{p-1}}))(x)$$

$$= A_p\left[\sum_{i=1}^{m}(\partial f/\partial x^i)dx^i \wedge dx^{i_1} \wedge \ldots \wedge dx^{i_{p-1}}\right](x)$$

$$= \left[\int_0^1 t^{p-1}\sum_i(\partial f/\partial x^i)_{tx}dt\right] dx^i \wedge dx^{i_1} \wedge \ldots \wedge dx^{i_{p-1}}(x)$$

$$= d\left(\int_0^1 t^{p-2}f(tx)dt\right) dx^{i_1} \wedge \ldots \wedge dx^{i_{p-1}}(x)$$

(since d and $\int$ commutes)

$$= (d \circ A_{p-1})(f\, dx^{i_1} \wedge \ldots \wedge dx^{i_{p-1}})(x).$$

Since $L_X = i_X d + di_X$, from (1.13) and (1.14), we obtain

$$id_{\wedge^p R^m} = A_p L_X = A_p i_X d + A_p di_X$$

$$= A_p i_X d + dA_{p-1} i_X.$$

Now we set

$$h_p = A_{p-1} \circ i_X$$

Thus we obtain

$$h_{p+1} \circ d + d \circ h_p = A_p \circ i_X \circ d + d \circ A_{p-1} \circ i_X = id_{\wedge^p R^m}. \square$$

Corollary 1.17.6 *If α is a p-form, $p \geq 1$, on R^m which is closed, then α is exact.*

Proof: We set

$$\beta = h_p \alpha.$$

Hence

$$d\beta = dh_p\alpha = \alpha - h_{p+1}(d\alpha) = \alpha. \square$$

Corollary 1.17.7 *The de Rham cohomology groups of R^m are all zero for $p \geq 1$.*

Since $H^0(R^m) = R$, we have computed all the de Rham cohomology groups of R^m. Moreover, if U is the open unit ball in R^m, since U and R^m are diffeomorphic, we deduce, from Corollary 1.17.3, that

$$H^p(U) = 0 \text{ for } p \geq 1.$$

From Corollary 1.17.6, we deduce that every closed p–form on a manifold M is locally exact. In fact, let α be a p–form on an m–dimensional manifold M such that $d\alpha = 0$. For each point $x \in M$ there exists an open neighborhood (U, φ) such that $\varphi(U)$ is the open unit ball in R^m. Then $(\varphi^{-1})^\star\alpha$ is a closed p–form on $\varphi(U)$ and hence $(\varphi^{-1})^\star\alpha = d\beta$, where β is a $(p-1)$–form on $\varphi(U)$. Therefore

$$\alpha = d(\varphi^\star\beta) \quad \text{on } U.$$

1.18 Linear connections. Riemannian connections

In this section we introduce the concept of linear connection. Further, we shall see that any Riemannian manifold possesses a unique linear connection satisfying certain conditions. In chapter 4, we shall generalize the notion of linear connection.

Definition 1.18.1 *A* **linear connection** *on an m–dimensional manifold M is a map that assigns to each pair of vector fields X and Y on M another vector field $\nabla_X Y$ such that:*
(1) $\nabla_{X_1+X_2}Y = \nabla_{X_1}Y + \nabla_{X_2}Y$,
(2) $\nabla_X(Y_1 + Y_2) = \nabla_X Y_1 + \nabla_X Y_2$,
(3) $\nabla_{fX}Y = f(\nabla_X Y)$,
(4) $\nabla_X(fY) = (Xf)Y + f(\nabla_X Y)$.

Remark 1.18.2 We notice that, if X_1, X_2, Y_1 and Y_2 are vector fields on M and if $X_1 = X_2$ and $Y_1 = Y_2$ in a neighborhood of a point $x \in M$, then $(\nabla_{X_1}Y_1)_x = (\nabla_{X_2}Y_2)_x$. This implies that ∇ induces a map $\chi(U) \times \chi(U) \longrightarrow \chi(U)$ satisfying (1)–(4), where U is an open set of M.

Now, let U be a coordinate neighborhood of M with local coordinates $(x^1, \ldots, x^m)$. Then we define m^3 functions Γ_{ij}^k on U by

$$\nabla_{\partial/\partial x^i}\partial/\partial x^j = \Gamma_{ij}^k(\partial/\partial x^k);$$

Γ_{ij}^k are called the **Christoffel components** of ∇.

Let $\bar{U}$ be another coordinate neighborhood with local coordinates $(\bar{x}^i)$. On $U \cap \bar{U}$ we have

$$\partial/\partial \bar{x}^i = (\partial x^j/\partial \bar{x}^i)(\partial/\partial x^j).$$

Hence, the transformation rule for the Γ's are:

$$\bar{\Gamma}_{ij}^k = (\partial x^p/\partial \bar{x}^i)(\partial x^q/\partial \bar{x}^j)(\partial \bar{x}^k/\partial x^r)\Gamma_{pq}^r$$

$$+(\partial \bar{x}^k/\partial x^p)(\partial^2 x^p/\partial \bar{x}^i \partial \bar{x}^j).$$

Then the Γ's **are not the components of a tensor field on** M. (This is a consequence of (4)).

Now, let $X, Y \in \chi(M)$. Then we locally have

$$\nabla_X Y = (X^j(\partial Y^i/\partial x^j) + \Gamma_{jk}^i X^j Y^k)(\partial/\partial x^i), \tag{1.15}$$

where $X = X^i(\partial/\partial x^i)$ and $Y = Y^i(\partial/\partial x^i)$.

Remark 1.18.3 From (1.15), we deduce that $(\nabla_X Y)(x)$ depends only on $X(x)$.

Definition 1.18.4 *Let $\sigma : R \longrightarrow M$ be a curve and X a vector field on M. We define the* **covariant derivative of X along σ** *by*

$$DX/dt = \nabla_{\dot{\sigma}(t)}X.$$

(From the Remark 1.18.3, we deduce that DX/dt is well-defined). We say that X is **parallel along** *σ if $DX/dt = 0$. We say that σ is a* **geodesic** *of ∇ if $\dot{\sigma}(t)$ is parallel along σ, i.e.,*

$$\nabla_{\dot{\sigma}}\dot{\sigma} = 0.$$

(As above, $\nabla_{\dot{\sigma}}\dot{\sigma}$ is well-defined, since DX/dt depends only on the values of X along σ; then it is sufficient to extend $\dot{\sigma}(t)$ to an arbitrary vector field on an open neighborhood of σ).

In local coordinates we have

$$DX/dt = ((dX^k(t)/dt) + \Gamma^k_{ij}(t)X^i(t)(dx^j/dt)(t))(\partial/\partial x^k),$$

and

$$\nabla_{\dot\sigma(t)}\dot\sigma(t) = ((d^2x^k/dt^2) + \Gamma^k_{ij}(t)(dx^i/dt)(dx^j/dt))(\partial/\partial x^k),$$

where $\sigma(t) = (x^i(t))$, $\dot\sigma(t) = (dx^i/dt)(\partial/\partial x^i)$. Hence, σ is a geodesic of ∇ if and only if it satisfies the following system of linear differential equations:

$$d^2x^k/dt^2 + \Gamma^k_{ij}(dx^i/dt)(dx^j/dt) = 0,\ 1 \leq k \leq m. \tag{1.16}$$

From the existence theorem for ordinary differential equations, we easily deduce the following.

Theorem 1.18.5 *For any point $x \in M$ and for any tangent vector $X \in T_xM$, there is a unique geodesic with initial condition (x, X), i.e., a unique geodesic $\sigma(t)$ defined on some interval $(-\epsilon, \epsilon)$, $\epsilon > 0$, such that $\sigma(0) = x$ and $\dot\sigma(0) = X$.*

Definition 1.18.6 *A linear connection ∇ on M is said to be* **geodesically complete** *if every geodesic $\sigma(t)$ can be extended so as to be defined for all $t \in R$.*

Example.- Define on R^m a linear connection ∇ by

$$\nabla_{\partial/\partial x^i}\partial/\partial x^j = 0,$$

for all $i, j = 1, \ldots, m$. Then $\Gamma^k_{ij} = 0$ and (1.16) becomes

$$d^2x^t/dt^2 = 0,\ 1 \leq k \leq m.$$

Thus the geodesics are straight lines and ∇ is complete.

Next, we define the covariant derivative of a tensor field K of type $(0, r)$ or $(1, r)$, $r \geq 1$, as follows. First, if K is a tensor field of type $(0, r)$ (resp. $(1, r)$) and X a vector field on M, we define a tensor field $\nabla_X K$ of the same type by

$$(\nabla_X K)(X_1, \ldots, X_r) = X(K(X_1, \ldots, X_r))$$

$$-\sum_{i=1}^{r} K(X_1,\ldots,\nabla_X X_i,\ldots,X_r)$$

(resp.

$$(\nabla_X K)(X_1,\ldots,X_r) = \nabla_X(K(X_1,\ldots,X_r))$$

$$-\sum_{i=1}^{r} K(X_1,\ldots,\nabla_X X_i,\ldots,X_r)).$$

$\nabla_X K$ is called the **covariant derivative of** K **along** X. Then we define ∇K to be the tensor field of type $(0, r+1)$ (resp. $(1, r+1)$) given by

$$(\nabla K)(X_1,\ldots,X_r,X) = (\nabla_X K)(X_1,\ldots,X_r).$$

∇K is called the **covariant derivative** of K. We say that K is **parallel** if $\nabla K = 0$.

Definition 1.18.7 *Let ∇ be a linear connection on M. Then the* **torsion tensor** *of ∇ is the tensor field T of type (1,2) on M defined by*

$$T(X,Y) = \nabla_X Y - \nabla_Y X - [X,Y].$$

Clearly, T is skew–symmetric, i.e., $T(X,Y) = -T(Y,X)$. ∇ is said to be **symmetric** if its torsion tensor T vanishes.

In local coordinates we have

$$T(\partial/\partial x^i, \partial/\partial x^j) = T^k_{ij}(\partial/\partial x^k),$$

where

$$T^k_{ij} = \Gamma^k_{ij} - \Gamma^k_{ji}.$$

Hence ∇ is symmetric if and only if $\Gamma^k_{ij} = \Gamma^k_{ji}$ in any coordinate neighborhood.

Definition 1.18.8 *Let ∇ be a linear connection on M. Then the* **curvature tensor** *of ∇ is the tensor field R of type (1,3) on M defined by*

$$R(X,Y,Z) = \nabla_X \nabla_Y Z - \nabla_Y \nabla_X Z - \nabla_{[X,Y]} Z.$$

Clearly, $R(X,Y,Z) = -R(Y,X,Z)$. ∇ is said to be **flat** if its curvature tensor R vanishes.

In local coordinates we have

$$R(\partial/\partial x^i, \partial/\partial x^j, \partial/\partial x^k) = R^{\ell}_{ijk}(\partial/\partial x^{\ell}).$$

and we easily obtain

$$R^{\ell}_{ijk} = ((\partial\Gamma^{\ell}_{jk}/\partial x^i) - (\partial\Gamma^{\ell}_{ik}/\partial k^j))$$

$$+(\Gamma^{r}_{jk}\Gamma^{\ell}_{ir} - \Gamma^{r}_{ik}\Gamma^{\ell}_{jr}).$$

Sometimes we shall consider the tensor field $R(X,Y)$ of type (1,1) defined by

$$R(X,Y)Z = R(X,Y,Z).$$

Theorem 1.18.9 (Bianchi's identities). *For a linear connection ∇ on M, we have*

Bianchi's 1st identity

$$\mathcal{G}\{R(X,Y,Z)\} = \mathcal{G}\{T(T(X,Y)Z) + (\nabla_X T)(Y,Z)\};$$

Bianchi's 2nd identity

$$\mathcal{G}\{(\nabla_X R)(Y,Z) + R(T(X,Y),Z)\} = 0;$$

where $\mathcal{G}$ denotes the cyclic sum with respect to X, Y and Z.

In particular, if $T = 0$, then

Bianchi's 1st identity: $\mathcal{G}\{R(X,Y,Z)\} = 0$;

Bianshi's 2nd identity: $\mathcal{G}\{(\nabla_X R)(Y,Z)\} = 0$.

(For a proof see Spivak [113] or Kobayashi and Nomizu [81]).

Next we shall obtain a specific connection for a Riemannian manifold.

Theorem 1.18.10 *Let M be a Riemannian manifold with Riemannian metric g. Then there is a unique symmetric linear connection ∇ on M such that $\nabla g = 0$. This connection ∇ is called the* **Riemannian connection** *(sometimes called the* **Levi–Civita connection***).*

Proof: EXISTENCE. Given two vector fields X and Y on M, we define $\nabla_X Y$ by the following equation:

$$2g(\nabla_X Y, Z) = Xg(Y,Z) + Yg(X,Z) - Zg(X,Y)$$

$$+ g([X,Y],Z) + g([Z,X],Y) + g([Z,Y],X), \qquad (1.17)$$

for all vector field Z on M. It is a straightforward computation that the map $(X,Y) \longrightarrow \nabla_X Y$, satisfies the four conditions of Definition 1.18.1. Similarly, we prove that ∇ has no torsion. To show that $\nabla g = 0$, it is sufficient to prove that

$$Xg(Y,Z) = g(\nabla_X Y, Z) + g(Y, \nabla_X Z),$$

which follows directly from (1.17).
UNIQUENESS. It is a straightforward verification that if $\bar{\nabla}$ satisfies $\bar{\nabla}_X Y - \bar{\nabla}_Y X - [X,Y] = 0$ and $\bar{\nabla} g = 0$, then it satisfies (1.17). Then $\bar{\nabla} = \nabla$.□

Remark 1.18.11 It is obvious that Theorem 1.18.10 holds for pseudo–Riemannian manifolds.

From (1.17), one easily deduces that the Christoffel components Γ_{ij}^k of the Riemannian connection ∇ are given by

$$\sum_{k=1}^{m} g_{hk}\Gamma_{ij}^k = \frac{1}{2}\left\{(\partial g_{hj}/\partial x^i) + (\partial g_{ih}/\partial x^j) - (\partial g_{ij}/\partial x^h)\right\}.$$

Let M be a Riemannian manifold with Riemannian metric g. Then the **arc length** of a C^∞ curve $\sigma : [a,b] \longrightarrow M$ is defined by

$$L(\sigma) = \int_a^b (g(\dot{\sigma}(t), \dot{\sigma}(t))^{1/2} dt.$$

In local coordinates we have

$$L(\sigma) = \int_a^b \left(\sum_{i,j} g_{ij}(dx^i/dt)(dx^j/dt)\right)^{1/2} dt,$$

where $\sigma(t) = (x^i(t))$.

This definition can be generalized to a piecewise C^∞ curve of the obvious manner.

Now, suppose that M is connected. Then we define the **distance function** $d(x, y)$ on M as follows. The distance $d(x, y)$ between two points x and y is the infimum of the lengths of all piecewise C^∞ curves joining x and y. Then we have:
(1) $d(x, y) \geq 0$;
(2) $d(x, y) = d(y, x)$;
(3) $d(x, y) \leq d(x, z) + d(z, y)$;
(4) $d(x, y) = 0$ if and only if $x = y$.
(See Boothby [9] for a proof of (4)). Thus d is a metric on M and we can prove that the **topology defined by** d **is the same as the manifold topology of** M (see Boothby [9] or Kobayashi and Nomizu [81]).

A Riemannian manifold admits **convex neighborhoods**. Precisely, for each $x \in M$, there is a neighborhood U of x such that any two points x_1 and x_2 of U may be joined by one and only one geodesic σ lying in U; this geodesic is **minimal** if for any other curve γ from x_1 to x_2, we have $L(\sigma) \leq L(\gamma)$. (See Kobayashi and Nomizu [81] for a proof).

Theorem 1.18.12 (Hopf–Rinow). *Let M be a connected Riemannian manifold. Then the following properties are equivalent:*
(1) M is geodesically complete.
(2) M is a complete metric space with respect to the distance function d.
(3) Every closed and bounded subset of M (with respect to d) is compact. □

We shall refer to Kobayashi and Nomizu [81] for a proof.

Since every compact metric space is complete, we have the following.

Corollary 1.18.13 *If a connected Riemannian manifold is compact, then any pair of points $x, y \in M$ may be joined by a geodesic whose length is precisely $d(x, y)$.* □

1.19 Lie groups

Lie groups are the most important special class of differentiable manifolds. Lie groups are differentiable manifolds which are also groups and in which the group operations are C^∞.

Definition 1.19.1 *A* **Lie group** *G is a differentiable manifold which is at the same time a group such that the map $G \times G \longrightarrow G$ defined by $(a, b) \longrightarrow ab^{-1}$ is C^∞.*

We use e to denote the identity element of a Lie group G.

Remark 1.19.2 Let G be a Lie group. Then the map $a \longrightarrow a^{-1}$ is C^∞ since it is the composition $a \longrightarrow (e, a) \longrightarrow a^{-1}$ of C^∞ maps. Also, the map $(a, b) \longrightarrow ab$ is C^∞ since it is the composition $(a, b) \longrightarrow (a, b^{-1}) \longrightarrow ab$ of C^∞ maps.

Examples of Lie groups.
(1) R^m is a Lie group under vector addition.
(2) As we have seen in Section 1.2, the general linear group $Gl(m, R)$ is an open submanifold of $gl(m, R)$. Moreover $Gl(m, R)$ is a group with respect to matrix multiplication. It is easy to check that the map $(A, B) \longrightarrow AB^{-1}$, $A, B \in Gl(m, R)$ is C^∞.
(3) Let $C^\star$ be the non–zero complex numbers. Then $C^\star$ is a group with respect to the multiplication of complex numbers. Moreover $C^\star$ is a (real) 2–dimensional manifold covered by a single coordinate neighborhood $U = C^\star = R^2 - \{(0,0)\}$ with coordinates

$$z = x + \sqrt{-1}y \longrightarrow (x, y).$$

One easily check that the map $(z_1, z_2) \longrightarrow z_1 z_2^{-1}$ is C^∞.
(4) The circle S^1 may be identified with the complex numbers of absolute value $+1$. Since $| z_1 z_2 |=| z_1 || z_2 |$, then S^1 is a group with respect to the multiplication of complex numbers. Moreover the map $(z_1, z_2) \longrightarrow z_1 z_2^{-1}$ is C^∞ (see Exercise 1.22.1).

Proposition 1.19.3 *If G_1 and G_2 are Lie groups, then the direct product $G_1 \times G_2$ of these groups (with the multiplication given by $(a_1, a_2)(b_1, b_2) = (a_1 b_1, a_2 b_2)$), with the C^∞ structure of product manifolds is a Lie group.*

The proof is left to the reader as an exercise.

Corollary 1.19.4 *The m–torus T^m is a Lie group.*

Definition 1.19.5 *A* **Lie algebra** *over R is a real vector space g with a bilinear map $[\ ,\ ] : g \times g \longrightarrow g$ (called the* **bracket***) such that:*
(1) $[X, Y] = -[Y, X]$,
*(2) $[[X, Y], Z] + [[Y, Z], X] + [[Z, X], Y] = 0$, (***Jacoby identity***), for all $X, Y, Z \in g$.*

Examples of Lie algebras

(1) Let V be an m–dimensional vector space. Then V becomes a Lie algebra by setting $[X, Y] = 0$, for all vector $X, Y \in V$. (A Lie algebra g such that $[X, Y] = 0$ for all $X, Y \in g$ is called **Abelian**).

(2) Let M be an m–dimensional manifold. Then the vector space $\chi(M)$ of all vector fields on M is a Lie algebra with respect to the Lie bracket of vector fields.

(3) The set $gl(m, R)$ is a real vector space of dimension m^2. If we set

$$[A, B] = AB - BA,$$

then $gl(m, R)$ becomes a Lie algebra.

Definition 1.19.6 *Let G be a Lie group and $a \in G$. We denote by ℓ_a (resp. r_a) the* **left translation** *(resp.* **right translation***) of G by a, i.e., $\ell_a(b) = ab$ (resp. $r_a(b) = ba$). Clearly ℓ_a is a diffeomorphism with inverse $\ell_{a^{-1}}$. A vector field X on G is called* **left invariant** *if it is invariant by all left invariant translations ℓ_a, i.e.,*

$$d\ell_a(b)X(b) = X(ab), \ a, b \in G.$$

A p–form ω on G is said to be **left invariant** if it is invariant by all ℓ_a, i.e.,

$$\ell_a^* \omega = \omega, \ a \in G.$$

Next we shall prove that there is a finite dimensional Lie algebra associated with each finite dimensional Lie group.

Let G be a Lie group and g its set of left invariant vector fields. Clearly g is a real vector space. Define a linear map

$$\alpha : g \longrightarrow T_e G$$

by

$$\alpha(X) = X(e).$$

Then α is a linear isomorphism. In fact, α is injective, since if $\alpha(X) = \alpha(Y)$ then for each $a \in G$ we have

$$X(a) = d\ell_a(e)X(e) = d\ell_a(e)Y(e) = Y(a).$$

Moreover, α is surjective, since, given $v \in T_eG$, we set $X(a) = d\ell_a(e)v$ and $X \in g$ because

$$d\ell_a(b)X(b) = d\ell_a(b)d\ell_b(e)v = d\ell_{ab}(e)v = X(ab).$$

Hence

$$dim\, g = dim\, T_eG = dim\, G.$$

Now, the Lie bracket $[X, Y]$ of two left invariant vector fields X and Y is a left invariant vector field (see Exercise 1.22.8). Then g is a Lie algebra which is called the **Lie algebra of** G.

Let X be a left invariant vector field on G. If φ_t is a local 1–parameter group of local transformations of G generated by X and $\varphi_t(e)$ is defined for $t \in (-\epsilon, \epsilon)$, $\epsilon > 0$, then $\varphi_t(a)$ can be defined for $t \in (-\epsilon, \epsilon)$ for every $a \in G$. Furthermore we have $\varphi_t(a) = \ell_a(\varphi_t(e))$ since φ_t commutes with ℓ_a for every $a \in G$ (see Exercise 1.22.4). Hence $\varphi_t(a)$ is defined for $t \in R$ (see Exercise 1.22.2), and X is complete. We set

$$\varphi_1(e) = \exp X.$$

Now $\psi_s = \varphi_{st}$ is the 1–parameter group of transformations generated by tX. Indeed, we have that the tangent vector to the curve $s \longrightarrow \psi_s(a)$ at $s = 0$ is equal to t times the tangent vector to the curve $s \longrightarrow \varphi_{st}(a)$ at $s = 0$. Hence we deduce

$$\exp tX = \psi_1(e) = \varphi_t(e), \text{ for all } t \in R.$$

The map $X \longrightarrow \exp X$ of g into G is called the **exponential map**. We have

$$\exp(s+t)X = (\exp sX)(\exp tX),$$

$$\exp(-tX) = (\exp tX)^{-1}.$$

We call $\exp tX$ the **1–parameter subgroup of** G **generated by** X.

Proposition 1.19.7 exp : $g \longrightarrow G$ *is* C^∞ *and* $d\exp(0) : T_0g \longrightarrow T_eG$ *is the identity map (with the usual identifications* $T_0g \cong g$ *and* $T_eG \cong g$*), so* exp *gives a diffeomorphism of a neighborhood of 0 in* g *onto a neighborhood of* e *in* G *(for a proof see Warner [123]).*

Example. Let $Gl(m,R)$ be the general linear group. Since $gl(m,R)$ is a vector space then we have a canonical identification

$$T_I(gl(m,R)) \xrightarrow{\beta} gl(m,R),$$

where I is the identity matrix. Moreover, we have a canonical identification $T_I(Gl(m,R)) \cong T_I(gl(m,R))$. Now, if g is the Lie algebra of $Gl(m,R)$ we define a canonical linear isomorphism

$$\gamma : g \longrightarrow gl(m,R)$$

by

$$\gamma(X) = \beta(X(I)).$$

One easily check that γ is in fact a Lie algebra isomorphism (i.e., $\gamma[X,Y] = [\gamma X, \gamma Y]$) and then $gl(m,R)$ **is the Lie algebra of** $Gl(m,R)$. It is known that the exponential map $\exp : gl(m,R) \longrightarrow Gl(m,R)$ coincides with the usual exponential map

$$\exp A = \sum_{i=0}^{\infty} A^i/i! = I + A + A^2/2! + \dots$$

(see Warner [123] for more details).

Remark 1.19.8 (1) Let V be an m–dimensional vector space. Let $End(V)$ denote the group of endomorphisms of V, and let $Aut(V)$ denote the vector space of automorphisms of V. Then $Aut(V)$ is a Lie group with Lie algebra $End(V)$. A basis of V determines a diffeomorphism of $End(V)$ with $gl(m,R)$ sending $Aut(V)$ onto $Gl(m,R)$.
(2) Let $Gl(m,C)$ be the **complex general linear group**. We can easily prove that $Gl(m,C)$ is a $2m^2$–dimensional Lie group with Lie algebra $gl(m,C)$. If V is a complex m– dimensional vector space, and if $End(V)$ denotes the group of complex linear transformations of V, and $Aut(V)$ denotes the set of complex automorphisms of V, then $Aut(V)$ is a $2m^2$–dimensional Lie group with Lie algebra $End(V)$.

Definition 1.19.9 *(1) Let G be a Lie group. By a* **Lie subgroup** *H of G we mean a subgroup which is at the same time a submanifold of G such that H itself is a Lie group with respect to this differentiable structure. (2) Let g be a Lie algebra. By a* **Lie subalgebra** *of g we mean a vector subspace h of g such that $[X,Y] \in h$ whenever $X, Y \in h$.*

Let H be a Lie subgroup of G. If $X \in h$ (Lie algebra of H) then X is determined by its value at e, $X(e)$, and $X(e) \in T_eH \subset T_eG$ determines a left invariant vector field $\tilde{X}$ on G, i.e., $\tilde{X} \in g$ (Lie algebra of G). It follows that h can be identified with a Lie subalgebra $\tilde{h}$ of g. Conversely, every subalgebra $\tilde{h}$ of g is the Lie algebra of a unique connected Lie subgroup H of G. To see this, we first define an involutive distribution $\mathcal{H}$ on G as follows:

$$\mathcal{H}(a) = \{X(a)/X \in \tilde{h}\}.$$

Then the maximal integral submanifold through e of $\mathcal{H}$ is a Lie subgroup H of G and the Lie algebra h of H can be identified to $\tilde{h}$. Hence we have

Proposition 1.19.10 *There is a 1-1 correspondence between connected Lie subgroups of G and Lie subalgebras of the Lie algebra g of G.*

Let H be a Lie subgroup of a Lie group G. Then H is an immersed submanifold of G, but H is not necessarily an embedded submanifold (see Example below). We have the following result about Lie subgroups which are embedded submanifolds.

Theorem 1.19.11 *A Lie subgroup H of a Lie group G is an embedded submanifold if and only if H is closed*

(See Warner [123] for a proof).
A further result is the following.

Theorem 1.19.12 (Cartan Theorem). *Let G be a Lie group, and let A be a closed abstract subgroup of G. Then A has a unique C^∞ structure which makes A into a Lie subgroup of G.*

We omit the proof (see Spivak [113], Warner [123]).
Examples
The following Lie groups are closed subgroups of the complex general linear group $Gl(m,C)$ (if $A \in Gl(m,C)$, then we denote by A^t the transpose of A and by $\bar{A}$ the complex conjugate of A):
(1) <u>**Unitary group**</u>

$$U(m) = \{A \in Gl(m,C)/\ \overline{A^t}A = I\}$$

Its Lie algebra is

$$u(m) = \{A \in gl(m, C) / \; A^t + \bar{A} = 0\}$$

A matrix $A \in U(m)$ is called **skew–Hermitian.**
(2) **Special linear group**

$$Sl(m, C) = \{A \in Gl(m, C) / \; \det A = 1\}$$

Its Lie algebra is

$$sl(m, C) = \{A \in gl(m, C) / \; \mathrm{trace}\, A = 0\}$$

(3) **Complex orthogonal group**

$$O(m, C) = \{A \in Gl(m, C) / \; A^t A = I\}$$

Its Lie algebra is

$$o(m, C) = \{A \in gl(m, C) / \; A^t + A = 0\}$$

A matrix $A \in O(m, C)$ is called **skew–symmetric.**
(4) **Special unitary group**

$$SU(m) = U(m) \cap Sl(m, C)$$

Its Lie algebra is

$$su(m) = u(m) \cap sl(m, C).$$

(5) **Real special linear group**

$$Sl(m, R) = Sl(m, C) \cap Gl(m, R) = \{A \in Gl(m, R) / \det A = 1\}$$

Its Lie algebra is

$$sl(m, R) = \{A \in gl(m, R) / \; \mathrm{trace}\, A = 0\}$$

(6) **Orthogonal group**

$$O(m) = U(m) \cap Gl(m, R) = \{A \in Gl(m, R) / \; A^t A = I\}$$

Its Lie algebra is

$$o(m) = o(m,C) \cap gl(m,R) = \{A \in gl(m,R)/\ A^t + A = 0\}.$$

(7) **Special orthogonal group**

$$SO(m) = O(m) \cap Gl(m,R).$$

Its Lie algebra is also $o(m)$.

The dimensions of these Lie groups are easily computed from their Lie algebras:

$$dim\ U(m) = m^2,$$

$$dim\ Sl(m,C) = 2m^2 - 2,$$

$$dim\ O(m,C) = m(m-1),$$

$$dim\ SU(m) = m^2 - 1,$$

$$dim\ Sl(m,R) = m^2 - 1,$$

$$dim\ O(m) = dim\ SO(m) = (m(m-1))/2.$$

Definition 1.19.13 *(1) A map $\varphi : G \longrightarrow H$ is a* **Lie group homomorphism** *if φ is both C^∞ and a group homomorphism. We call φ a* **Lie group isomorphism** *if, in addition, φ is a diffeomorphism. A Lie group isomorphism of G into itself is called a* **Lie group automorphism**. *If $H = Aut(V)$ for some vector space V, or if $H = Gl(m,C)$ or $Gl(m,R)$, then a homomorphism $\varphi : G \longrightarrow H$ is called a* **representation of the Lie group** *G.*
(2) A linear map $\psi : g \longrightarrow h$ is a **Lie algebra homomorphism** *if $\psi[X,Y] = [\psi X, \psi Y]$, for all $X,Y \in g$. If ψ is, in addition, a linear isomorphism, then ψ is calledd a* **Lie algebra isomorphism**. *An isomorphism of g into itself is called an* **automorphism**. *If $h = End(V)$ for some vector space V, or if $h = gl(m,C)$ or $gl(m,R)$, then a homomorphism $\psi : g \longrightarrow h$ is called a* **representation of the Lie algebra** *g.*

Example
Let $T^2 = S^1 \times S^2$ and let $\varphi : R^2 \longrightarrow T^2$ be given by

$$\varphi(x_1, x_2) = (e^{2\pi i x_1}, e^{2\pi i x_2}).$$

Then φ is a C^∞ map of rank 2 everywhere and is a Lie group homomorphism. Now, let α be an irrational number and define $F : R \longrightarrow R^2$ by $F(t) = (t, \alpha t)$. Obviously F is an embedding with image being the line through (0,0) of slope α. Hence $\varphi' = \varphi \circ F$ is an injective immersion of R into T^2 and $\varphi'(R)$ is an immersed submanifold of T^2. Furthermore φ' is a Lie group homomorphism so that $\varphi'(R)$ is a Lie subgroup of T^2. But $\varphi'(R)$ is dense in T^2 and, thus, it is not an embedded submanifold.

Now, let $\varphi : G \longrightarrow H$ be a Lie group homomorphism. Since φ maps the identity of G to the identity of H, then we have

$$d\varphi(e) : T_e G \longrightarrow T_e H$$

and hence we can define a Lie algebra homomorphism $\tilde{\varphi} : g \longrightarrow h$ of g (Lie algebra of G) into h (Lie algebra of H) as follows: If $X \in g$, then $\tilde{\varphi}(X)$ is the unique left invariant vector field on H such that

$$\tilde{\varphi}(X)(e) = d\varphi(e)(X(e)).$$

In particular, for every $a \in G$, we can define an **inner automorphism** $i_a : G \longrightarrow G$ by $i_a(b) = aba^{-1}$. Then i_a induces an automorphism of g (Lie algebra of G), denoted by $Ad(a)$. The representation

$$Ad : G \longrightarrow Aut(g)$$

given by

$$Ad : a \longrightarrow Ad(a)$$

is called the **adjoint representation of** G. For every $a \in G$ and $X \in g$ we have

$$(Ad(a))(X)(e) = di_a(e)(X(e))$$

$$= dr_{a^{-1}}(a) d\ell_a(e)(X(e))$$

$$= (dr_{a^{-1}}(a))X(a)$$

$$= (Tr_{a^{-1}})(X)(e),$$

since X is left invariant and $i_a = r_{a^{-1}} \circ \ell_a$. Thus

$$(Ad(a))X = (Tr_{a^{-1}})X.$$

Proposition 1.19.14 *Let $X, Y \in g$. Then*

$$[Y, X] = \lim_{t \to 0}(1/t)[Ad(\exp(-tX))Y - Y]$$

Proof: Let $X, Y \in g$ and φ_t the 1–parameter group of transformations of G generated by X. Since $\varphi_t(a) = \ell_a(\varphi_t(e)) = \ell_a(\exp tX) = a(\exp tX) = r_{\exp tX}(a)$, we deduce that $\varphi_t = r_{\exp tX}$.

By Exercise 1.22.7, we have

$$[Y, X] = \lim_{t \to 0}(1/t)[(T\varphi_t)Y - Y]$$

$$= \lim_{t \to 0}(1/t)[(Tr_{\exp tX})Y - Y]$$

$$= \lim_{t \to 0}(1/t)[Ad(\exp(-tX))Y - Y]. \square$$

Now, the adjoint representation $Ad : G \longrightarrow Aut(g)$ induces a Lie algebra homomorphism $ad : g \longrightarrow End(g)$. It is not hard to prove that

$$ad(X)(Y) = ad_X Y = [X, Y]$$

(see Spivak [113], Warner [123]). The representation $ad : g \longrightarrow End(g)$ is called the **adjoint representation of** g.

Let G be a Lie group. If ω is a left invariant 1–form and X a left invariant vector field on G, then

$$\omega(X)(a) = \omega(a)(X(a)) = \omega(a)(d\ell_a(e)X(e))$$

$$= (\ell_a^* \omega(a))(X(e)) = \omega(e)X(e) = \omega(X)(e)$$

and thus $\omega(X)$ is constant on G. Furthermore, if ω is a left invariant form, then so is $d\omega$, since d commutes with ℓ_a^*, for every $a \in G$. Hence, if ω is a left invariant 1–form and $X, Y \in g$, we obtain

$$d\omega(X,Y) = -\omega[X,Y], \tag{1.18}$$

which is called the **Maurer–Cartan equation**.

Clearly, the vector space of left invariant 1–forms on G can be identified with the dual space g^* of $G : \omega \longrightarrow \tilde{\omega}$, where $\tilde{\omega}(X) = \omega(e)X(e)$. Thus if $X_1, \ldots, X_m$ is a basis for g and $\theta^1, \ldots, \theta^m$ is the dual basis for g^*, then $\theta^1, \ldots, \theta^m$ are left invariant 1–forms on G. We set

$$[X_i, X_j] = C_{ij}^k X_k,$$

where the C_{ij}^k's are called the **structure constants** of G with respect to the basis $X_1, \ldots, X_m$. From (1.18) it is easy to check that the Maurer–Cartan equation is given by

$$d\theta^k = -1/2 \sum_{i,j=1}^{m} C_{ij}^k \theta^i \wedge \theta^j, \; 1 \leq k \leq m,$$

since $C_{ij}^k = -C_{ji}^k$.

1.20 Principal bundles. Frame bundles

Before introducing principal bundles, we need some generalities about Lie groups acting on manifolds.

Definition 1.20.1 *Let M be a differentiable manifold and G a Lie group. A C^∞ map $\phi : M \times G \longrightarrow M$ such that*
(1) $\phi(x, ab) = \phi(\phi(x,a), b)$,
(2) $\phi(x, e) = x$,
for all $a, b \in G$ and $x \in M$ is called an **action of G on M on the right**.

If G acts on M on the right, then for a fixed $a \in G$, the map $\phi_a : M \longrightarrow M$, $\phi_a(x) = \phi(x, a)$ is a transformation of M (for this reason we call G a **Lie transformation group**). For a fixed $x \in M$, we denote by $\phi_x : G \longrightarrow M$ the C^∞ map defined by $\phi_x(a) = \phi(x, a)$.

We say that G acts **effectively** (resp. **freely**) if $\phi_a(x) = x$ for all $x \in M$ (resp. for some $x \in M$) implies that $a = e$, i.e., $\phi_a = id_M$.

For the sake of simplicity we sometimes set $\phi(x, a) = xa$.

Remark 1.20.2 Similarly, we can define an **action of G on M on the left** as a C^∞ map $\phi : G \times M \longrightarrow M$ such that
(1) $\phi(ab, x) = \phi(a, \phi(b, x))$,
(2) $\phi(e, x) = x$,
for all $a, b \in G$ and $x \in M$.

Now, let us suppose that G acts on the right on M and let g be the Lie algebra of G. Then to each $A \in g$ we assign a vector field λA on M as follows: λA is the infinitesimal generator of the 1–parameter group of transformations of M given by

$$(t, x) \in R \times M \longrightarrow \phi(x, exp\, tA)$$

Proposition 1.20.3 *We have*

$$(\lambda A)(x) = d\phi_x(e)A_e$$

Proof: In fact,

$$d\phi_x(e)A_e = d\phi_x(0)(d\,exp_A(0))(d/dt(0))$$

(see Exercise 1.22.23).

$$= d(\phi_x \circ exp_A)(0)(d/dt(0))$$

$$= (\lambda A)(x),$$

since

$$(\phi_x \circ exp_A)(t) = \phi_x(exp\, tA) = \phi(x, exp\, tA). \square$$

Proposition 1.20.4 *Assume that G acts on M on the right. Then*
(1) The map $\lambda : g \longrightarrow \chi(M)$ is linear.
(2) $\lambda[A, B] = [\lambda A, \lambda B]$.
(3) If G acts effectively then λ is injective.
(4) If G acts freely, then for each non–zero $A \in g$, λA never vanishes.

Proof: (1) follows from Proposition 1.20.3. To prove (2) we note that

$$[A,B] = \lim_{t\to 0}(1/t)[B - Ad(exp(-tA))],$$

by Proposition 1.19.14.

On the other hand, we have

$$\phi_{exp\,tA} \circ \phi_{x(exp\,(-tA))}(a) = x[exp\,(-tA)a(exp\,tA)],$$

for all $a \in G$. Since $\phi(x, exp\,tA) = x(exp\,tA)$ is the 1–parameter group of transformations generated by λA, we have

$$[\lambda A, \lambda B](x) = \lim_{t\to 0}(1/t)[(\lambda B)(x) - (T\phi_{exp\,tA})(\lambda B)(x)]$$

$$= \lim_{t\to 0}(1/t)[(T\phi_x)B_e - (T\phi_{exp\,tA}) \circ (T\phi_{x\,exp(-tA)})B_e]$$

$$= \lim_{t\to 0}(1/t)[(T\phi_x)B_e - (T\phi_x)(Ad(exp(-tA))B_e)]$$

$$= (T\phi_x)(\lim_{t\to 0}(1/t)[B_e - Ad(exp(-tA))B_e])$$

$$= (T\phi_x)([A,B]_e) = (\lambda[A,B])(x),$$

from Proposition 1.19.14.

Hence $\lambda : g \longrightarrow \chi(M)$ is a Lie algebra homomorphism.

To prove (3), suppose that $\lambda A = 0$ everywhere in M. This implies that, for every $x \in M$, the 1–parameter group of transformations $x(exp\,tA)$ must be trivial, i.e., $x(exp\,tA) = x$. If G acts effectively, then $exp\,tA = 0$ for every t and hence $A = 0$.

To prove (4), suppose $(\lambda A)(x) = 0$ for some $x \in M$. Then $x(exp\,tA) = x$ for every t. If G acts freely, then $exp\,tA = 0$ for every t and, hence $A = 0$. □

Definition 1.20.5 *Let M be a differentiable manifold and G a Lie group. A* **principal bundle over** *M*, **with group** *G, consists of a fibred manifold P over M with* **projection** $\pi : P \longrightarrow M$ *and an action of G on P on the right satisfying the following conditions:*

(1) $\pi(ua) = \pi(a)$ *for all* $u \in P$ *and* $a \in G$;

(2) P is locally trivial, i.e., for each $x \in M$ there is a neighborhood U of x and a diffeomorphism $\psi : \pi^{-1}(U) \longrightarrow U \times G$ such that $\psi(u) = (\pi(u), \varphi(u))$ where $\varphi : \pi^{-1}(U) \longrightarrow G$ is a map satisfying $\varphi(ua) = (\varphi(u))a$ for all $u \in \pi^{-1}(U)$ and $a \in G$.

A principal bundle will be denoted by $P(M, G, \pi)$, $P(M, G)$, or simply P. We call P the **total space** *or the* **bundle space**, *M the* **base space** *and G the* **structure group**.

Since $\pi(ua) = \pi(u)$ we deduce that $\{ua/a \in G\} \subset \pi^{-1}(\pi(u))$. Moreover, since $\varphi(ua) = (\varphi(u))a$, we easily see that $\{ua/a \in G\} = (\pi^{-1}(\pi(u))$. Notice also that if $ua = u$ for some $u \in P$, then $a = e$. Hence G acts freely on P.

Each fibre $\pi^{-1}(x)$, $x \in M$, of P is diffeomorphic to G. If $x = \pi(u)$ for some $u \in P$, then $\pi^{-1}(x)$ is an embedded submanifold of P and the tangent space to $\pi^{-1}(x)$ at u is a vector subspace V_u of T_uP which is called the **vertical subspace at u**. Clearly we have

$$V_u = Ker\{d\pi(u) : T_uP \longrightarrow T_xM\}.$$

A tangent vector in V_u is called **vertical**.

Examples of principal bundles

(1) <u>Trivial principal bundles</u>

Let M be a differentiable manifold and G a Lie group. We set $P = M \times G$ and define an action $P \times G \longrightarrow P$ by

$$(x, a)b = (x, ab).$$

If we define $\pi : M \times G \longrightarrow M$ by $\pi(x, a) = x$, then P is a principal bundle over M with structure group G and projection π.

(2) <u>Frame bundles</u>

Let M be an m–dimensional manifold. A **linear frame** u at a point $x \in M$ is an ordered basis $X_1, \ldots, X_m$ of the tangent space T_xM. Let FM be the set of all linear frames at all points of M and let π be the map of FM onto M which maps a linear frame u at x into x. We define an action of the general linear group $Gl(m, R)$ on FM on the right as follows. If $a = (a^i_j) \in Gl(m, R)$ and $u = (X_1, \ldots, X_m)$ is a linear frame at x, then ua is the linear frame $(Y_1, \ldots, Y_m)$ at x defined by

$$Y_i = \sum_{j=1}^{m} a^j_i X_j.$$

In order to introduce a differentiable structure in FM we proceed as follows. Let $(x^1, \ldots, x^m)$ be a local coordinate system in a coordinate neighborhood U of M. If $x \in U$ then $(\partial/\partial x^1)_x, \ldots, (\partial/\partial x^m)_x$ is a linear frame at x. Hence every linear frame u at x can be expressed uniquely in the form $u = (X_1, \ldots, X_m)$ with

$$X_i = \sum_{j=1}^{m} X_i^j (\partial/\partial x^j),$$

where $(X_i^j) \in Gl(m, R)$. Thus we have a bijective map

$$\psi : \pi^{-1}(U) \longrightarrow U \times Gl(m, R)$$

given by

$$\psi(u) = (x, (X_i^j)).$$

We can make FM into a differentiable manifold by taking (x^i, X_i^j) as a local coordinate system in $\pi^{-1}(U)$. It is clear that the action $FM \times Gl(m, R) \longrightarrow FM$, $u \longrightarrow ua$, is C^∞. Thus, if we set $\varphi(u) = (X_i^j) \in Gl(m, R)$, where $u = (X_1, \ldots, X_m)$, then it is easy to check that FM is a principal bundle over M with structure group $Gl(m, R)$ and projection π. We call FM the **frame bundle** (or **bundle of linear frames**) of M. We notice that if $\{e_1, \ldots, e_m\}$ is the canonical basis for R^m and $u = (X_1, \ldots, X_m)$ is a linear frame at x, then u may be considered as a linear isomorphism $u : R^m \longrightarrow T_xM$ defined by $u(e_i) = X_i$, $1 \leq i \leq m$. On the other hand, a non–singular matrix $a \in Gl(m, R)$ may be considered as the automorphism $a : R^m \longrightarrow R^m$ given by $a(e_i) = a_i^j e_j$. Hence the action of $Gl(m, R)$ in FM is given by the composition $R^m \xrightarrow{a} R^m \xrightarrow{u} T_xM$.

Given a principal bundle $P(M, G)$, the action of G on P induces a Lie algebra homomorphism λ of the Lie algebra g of G into $\chi(P)$ by Proposition 1.20.4. For each $A \in g$, λA is called the **fundamental vector field corresponding to** A. Since the action of G sends each fibre into itself, $(\lambda A)_u \in V_u$. As G acts freely on P, if $A \neq 0$ then λA never vanishes. Then the linear map

$$g \longrightarrow V_u$$

defined by

$$A \longrightarrow (\lambda A)_u$$

is injective. Since $dim\, g = dim\, V_u$, we deduce that this map is a linear isomorphism.

For each $a \in G$, let $R_a : P \longrightarrow P$ be the diffeomorphism defined by $R_a(u) = ua$. We have the following.

Proposition 1.20.6 *If λA is the fundamental vector field corresponding to $A \in g$, then $(TR_a)(\lambda A)$ is the fundamental vector field corresponding to $(Ad(a^{-1}))A \in g$.*

Proof: Since λA is induced by the 1–parameter group of transformations $R_{exp\, tA}$ of P, then the vector field $(TR_a)(\lambda A)$ is induced by the 1–parameter group of transformations $R_a R_{exp\, tA} R_{a^{-1}} = R_{a^{-1}(\exp\, tA)a}$ (see Exercise 1.22.4). But $a^{-1}(exp\, tA)a$ is the 1– parameter group of transformations of G generated by $Ad(a^{-1})A$. Thus $R_{a^{-1}(\exp\, tA)a}$ is generated by $\lambda(Ad(a^{-1})A)$.$\square$

Definition 1.20.7 *Let $P(M,G)$ be a principal bundle with projection π. A* **section** *s of P over an open set $U \subset M$ is a section of the fibred manifold $\pi : P \longrightarrow M$, i.e., a C^∞ map $s : U \longrightarrow P$ such that $\pi \circ s = id_U$. If P is the frame bundle FM of a manifold M, then a section s of FM over U is called a* **local frame field** *of U. If $x \in U$, hence $s(x) = (s_1(x), \ldots, s_m(x))$, $m = dim\, M$, is a basis of T_xM. Thus a local frame field s on U determines m vector fields $s_1, \ldots, s_m$ on U which are linearly independent at each point of U.*

Remark 1.20.8 Given a principal bundle $P(M,G)$ with projection π, there always exist local sections. In fact, since P is locally trivial, for each $x \in M$, there exists an open set $U \subset M$ and a diffeomorphism $\psi : \pi^{-1}(U) \longrightarrow U \times G$ such that $\psi(u) = (\pi(u), \varphi(u))$, where $\varphi : \pi^{-1}(u) \longrightarrow G$ satisfies $\varphi(ua) = (\varphi(u))a$. Hence we can define a local section $s : U \longrightarrow P$ by

$$s(y) = \psi^{-1}(y, e).$$

Definition 1.20.9 *(1)* **A principal bundle homomorphism** *$f : P'(M', G') \longrightarrow P(M,G)$ consists of a map $f_1 : P' \longrightarrow P$ and a Lie group homomorphism $f_2 : G' \longrightarrow G$ such that $f_1(u'a') = f_1(u')f_2(a')$ for all $u' \in P'$ and $a' \in$*

G'. Thus f_1 maps each fibre of P' into a fibre of P and then we can define a map $f_3 : M' \longrightarrow M$ by $f_3(x') = \pi(f_1(u'))$, $x' \in M'$, $u' \in P'$, $\pi'(u') = x'$ where $\pi' : P' \longrightarrow M'$ and $\pi : P \longrightarrow M$ are the corresponding projections. Hence the following diagram

$$\begin{array}{ccc} P' & \xrightarrow{f_1} & P \\ {\scriptstyle \pi'}\downarrow & & \downarrow{\scriptstyle \pi} \\ M' & \xrightarrow{f_3} & M \end{array}$$

is commutative. For the sake of simplicity, we shall denote f_1, f_2 and f_3 by f.
(2) f is called an **embedding** *if $f : P' \longrightarrow P$ is an embedding and $f : G' \longrightarrow G$ is injective. By identifying P' with $f(P')$, G' with $f(G')$ and M' with $f(M')$, we say that $P'(M', G')$ is a* **principal subbundle** *of $P(M, G)$.*
(3) If, moreover, $M' = M$ and the induced map $f : M \longrightarrow M$ is the identity transformation of M, $f : P'(M, G') \longrightarrow P(M, G)$ is called a **reduction** *of G to G'. The subbundle $P'(M, G')$ is called a* **reduced bundle** *and we say that the structure group of G is* **reducible** *to the Lie subgroup G'.*

Let $P(M, G)$ be a principal bundle with projection $\pi : P \longrightarrow M$. Since P is locally trivial, it is possible to choose an open covering $\{U_\alpha\}$ of M and a diffeomorphism $\psi_\alpha : \pi^{-1}(U_\alpha) \longrightarrow U_\alpha \times G$ for each α such that $\psi_\alpha(u) = (\pi(u),\ \varphi_\alpha(u))$, where $\varphi_\alpha : \pi^{-1}(u_\alpha) \longrightarrow G$ is a map satisfying $\varphi_\alpha(ua) = (\varphi_\alpha(u))a$ for all $u \in \pi^{-1}(u_\alpha)$ and $a \in G$. Hence we can define a map $\psi_{\beta\alpha} : U_\alpha \cap U_\beta \longrightarrow G$ by

$$\psi_{\beta\alpha}(x) = \varphi_\beta(u)(\varphi_\alpha(u))^{-1},\ u \in \pi^{-1}(x). \tag{1.19}$$

In fact, (1.19) depends only on $\pi(u)$ not on u, since if $a \in G$, then we have

$$\varphi_\beta(ua)(\varphi_\alpha(ua))^{-1} = \varphi_\beta(u)a(\varphi_\alpha(u)a)^{-1} = \varphi_\beta(u)(\varphi_\alpha(u))^{-1}.$$

Definition 1.20.10 *The family of maps $\psi_{\beta\alpha}$ are called* **transition functions** *of P corresponding to the open covering $\{U_\alpha\}$.*

Proposition 1.20.11 *The transition functions of P verify*
(1) $\psi_{\alpha\alpha}(x) = e$; for $x \in U_\alpha$;
(2) $\psi_{\beta\alpha}(x) = (\psi_{\alpha\beta}(x))^{-1}$, for $x \in U_\alpha \cap U_\beta$;
(3) $\psi_{\gamma\alpha}(x) = \psi_{\gamma\beta}(x) \circ \psi_{\beta\alpha}(x)$, for $x \in U_\alpha \cap U_\beta \cap U_\gamma$.

The principal bundle P can be reconstructed from the transition functions:

Proposition 1.20.12 *Let M be a manifold, $\{U_\alpha\}$ an open covering of M and G a Lie group. Assume that there exist a family of functions $\psi_{\beta\alpha} : U_\alpha \cap U_\beta \longrightarrow G$ satisfying (1)–(3) of Proposition 1.20.11. Then there is a principal bundle $P(M, G)$ with transition functions $\psi_{\beta\alpha}$.*

Proof: We set

$$X = \bigcup_\alpha (\{\alpha\} \times U_\alpha \times G).$$

Then each element of X is a triple (α, x, a) where α is some index, $x \in U_\alpha$ and $a \in G$. Clearly X is a differentiable manifold. We introduce an equivalence relation $\sim$ on X as follows. We say that (α, x, a) is equivalent to (β, y, b) if and only if $x = y \in U_\alpha \cap U_\beta$ and $b = \psi_{\beta\alpha}(x)a$. Then $P = X/\sim$. The projection $\pi : P \longrightarrow M$ is defined by $\pi[\alpha, x, a] = x$, where $[\alpha, x, a]$ denotes the equivalence class of (α, x, a). The action of G on P is given by

$$[\alpha, x, a]b = [\alpha, x, ab].$$

In order to make P into a differentiable manifold we note that, by the natural projection $X \longrightarrow P = X/\sim$, each $\{\alpha\} \times U_\alpha \times G$ is mapped one–to–one onto $\pi^{-1}(U_\alpha)$. Hence we introduce a differentiable structure on P by requiring that the map $X \longrightarrow P$ induces a diffeomorphism of $\{\alpha\} \times U_\alpha \times G$ onto $\pi^{-1}(U_\alpha)$. Now we can easily check that $P(M, G)$ is a principal bundle over M with structure group G and projection π. Moreover if we define

$$\psi_\alpha : \pi^{-1}(U_\alpha) \longrightarrow U_\alpha \times G$$

by

$$\psi_\alpha[\alpha, x, a] = (x, a),$$

then the transition functions $\psi'_{\beta\alpha} : U_\alpha \cap U_\beta \longrightarrow G$ corresponding to $\{U_\alpha\}$ are precisely $\psi_{\beta\alpha}$.□

Theorem 1.20.13 *The structure group G of $P(M, G)$ is reducible to a subgroup G' if and only if there exists an open covering $\{U_\alpha\}$ of M with transition functions $\psi_{\beta\alpha}$ which take their values in G'.*

Proof: Suppose that G is reducible to G'. Then there exists a reduced bundle $P'(M,G')$ with projection $\pi' = \pi/P'$, where $\pi : P \longrightarrow M$ is the projection of P over M. Let $\{U_\alpha\}$ be an open covering of M with diffeomorphisms $\psi'_\alpha : (\pi')^{-1}(U_\alpha) \longrightarrow U_\alpha \times G'$, and let $\psi'_{\beta\alpha} : U_\alpha \cap U_\beta \longrightarrow G'$ be the corresponding transition functions. We extend ψ'_α to $\pi^{-1}(U_\alpha)$ as follows. If by $u \in \pi^{-1}(U_\alpha)$, then $u = u'a$, where $u' \in (\pi')^{-1}(U_\alpha)$ and $a \in G$. Then we define

$$\varphi_\alpha : \pi^{-1}(U_\alpha) \longrightarrow G$$

by

$$\varphi_\alpha(u) = \varphi_\alpha(u'a) = \varphi'_\alpha(u')a$$

and

$$\psi_\alpha : \pi^{-1}(U_\alpha) \longrightarrow U_\alpha \times G$$

by

$$\psi_\alpha(u) = (\pi(u),\ \varphi_\alpha(u)).$$

Then the corresponding transition funtions $\psi_{\beta\alpha} : U_\alpha \cap U_\beta \longrightarrow G$ are given by $\psi_{\beta\alpha}(x) = \psi'_{\beta\alpha}(x) \in G'$, for all $x \in U_\alpha \cap U_\beta$.

Conversely, suppose that there is an open covering $\{U_\alpha\}$ of M with transition functions $\psi_{\beta\alpha}$ taking values in a Lie subgroup G' of G. From Exercise 1.22.20, the map $\psi_{\beta\alpha} : U_\alpha \cap U_\beta \longrightarrow G'$ is C^∞ with respect to the differentiable structure of G'. Hence, proceeding as in Proposition 1.20.12, we construct a principal bundle $P'(M,G')$ with transition functions $\psi_{\beta\alpha}$. Now, it is easy to prove that P' is a subbundle of P (details are left to the reader as an exercise). □

Example

Let FM be the frame bundle of an m–dimensional manifold M. Let $\{U_\alpha\}$ be an open covering of M by coordinate neighborhoods U_α (i.e., an atlas of M) with coordinate functions $x^1_\alpha, \ldots, x^m_\alpha$. Then the transition functions $\psi_{\beta\alpha}$ of FM corresponding to $\{U_\alpha\}$ are given by

$$\psi_{\beta\alpha}(x) = (\partial x^j_\beta / \partial x^i_\alpha)(x),\ x \in U_\alpha \cap U_\beta. \tag{1.20}$$

In fact,

$$\psi_{\beta\alpha}(x) = \varphi_\beta(u)(\varphi_\alpha(u))^{-1},$$

where u is an arbitrary linear frame at x. If we take $u = (\partial/\partial x^1_\alpha, \ldots, \partial/\partial x^m_\alpha)$, then we obtain (1.20).

1.21 G–structures

This section concerns to G–structures. G–structures on manifolds has been introduced by Chern in 1953. For a general theory of G–structures see Bernard [7], Chern [14] and Fujimoto [57].

Definition 1.21.1 *Let M be a differentiable manifold of dimension m and G a Lie subgroup of $Gl(m, R)$. A G–**structure** on M is a reduction of the frame bundle FM to G.*

Thus, a G–structure on M is defined by a principal bundle $B_G(M)$ over M with structure group G satisfying the following conditions:

1. $B_G(M)$ is a submanifold of FM;
2. The projection $\pi' : B_G(M) \longrightarrow M$ is the restriction of the projection $\pi : FM \longrightarrow M$;
3. The map $R'_a : B_G(M) \longrightarrow B_G(M)$ is the restriction of the map $R_a : FM \longrightarrow FM$ for all $a \in G$.

In the sequel we will denote π' and R'_a by π and R_a, respectively.
From Theorem 1.20.13, we have,

Proposition 1.21.2 *A manifold M of dimension m admites a G–structure if and only if there is an atlas $\{U_\alpha\}$ of M with coordinate functions $x^1_\alpha, \ldots, x^m_\alpha$, such that the Jacobian matrices $(\partial x^j_\beta / \partial x^i_\alpha)(x)$, for all $x \in U_\alpha \cap U_\beta$, belong to G.*

The following result will be very useful in the next chapters.

Proposition 1.21.3 *Let M be an m–dimensional manifold, G a Lie subgroup of $Gl(m, R)$, and B a subset of FM. Then B is a G–structure on M if and only if the following conditions are satisfied:*
(1) $\pi(B) = M$;
(2) $(\pi/B)^{-1}(x) = uG = \{ua/a \in G\}$, $x \in M$, $x = \pi(u)$;
(3) for each $x \in M$ there exists a neighborhood U of x and a C^∞ section $s : U \longrightarrow FM$ of FM over U such that $s(U) \subset B$.

Proof: It is clear that if B is a G–structure on M, then (1), (2) and (3) hold. Let us prove the converse. From (3) we can choose an open covering

$\{U_\alpha\}$ of M such that, for each U_α, there exists a local section s_α of FM over U_α taking values in B. Since $\pi(s_\alpha(x)) = \pi(s_\beta(x))$, for each $x \in U_\alpha \cap U_\beta$, then, from (2), there exists a unique element $\psi_{\beta\alpha}(x) \in G$ such that

$$s_\alpha(x) = s_\beta(x)\psi_{\beta\alpha}(x).$$

Hence $\psi_{\beta\alpha} : U_\alpha \cap U_\beta \longrightarrow Gl(m, R)$ is a C^∞ map taking values in G. By Exercise 1. , it follows that $\psi_{\beta\alpha} : U_\alpha \cap U_\beta \longrightarrow G$ is C^∞ with respect to the differentiable structure on G. Now, we define a map

$$\Phi_\alpha : U_\alpha \times G \longrightarrow (\pi/B)^{-1}(U_\alpha)$$

by

$$\Phi_\alpha(x, a) = s_\alpha(x)a,$$

for all $x \in U_\alpha$ and $a \in G$. We can easily check that Φ_α is bijective. Moreover, we have

$$(\Phi_\beta^{-1} \circ \Phi_\alpha)(x, a) = (x, \psi_{\beta\alpha}(x)a). \tag{1.21}$$

We set $\psi_\alpha = \Phi_\alpha^{-1}$. Then (1.21) becomes

$$(\psi_\beta \circ \psi_\alpha^{-1})(x, a) = (x, \psi_{\beta\alpha}(x)a). \tag{1.22}$$

Hence we introduce a differentiable structure on B such that ψ_α becomes a diffeomorphism. (From (1.22) this differentiable structure is well–defined). Moreover, if we define the action of G on B by the restriction of the action of $Gl(m, R)$ on FM, B becomes a principal bundle over M with structure group G and projection π/B. It easily follows that the inclusion $i : B \longrightarrow FM$ is an embedding and thus B is a G– structure. □

Remark 1.21.4 We notice that the function $\psi_{\beta\alpha} : U_\alpha \cap U_\beta \longrightarrow G$ are precisely the transition functions of B corresponding to the open covering $\{U_\alpha\}$.

Definition 1.21.5 *Let $B_G(M)$ be a G–structure on a differentiable manifold M. A linear frame u at $x \in M$ which belongs to $B_G(M)$ is called an* **adapted frame** *at x. A local section s of FM taking values in $B_G(M)$ is called an* **adapted frame field**.

From Proposition 1.21.3 we easily deduce the following.

Corollary 1.21.6 *M possesses a G–structure if and only if there exists an open covering $\{U_\alpha\}$ of M and local frame fields s_α on U_α such that if we set for $x \in U_\alpha \cap U_\beta$*

$$s_\alpha(x) = s_\beta(x)\psi_{\beta\alpha}(x),$$

then $\psi_{\beta\alpha}(x) \in G$.

Now we set $s_\alpha(x) = (X_1^\alpha(x), \ldots, X_m^\alpha)$ and define the m 1–forms θ_α^i on U_α by

$$\theta_\alpha^i(x)(X_j^\alpha(x)) = \delta_j^i.$$

Then $\{\theta_\alpha^1, \ldots, \theta_\alpha^m\}$ is called the **(adapted) coframe field** dual to $\{X_1^\alpha, \ldots, X_m^\alpha\}$.

Examples of G–structures

(1) $\{e\}$–structures

Let e be the identity matrix of $Gl(m, R)$. From Proposition 1.21.3, it follows that M possesses an $\{e\}$–structure if and only if there exists a (global) section $s : M \longrightarrow FM$, i.e., a global frame field. Hence giving an $\{e\}$–structure is the same as giving a global frame field. Then M **possesses an $\{e\}$–structure if and only if M is a parallelizable manifold.**

(2) $O(m)$–structures (Riemannian structures)

Let $< , >$ be the natural inner product in R^m for which the canonical basis $\{e_1, \ldots, e_m\}$ is orthonormal. Then the orthogonal group $O(m)$ may be described as follows:

$$O(m) = \{A \in Gl(m, R) / < A\xi, A\eta > = < \xi, \eta >, \quad \text{for all } \xi, \eta \in R^m\}$$

Suppose that M possesses an $O(m)$–structure $B_{O(m)}(M)$. Then we can define a Riemannian metric g on M as follows. For each $x \in M$ we set

$$g_x(X, Y) = < u^{-1}X, u^{-1}Y >, \ X, Y \in T_xM,$$

where u is a linear frame at x, $u \in B_{O(m)}(M)$ (here u is considered as a linear isomorphism $u : R^m \longrightarrow T_xM$). The invariance of $< , >$ by $O(m)$ implies that $g_x(X, Y)$ is independent of the choice of $u \in B_{O(m)}(M)$. To prove that g is C^∞ it is sufficient to consider local sections of $B_{O(m)}(M)$.

Conversely, let M be a Riemannian manifold with Riemannian metric g. We set

$$O(M) = \{u \in FM / < \xi, \eta >= g_x(u\xi, u\eta), \ \ x = \pi(u), \ \xi, \eta \in R^m\}.$$

We notice that a linear frame $u = (X_1, \ldots, X_m)$ at x belongs to $O(M)$ if and only if $\{X_1, \ldots, X_m\}$ is an orthonormal basis of $T_x M$ with respect to g_x. It easily follows that $\pi(O(M)) = M$ and $(\pi/_{O(M)})^{-1}(x) = u\,O(m)$, $x \in M$, $x = \pi(u)$. Moreover, for each $x \in M$, we can choose a neighborhood U of x and a frame field $s = (X_1, \ldots, X_m)$ on U such that $\{X_1(y), \ldots, X_m(y)\}$ is an orthonormal basis of $T_y M$ for all $y \in U$. In fact, we start with an arbitrary frame field $\{Y_1, \ldots, Y_m\}$ on a neighborhood W of x and, by the usual Gramm–Schmidt argument, we obtain $\{X_1, \ldots, X_m\}$ on U, $U \subset W$. From Proposition 1.21.3 we deduce that $O(M)$ is an $O(m)$–structure on M. $O(M)$ is called the **orthonormal frame bundle** of M and an element $u \in O(M)$ is called an **orthonormal frame.**

Thus **giving an $O(m)$–structure on M is the same as giving a Riemannian metric on M.**

(3) In the next chapters we consider more examples of G–structures: Almost tangent structures, almost product structures, almost Hermitian structures, almost contact structures and almost symplectic structures.

Definition 1.21.7 *(1) Let $f : M \longrightarrow M'$ be a local diffeomorphism. Then f induces a map $Ff : FM \longrightarrow FM'$ as follows. If $u = (X_1, \ldots, X_m)$, dim M = dim $M' = m$, is a linear frame at $x \in M$, then $Ff(u)$ is the linear frame at $f(x) \in M'$ given by $Ff(u) = (df(x)X_1, \ldots, df(x)X_m)$. Ff is called the* **natural lift** *of f. One can easily checks that Ff is a principal bundle homomorphism.*
(2) Let $B_G(M)$ and $B_G(M')$ be G–structures on M and M', respectively. Let $f : M \longrightarrow M'$ be a diffeomorphism. We say that f is an **isomorphism** *of $B_G(M)$ onto $B_G(M')$ if*

$$(Ff)(B_G(M)) = B_G(M').$$

If $M = M'$ and $B_G(M) = B_G(M')$, then f is called an **automorphism** *of $B_G(M)$.*
(3) Let $B_G(M)$ and $B_G(M')$ be G–structures on M and M', respectively. We say that $B_G(M)$ and $B_G(M')$ are **locally isomorphic** *if for each pair $(x, x') \in M \times M'$, there are open neighborhoods U of x and U' of x' and a local diffeomorphism $f : U \longrightarrow U'$ such that $(Ff)(B_G(M)/_U) = B_G(M')/_{U'}$.*

f is called a **local isomorphism** *of $B_G(M)$ onto $B_G(M')$. If $M = M'$ and $B_G(M) = B_G(M')$, then f is called a* **local automorphism**.

Examples:
(1) If $G = O(m)$, then a diffeomorphism $f : M \longrightarrow M'$ is an isomorphism if and only if

$$g'_{f(x)}(df(x)X, df(x)Y) = g_x(X,Y),$$

$X, Y \in T_xM$, where g and g' are the corresponding Riemannian metrics on M and M', respectively. f is called an **isometry**.
(2) If $G = Sp(m)$, then a diffeomorphism $f : M \longrightarrow M'$ is an isomorphism if and only if

$$f^* \omega' = \omega,$$

where ω and ω' are the almost symplectic forms on M and M', respectively. If ω and ω' are symplectic, then f is an isomorphism if and only if f is a **symplectomorphism** (see Section 5.2).

Now, let FR^m be the frame bundle of the Euclidean space R^m and $(x^1, \ldots, x^m)$ the canonical coordinate system of R^m. Hence R^m possesses an $\{e\}$–structure given by the global frame field

$$s : x \in R^m \longrightarrow s(x) = ((\partial/\partial x^1)_x, \ldots, (\partial/\partial x^m)_x).$$

Moreover, we obtain a principal bundle isomorphism

$$FR^m \xrightarrow{\psi} R^m \times Gl(m, R)$$

defined by

$$\psi(u) = (x, (X_i^j)),$$

where $u = (X_1, \ldots, X_m)$ is a linear frame at x and $X_i = X_i^j(\partial/\partial x^j)_x$. Thus if G is an arbitrary subgroup of $Gl(m, R)$ we obtain a G–structure $B_G(R^m)$ on R^m by setting

$$B_G(R^m) = \psi^{-1}(R^m \times G).$$

In fact, $B_G(R^m)$ is obtained by the group enlarging $\{e\} \longrightarrow G$, i.e.,

$$B_G(R^m) = \{s(x)a / \ x \in R^m, \ a \in G\}.$$

This G–structure $B_G(R^m)$ is called the **standard G–structure** on R^m.

Definition 1.21.8 *A G–structure $B_G(M)$ on an m–dimensional manifold M is said to be* **integrable** *if it is locally isomorphic to the standard G–structure $B_G(R^m)$ on R^m.*

It is easy to check the following.

Proposition 1.21.9 *A G–structure $B_G(M)$ on M is integrable if and only if there is an atlas $\{(U_\alpha, x^1_\alpha, \ldots, x^m_\alpha)\}$ such that for each U_α the local frame field $x \longrightarrow ((\partial/\partial x^1_\alpha)_x, \ldots, (\partial/\partial x^m_\alpha)_x)$ takes its values in $B_G(M)$.*

Examples
(1) If a Riemannian structure is integrable then the Riemannian connection is flat. The converse is also true (see Fujimoto [57]).
(2) In the next chapters we obtain necessary and sufficient conditions for integrability of many examples of G–structures.

1.22 Exercises

1.22.1 Let $F : N \longrightarrow M$ be a C^∞ map and suppose that $F(N) \subset A$, A being an embedded submanifold of M. Prove that F is C^∞ as a map from N to A.
1.22.2 Let X be a vector field on a manifold M and φ_t a 1–parameter group of local transformations generated by X. Prove that if $\varphi_t(x)$ is defined on $(-\epsilon, \epsilon) \times M$ for some $\epsilon > 0$, then X is complete.
1.22.3 Define a (global) 1–parameter group of transformations φ_t on R^2 by

$$\varphi_t(x, y) = (x\, e^t, y\, e^{-t}),\ t \in R.$$

Determine the infinitesimal generator.
1.22.4 (1) Let φ be a transformation of M and $T\varphi : TM \longrightarrow TM$ the vector bundle isomorphism defined by $T\varphi(v) = d\varphi(x)(v)$, $v \in T_xM$. Show that if X is a vector field on M, then $T\varphi(X)$ defined by $T\varphi(X)(\varphi(x)) = d\varphi(x)(X(x))$ is a vector field on M.
(2) Suppose that X generates a local 1–parameter group of local transformations φ_t. Prove that the vector field $T\varphi(X)$ generates $\varphi \circ \varphi_t \circ \varphi^{-1}$.
(3) Prove that X is invariant by φ, i.e., $T\varphi(X) = X$, if and only if φ commutes with φ_t.
1.22.5 Let $p : E \longrightarrow M$ be a fibred manifold. Prove that there exists a global section s of E over M (*Hint*: use partitions of unity).

1.22.6 Let $p : E \longrightarrow M$ be a vector bundle. A **metric** g in E is an assignement of an inner product g_x on each fiber E_x, $x \in M$, such that, if s_1 and s_2 are local sections over an open set U of M, then the function $g(s_1, s_2) : U \longrightarrow R$ defined by

$$g(s_1, s_2)(x) = g_x(s_1(x), s_2(x))$$

is differentiable. Prove that a metric in TM induces a Riemannian metric on M, and conversely.

1.22.7 Let X and Y be two vector fields on a manifold M and φ_t the 1–parameter group of local transformations generated by X. Show that

$$[X, Y](x) = \lim_{t \to 0} (1/t)[Y(x) - (T\varphi_t)Y))(x)],$$

for all $x \in M$. (The right–hand of this formula is called the **Lie derivative of** Y **with respect to** X and denoted by $L_X Y$).

1.22.8 Let $F : M \longrightarrow N$ be a C^∞ map. Two vector fields X on M and Y on N are called F**–related** if $dF(x)\, X(x) = Y(F(x))$ for all $x \in M$. Prove that if X_1 and X_2 are vector fields on M F–related to vector fields Y_1 and Y_2 on N, respectively, then $[X_1, X_2]$ and $[Y_1, Y_2]$ are F–related.

1.22.9 Let X be a vector field on a manifold M of dimension m and $x \in M$. Prove that if $X(x) \neq 0$, then there exists a coordinate neighborhood U of x with local coordinates $x^1, \ldots, x^m$ such that $X = \partial / \partial x^1$ on U.

1.22.10 Let M be a connected m–dimensional manifold and f a function on M. Prove that $df = 0$ if and only if f is constant on M.

1.22.11 Consider the product manifold $M_1 \times M_2$ with the canonical projections $\pi_1 : M_1 \times M_2 \longrightarrow M_1$ and $\pi_2 : M_1 \times M_2 \longrightarrow M_2$. Prove that the map

$$T_{(x_1, x_2)}(M_1 \times M_2) \longrightarrow T_{x_1} M_1 \oplus T_{x_2} M_2$$

defined by $v \longrightarrow (d\pi_1(x_1, x_2)v,\ d\pi_2(x_1, x_2)v)$ is a linear isomorphism.

1.22.12 Let α and β closed forms on a manifold M. Prove that $\alpha \wedge \beta$ is closed. If, in addition, β is exact, prove that $\alpha \wedge \beta$ is exact.

1.22.13 Let

$$\alpha = \frac{1}{2\pi} \frac{x\, dy - y\, dx}{x^2 + y^2}.$$

Prove that α is a closed 1–form on $R^2 - \{(0,0)\}$. Compute the integral of α over S^1 and prove that α is not exact.

<u>**1.22.14**</u> (1) Let V be an oriented vector space of dimension m and $F : V \longrightarrow V$ a linear map. Prove that there is a unique constant det F, called **determinant** of F such that $F^\star\omega = (det\, F)\omega$, for all $\omega \in \wedge^m V$.
(2) Show that this definition of determinant is the usual one in Linear Algebra.
(3) Let M_1 and M_2 be two orientable m–dimensional manifolds with volume forms ω_1 and ω_2, respectively. Show that if $F : M_1 \longrightarrow M_2$ is a C^∞ map, then there exists a unique C^∞ function det F on M_1, called **determinant** of F (with respect to ω_1 and ω_2) such that $F^\star\omega_2 = (det\, F)\omega_1$.
(4) Let M be an orientable m–dimensional manifold with volume form ω. Prove the following assertions:
(i) If $F, G : M \longrightarrow M$ are C^∞ maps, then

$$det(F \circ G) = [(det\, F) \circ G](det\, G).$$

(ii) If $H = id_M$, then det $H = 1$.
(iii) If F is a diffeomorphism, then

$$det\, F^{-1} = \frac{1}{(det\, F) \circ F^{-1}}.$$

(Here, all the determinants are defined with respect to ω).
<u>**1.22.15**</u> Prove that the de Rham cohomology groups of S^1 are:

$$H^0(S^1) = R,\ H^1(S^1) = R \text{ and } H^p(S^1) = 0 \text{ for } p > 1.$$

<u>**1.22.16**</u> Let M be a compact orientable m– dimensional manifold without boundary. Prove that $H^m(M) \neq 0$.
<u>**1.22.17**</u> Let M be a Riemannian manifold with Riemannian metric g. Prove that an arbitrary submanifold N of M becomes a Riemannian manifold with the **induced Riemannian metric** g' defined by $g'_x(X,Y) = g_x(X,Y)$, for all $x \in N$ and $X, Y \in T_xN \subset T_xM$. (Thus S^m is a Riemannian manifold with the induced Riemannian metric from R^{m+1}).
<u>**1.22.18**</u> Let M be an m–dimensional Riemannian manifold with Riemannian metric g and curvature tensor R. The **Riemannian curvature tensor**, denoted also by R, is the tensor field of type (0,4) on M defined by

$$R(X_1, X_2, X_3, X_4) = g(R(X_3, X_4, X_2), X_1).$$

Prove that R satisfies the following identities:
(1) $R(X_1, X_2, X_3, X_4) = -R(X_2, X_1, X_3, X_4)$.

(2) $R(X_1, X_2, X_3, X_4) = -R(X_1, X_2, X_4, X_3)$.
(3) $R(X_1, X_2, X_3, X_4) + R(X_1, X_3, X_4, X_2) + R(X_1, X_4, X_2, X_3) = 0$.
(4) $R(X_1, X_2, X_3, X_4) = R(X_3, X_4, X_1, X_2)$.

Now, let π be a plane in the tangent space T_xM, i.e., π is a 2–dimensional vector subspace of T_xM. We define the **sectional curvature** $K(\pi)$ of π by

$$K(\pi) = R(X_1, X_2, X_1, X_2),$$

where $\{X_1, X_2\}$ is an orthonormal basis of π. Proves that $K(\pi)$ is independent of the choice of this orthonormal basis, and that the set of values of $K(\pi)$ for all planes π in T_xM determine the Riemannian curvature tensor at x.

1.22.19 Let M be a Riemannian manifold with Riemannian metric g and curvature tensor R. We define the **Ricci tensor** S of M by

$$S(X, Y) = \sum_{i=1}^{m} R(e_i, Y, e_i, X),$$

where $\{e_1, \ldots, e_m\}$ is an orthonormal frame at x.

(1) Prove that $S(X, Y)$ does not depends on the choice of the orthonormal frame $\{e_i\}$.

(2) Prove that S is a symmetric tensor field of type (0,2) on M.

(3) Prove that in local coordinates we have

$$R_{ij} = \sum_{k=1}^{m} R^k_{kij},$$

where R_{ij} denotes the components of S.

(4) Prove that $\gamma(x) = S(e_1, e_1) + \ldots + S(e_m, e_m)$ does not depends on the choice of the orthonormal frame $\{e_i\}$ at x; $\gamma(x)$ is called the **scalar curvature** at x.

(5) Prove that in local coordinates we have

$$r = \sum_{i,j} g^{ij} R_{ij},$$

where (g^{ij}) is the inverse matrix of (g_{ij}).

1.22.20 Let G' be a Lie subgroup of a Lie group G and $F : M \longrightarrow G$ a C^∞ map such that F take values in G'. Prove that $F : M \longrightarrow G'$ is also C^∞.

1.22.21 Let G be a Lie group with Lie algebra g. We define a canonical g–valued 1–form θ on G by

$$\theta_a(X) = \alpha^{-1}(Tl_{a^{-1}}(X)),$$

for all $X \in T_aG$. Prove that $\theta(A) = A$ for all $A \in g$.
1.22.22 (1) Show that each element of $Gl(m, R)$ has a **polar decomposition**, that is, each matrix $K \in Gl(m, R)$ can be expressed in the form

$$K = RJ,$$

where R is a positive definite symmetric matrix and $J \in O(m)$. (Recall that a symmetric matrix R is **positive definite** if each of its (real) eigenvalues is strictly positive).
(2) Let E be a real vector space of dimension m and $<,>$ any inner product on E. For each $R \in Aut(E)$ we define the transpose R^t of R by $< R^t x, y >=< x, Ry >$. We say that R is **symmetric** if $R^t = R$. If R is symmetric then all the eigenvalues of R are real, and that R is **positive definite** if each of its eigenvalues is strictly positive. An element J of $Aut(E)$ is **orthogonal** if $< Jx, Jy >=< x, y >$. (Equivalently, if $\{e_1, \ldots, e_m\}$ is a basis for E and (R^i_j) is the matrix given by $Re_j = R^i_j e_i$, then R is symmetric (resp., positive definite, orthogonal) if and only if (R^i_j) is symmetric (resp., positive definite, orthogonal)). Prove that each element K of $Aut(E)$ has a polar decomposition $K = RJ$, where R is positive definite symmetric and J is orthogonal.
1.22.23 Let G be a Lie group with Lie algebra g. For a fixed $A \in g$, define $\exp_A : R \longrightarrow G$ by $\exp_A(t) = \exp tA$. Prove that $\exp_A$ is a Lie group homomorphism such that the tangent vector to the curve $\exp_A$ at $t = 0$ is precisely $A(e)$.
1.22.24 Let G be a Lie group with Lie algebra g and $X, Y \in g$. Prove that

$$(\exp tX)(\exp tY) = \exp\{t(X + Y) + \frac{t^2}{2}[X, Y] + O(t^3)\}.$$

1.22.25 Prove that the unitary group $U(m)$ is compact (*Hint* : $U(m)$ is closed and also it is bounded in $Gl(m, R)$). It follows that $SU(m)$, $O(m)$ and $SO(m)$ are also compact.
1.22.26 Prove that if $F : G \longrightarrow H$ is a Lie group homomorphism then: (i) F has constant rank; (ii) the kernel of F is a Lie subgroup; and (iii) dim $Ker\, F = dim\, G_1 - rank\, F$.
1.22.27 Let $P(M, G)$ be a principal bundle over M with structure group G and projection π. Assume that G acts on R^n on the left. Then G acts on the product manifold $P \times R^n$ on the right as follows:

$$(u, \xi)a = (ua, a^{-1}\xi),\ u \in P,\ \xi \in R^n,\ a \in G.$$

This action determines an equivalence relation on $P \times F$. We denote by E the quotient space and by $p : E \longrightarrow M$ the map defined by $p[u, \xi] = \pi(u)$.

(1) Prove that $p : E \longrightarrow M$ is a vector bundle of rank n which is said to be **associated** with P(*Hint* : If U is an open set of M such that $\psi : \pi^{-1}(U) \longrightarrow U \times G$ is a local trivialization of P, then we define $\tilde{\psi} : p^{-1}(U) \longrightarrow U \times R^n$ as follows. Given $[u, \xi] \in p^{-1}(U)$, then we have $\psi(u) = (x, a)$. Thus we define $\tilde{\psi}[u, \xi] = (x, a\xi)$ and $\tilde{\psi}^{-1} : U \times R^n \longrightarrow p^{-1}(U)$ is a local trivialization of E).

(2) Let $\{U_\alpha\}$ be an open covering of M and, for each U_α, let $\psi_\alpha : \pi^{-1}(U_\alpha) \longrightarrow U_\alpha \times G$ be a local trivialization of P. If $\{\psi_{\beta\alpha}\}$ are the transition functions corresponding to $\{u_\alpha\}$, prove that

$$(\tilde{\psi}_\beta \circ \tilde{\psi}_\alpha^{-1})(x, \xi) = (x, \psi_{\beta\alpha}(x)\xi).$$

(The maps $\psi_{\beta\alpha}$ are also called the **transition functions** of E corresponding to $\{U_\alpha\}$).

(3) If $P = FM$ and the action of $Gl(m, R)$, $m = dim\, M$, on R^m is the natural one, i.e., $a\ \xi \in R^m$ is the image of $\xi \in R^m$ by the linear isomorphism $a : R^m \longrightarrow R^m$, then prove that the associated vector bundle with FM is precisely TM, the tangent bundle of M.

(4) Conversely, let $p : E \longrightarrow M$ be a vector bundle of M. Let P be the set of all linear isomorphisms $u : R^n \longrightarrow E_x$ in all points x of M. Define $\pi : P \longrightarrow M$ by $\pi(u) = x$. Prove that: (i) P is a principal bundle over M with structure group $Gl(n, R)$ and projection π; and (ii) E is associated with P.

1.22.28 Let G' be a Lie subgroup of a Lie group G. Prove that if M possesses a G'–structure then it possesses also a G–structure.

1.22.29 Let M be a Riemannian manifold with Riemannian metric g. A vector field X on M is called a **Killing vector field** if $L_X g = 0$. Prove that X is a Killing vector field if and only if the local 1–parameter group of local transformations generated by X consists of local isometries.

1.22.30 Let M be an m–dimensional manifold. Prove that giving an $Sl(m, R)$–structure on M is the same as giving a volume form on M. Hence if M possesses an $Sl(m, R)$–structure, then M is orientable.æ

Chapter 2

Almost tangent structures and tangent bundles

2.1 Almost tangent structures on manifolds

In this section we introduce a geometric structure which is essential in the Lagrangian formulation of Classical Mechanics.

Definition 2.1.1 *Let M be a differentiable manifold of dimension 2n. An* **almost tangent structure** *J on M is a tensor field J of type (1,1) on M with constant rank n and satisfying $J^2 = 0$. In this case, M is called an* **almost tangent manifold**.

Let $x \in M$. Then

$$J_x : T_xM \longrightarrow T_xM$$

is a linear endomorphism. Since $J_x^2 = 0$, we have $ImJ_x \subset KerJ_x$. Furthermore, because rank $J_x = n$, we deduce that $ImJ_x = KerJ_x$. Then

$$ImJ = \bigcup_{x \in M} ImJ_x, KerJ = \bigcup_{x \in M} KerJ_x$$

are vector subbundles of TM of rank n.

Now, let H_x be a complement in T_xM of $KerJ_x$. Then

$$J_x : H_x \longrightarrow KerJ_x = ImJ_x$$

is a linear isomorphism. Hence, if $\{e_i\}$ is a basis of H_x, then $\{e_i, \bar{e}_i = Je_i\}$ is a basis of T_xM, that is, a linear frame at x, which is called an **adapted frame** to J. Let H'_x be another complement to $Ker J_x$ and $\{e'_i\}$ a basis of H'_x. Therefore, we have

$$e'_i = A_i^j e_j + B_i^j e_j,$$

$$\bar{e}'_i = Je'_i = A_i \bar{e}_j,$$

where A, B are $n \times n$ matrices, with A non-singular. Then the two adapted frames are related by the $2n \times 2n$ matrix

$$\begin{bmatrix} A & 0 \\ B & A \end{bmatrix},$$

where $A \in Gl(n, R)$.

Now, let G be the set of such matrices; G is a closed subgroup of $Gl(2n, R)$ and therefore a Lie subgroup of $Gl(2n, R)$. We put

$$B_G = \{\text{adapted frames at all points of } M\}.$$

We shall prove that B_G defines a G-structure on M. To do this, it is sufficient to find, for each $x \in M$, a local section $\sigma : U \longrightarrow FM$ of FM over a neighborhood U of x such that $\sigma(U) \subset B_G$. From the local triviality of $Ker\ J$ and TM, there exists a neighborhood U of x and a frame local field $\{X_1, \ldots, X_n, \bar{X}_1, \ldots, \bar{X}_n\}$ on U such that $\{X_i(y), \bar{X}_i(y)\}$ is an adapted frame at $y, y \in U$, that is, $\bar{X}_i(y) = J_y X_i(y)$. If we define

$$\sigma(y) = \{X_i(y), \bar{X}_i(y)\}, y \in U,$$

then σ is the required local section.

We remark that, with respect to an adapted frame, J is represented by the matrix

$$J_0 = \begin{bmatrix} 0 & 0 \\ I_n & 0 \end{bmatrix},$$

where I_n is the $n \times n$ identity matrix. In fact, the group G can be described as the invariance group of the matrix J_0, that is, $\alpha \in G$ if and only if $\alpha J_0 \alpha^{-1} = J_0$.

Suppose now given a G–structure B_G on M. Then we may define a tensor field J of type (1,1) on M as follows. We set

$$J_x(X) = p(J_0(p^{-1}(X))),$$

where $X \in T_xM, x \in M$ and $p \in B_G$ is a linear frame at x. From the definition of G, $J_x(X)$ is independent of the choice of p. In other words, J_x is defined as the linear endomorphism of T_xM which has at x the matrix representation J_0 with respect to one of the linear frames determined at x by B_G, and hence with respect to any other. Obviously, we have

$$\operatorname{rank} J = n, J^2 = 0.$$

Thus, J is an almost tangent structure on M.

Summing up, we have proved the following.

Proposition 2.1.2 *Giving an almost tangent structure is the same as giving a G– structure on M.*

Now, let g be a Riemannian metric on M. Then g_x determines an inner product on each tangent space T_xM. Let H_x be an orthogonal complement in T_xM to $KerJ_x$ with respect to g_x. If $\{e_i\}$ is an orthonormal basis of H_x, then $\{e_i, \bar{e}_i = J_x e_i\}$ is an orthonormal basis of T_xM, that is, an orthonormal frame at x, since

$$J_x : H_x :\longrightarrow KerJ_x$$

is an isometry. If $\{e'_i, \bar{e}'_i\}$ is another orthonormal frame at x obtained from a different orthonormal basis $\{e'_1\}$ of H_x, then the two orthonormal frames are related by the $(2n) \times (2n)$ matrix

$$\begin{bmatrix} A & 0 \\ 0 & A \end{bmatrix},$$

where $e'_i = A^j_i e_j, A \in O(n)$. Let B be the set of all orthonormal frames obtained as above in all points of M. Then B defines a $(O(n) \times O(n))$–structure on M. In fact, given a local frame field on a neighborhood of each point $x \in M$, we obtain a local section of FM taking values in B by the usual Gramm–Schmidt argument. Conversely, given a $(O(n) \times O(n))$–structure B on M, we obtain an almost tangent structure on M, since $(O(n) \times O(n)) \subset G$.

Summing up, we have the following.

Proposition 2.1.3 *Giving an almost tangent structure is the same as giving a $(O(n) \times O(n))$–structure on M.*

Since $O(n) \times O(n) \subset SO(n)$, we have

Proposition 2.1.4 *Every almost tangent manifold is orientable.*

2.2 Examples. The canonical almost tangent structure of the tangent bundle

In this section we shall prove that the tangent bundle of any manifold carries a canonically defined almost tangent structure (hence the name).

Let N be an n–dimensional differentiable manifold and TN its tangent bundle. We denote by $\tau_N : TN \longrightarrow N$ the canonical projection. For each $y \in T_xN$, let

$$V_y = Ker\{d\tau_N(y) : T_y(TN) \longrightarrow T_xN\}.$$

Then V_y is an n–dimensional vector subspace of $T_y(TN)$ and

$$V = \bigcup_{y \in TN} V_y$$

is a vector bundle over TN of rank n (in fact, a vector subbundle of $\tau_{TN} : TTN \longrightarrow TN$). V (sometimes denoted by $V(TN)$) is called the **vertical bundle.** A tangent vector v of TN at y such that $v \in V_y$ is called **vertical.** A **vertical vector field** X is a vector field X on TN such that $X(y) \in V_y$, for each $y \in TN$ (that is, X is a section of V). We remark that the vertical tangent vectors are tangent to the fibres of the projection τ_N.

Now, let $y \in T_xN, x \in N$. Then we may define a linear map

$$T_xN \longrightarrow V_y,$$

called the **vertical lift** as follows: for $u \in T_xN$, its vertical lift u^v to TN at y is the tangent vector at $t = 0$ to the curve $t \longrightarrow y + tv$. Furthermore, if X is a vector field on N, then we may define its vertical lift as the vector field X^v on TN such that

$$X^v(y) = (X(\tau_N(y)))^v.$$

If X is locally given by $X = X^i(\partial/\partial x^i)$ in a coordinate neighborhood U with local coordinates (x^i), then X^v is locally given by

$$X^v = X^i(\partial/\partial v^i)$$

with respect to the induced coordinates (x^i, v^i) on TU.

Next, we define a tensor field J of type (1,1) on TN as follows: for each $y \in TN$, J is given by

$$J_y(\bar{X}) = ((\tau_N)_*\bar{X})^v, \bar{X} \in T_y(TN).$$

Then J is locally given by

$$J(\partial/\partial x^i) = \partial/\partial v^i, J(\partial/\partial v^i) = 0,$$

or, equivalently,

$$J = (\partial/\partial v^i) \otimes (dx^i).$$

Consequently, J has constant rank n and $J^2 = 0$. Thus, J is an almost tangent structure on TN which is called the **canonical almost tangent structure** on TN (in Section 2.5, we shall give an alternative definition of J).

We can easily prove that

$$KerJ = ImJ = V.$$

To end this section we describe a family of almost tangent structures on the 2–torus T^2. The 2–torus $T^2 = S^1 \times S^1$ may be considered as a quotient manifold R^2/Z^2, where Z^2 is the **integral lattice** of R^2. So, the canonical global coordinates (x, y) of R^2 may be taking as local coordinates on T^2. Let α be any real number then

$$J_\alpha = \cos\alpha \sin\alpha(\partial/\partial x) \otimes (dx) - \cos^2\alpha(\partial/\partial x) \otimes (dy)$$

$$+ \sin^2\alpha(\partial/\partial y) \otimes (dx) - \cos\alpha \sin\alpha(\partial/\partial y) \otimes (dy)$$

determines an almost tangent structure on T^2. The vertical distribution V_α is tangent to the spiral which is the image of the line $y = xtg\alpha$ under the canonical projection $R^2 \longrightarrow T^2$.

2.3 Integrability

The fundamental problem of the theory of G–structures is to decide whether a given G–structure is equivalent to the standard G–structure on R^{2n}. In this section we establish a necessary and sufficient condition for an almost tangent structure J on a $2n$–dimensional manifold to be **integrable**, that is, locally equivalent to the standard almost tangent structure on R^{2n} (see Section 1.23).

Definition 2.3.1 *Let J be an almost tangent structure on a $2n$–dimensional manifold M. The* **Nijenhuis tensor** *N_J of J is a tensor field of type (1,2) given by*

$$N_J(X,Y) = [JX, JY] - J[JX,Y] - J[X,JY], X,Y \in \chi(M).$$

Now, let J_0 be the standard almost tangent structure on R^{2n}. Then J_0 is given by

$$J_0(\partial/\partial x^i) = \partial/\partial y^i, J_0(\partial/\partial y^i) = 0, \tag{2.1}$$

where (x^i, y^i) are the canonical coordinates on $R^{2n}, 1 \leq i \leq n$. If J is integrable, then there are local coordinates (x^i, y^i) on a neighborhood U of each point x of M such that J is locally given by (2.1). Hence, if J is integrable, then the Nijenhuis tensor N_J vanishes.

Next, we prove the converse. Suppose that $N_J = 0$. Therefore, we have

$$[JX, JY] = J[JX,Y] + J[X,JY].$$

Thus, the distribution $V = ImJ = KerJ$ is integrable. From the Frobenius theorem, we may find local coordinates (x^i, z^i) on a neighborhood of each point of M such that the leaves of the corresponding foliation are given by $x^i =$ constant, $1 \leq i \leq n$. Then the local vector fields

$$\{\partial/\partial z^i, 1 \leq i \leq n\}$$

determine a basis of V. Hence we have

$$J(\partial/\partial x^i) = A_i^j(\partial/\partial z^j), J(\partial/\partial z^i) = 0,$$

where (A_i^j) is a non–singular matrix of functions, since J has rank n. Let H be a complement of V in TM, that is,

$$TM = H \oplus V (\text{Whitney sum})$$

Then $J : H \longrightarrow V$ is a vector bundle isomorphism. Thus, there exists a local basis $\{\bar{Z}_i; 1 \le i \le n\}$ of H such that

$$J\bar{Z}_i = \partial/\partial z^i.$$

We have

$$\bar{Z}_i = \alpha_i^j(\partial/\partial x^j) + \beta_i^j(\partial/\partial z^j).$$

We set

$$Z_i = \alpha_i^j(\partial/\partial x^j).$$

Then $\{Z_i; 1 \le i \le n\}$ is a set of linearly independent local vector fields on M such that

$$JZ_i = \partial/\partial z^i.$$

Since

$$JZ_i = \alpha_i^k J(\partial/\partial x^k) = \alpha_i^k A_k^j(\partial/\partial z^j) = \partial/\partial z^i,$$

we deduce that

$$\alpha_i^k A_k^j = \delta_i^j.$$

Hence (α_i^j) is the inverse matrix of (A_i^j). Because $N_J = 0$, we have

$$0 = N_J(Z_i, Z_j) = -J[\partial/\partial z^i, Z_j] - J[Z_i, \partial/\partial z^j]$$

$$= -J((\partial \alpha_j^k/\partial z^i)(\partial/\partial x^k)) - J(-(\partial \alpha_i^k/\partial z^j)(\partial/\partial x^k))$$

$$= ((\partial \alpha_i^k/\partial z^j) - (\partial \alpha_j^k/\partial z^i))A_k^l(\partial/\partial z^l).$$

Since (A_k^l) is a non–singular matrix, we obtain

$$\partial \alpha_i^k/\partial z^j = \partial \alpha_j^k/\partial z^i \tag{2.2}$$

From the compatibility conditions (2.2), we deduce that there exist local functions $f^k = f^k(x^i, z^i)$ such that

$$\alpha_i^k = \partial f^k / \partial z^i.$$

Now, we make the following coordinate transformation:

$$x^i = x^i, y^i = f^i(x^j, z^j), 1 \leq i \leq n.$$

Then we have

$$\partial / \partial x^i = \partial / \partial x^i + (\partial f^j / \partial x^i)(\partial / \partial y^j)$$

$$\partial / \partial z^i = (\partial f^j / \partial z^i)(\partial / \partial y^j) = \alpha_i^j (\partial / \partial y^j)$$

Thus, we deduce

$$J(\partial / \partial x^i) = \partial / \partial y^i, J(\partial / \partial y^i) = 0.$$

Hence J is integrable.

Summing up, we have proved the following theorem due to J. Lehman–Lejeune [86]).

Theorem 2.3.2 *An almost tangent structure J is integrable if and only if its Nijenhuis tensor N_J vanishes identically.*

Corollary 2.3.3 *The canonical almost tangent structure on the tangent manifold TN of any manifold N is integrable.*

Corollary 2.3.4 *Every almost tangent structure on a 2–dimensional manifold M is integrable.*

Proof: Let H be an orthogonal complement of V with respect to any Riemannian metric on M. Let $\{X, JX\}$ be an adapted local field frame with $X \in H$. Then we have

$$N_J(X, JX) = 0.$$

Hence $N_J = 0$ vanishes. □

Remark 2.3.5 From Corollary 2.3.4, we deduce that the almost tangent structures J_α defined on the 2–torus T^2 are always integrable. Since the 2–torus is compact then in neither case J_α is globally equivalent to the standard almost tangent structure on R^2. Thus, the integrability of an almost tangent structure is a purely local matter.

2.4 Almost tangent connections

In this section we give a characterization of the integrability of an almost tangent structure in terms of a symmetric linear connection.

Definition 2.4.1 *Let J be an almost tangent structure on M. A linear connection ∇ on M is said to be an* **almost tangent connection** *if $\nabla J = 0$, that is,*

$$\nabla_X(JY) = J(\nabla_X Y), X, Y \in \chi(M).$$

Proposition 2.4.2 *On each almost tangent manifold there always exists an almost tangent connection.*

Proof: Let ∇ be any symmetric connection on M, for instance, the Riemannian connection of some Riemannian metric on M. Choose an orthogonal complement H in TM of $V = KerJ = ImJ$. Now, we define a tensor field Q of type (1,2) on M as follows:

$$Q(X,Y) = 0,$$

$$Q(JX,Y) = (\nabla_Y J)(X),$$

$$Q(X,JY) = (\nabla_X J)(Y),$$

$$Q(JX,JY) = (\nabla_{JX} J)(Y) + (J(\nabla_Y J)(X)),$$

for $X, Y \in H$. Let $\bar{\nabla} = \nabla - Q$. Then $\bar{\nabla}$ is a linear connection on M such that $\bar{\nabla} J = 0$. □

Now, let ∇ be a symmetric linear connection on an almost tangent manifold M with almost tangent structure J. Since ∇ is symmetric, we have

$$[X,Y] = \nabla_X Y - \nabla_Y X.$$

Hence the Nijenhuis tensor N_J of J may be written as follows:

$$N_J(X,Y) = (\nabla_{JX} J)(Y) - (\nabla_{JY} J)(X) + (J(\nabla_Y J))(X) - J((\nabla_X J))(Y).$$

Therefore, we have

Proposition 2.4.3 *If there is a symmetric almost tangent connection ∇ on M then J is integrable.*

Proof: In fact, suppose that ∇ is a symmetric almost tangent connection on M. Then $\nabla J = 0$, and consequently, we deduce $N_J = 0$. Thus, J is integrable. □

We now prove the converse.

Proposition 2.4.4 *If J is integrable, then there exists a symmetric almost tangent connection on M.*

Proof: Let $\bar{\nabla}$ be the linear connection on M constructed as in Proposition 2.4.2. Since J is integrable and ∇ is symmetric, we have

$$0 = N_J(X,Y) = (\nabla_{JX}J)(Y) - (\nabla_{JY}J)(X) + J((\nabla_Y J)(X)) - J((\nabla_X J)(Y)).$$

Consequently, Q is symmetric. Then we have

$$\bar{T}(X,Y) = T(X,Y) - Q(X,Y) + Q(Y,X)$$

$$= -Q(X,Y) + Q(Y,X) = 0,$$

where $\bar{T}$ denotes the torsion tensor of $\bar{\nabla}$. Then is the required symmetric almost tangent connection. □

2.5 Vertical and complete lifts of tensor fields to the tangent bundle

In Section 2.2, we have introduced the vertical lifts of vector fields on a manifold N to its tangent bundle TN. Next, we shall define the vertical and complete lifts of tensor fields.

Let F be a tensor field of type $(1,r)$, $r \geq 1$, on N. Then the **vertical lift** of F to TN is the tensor field of type $(1,r)$ on TN defined by

$$F^v_y(\bar{X}_1, \ldots, \bar{X}_r) = (F_x((\tau_N)_* \bar{X}_1, \ldots, (\tau_N)_* \bar{X}_r)^v,$$

where $\bar{X}_1, \ldots, \bar{X}_r \in T_y(TN), y \in T_xN, x \in N$. If

$$F = F^j_{i_1 \ldots i_r}(\partial/\partial x^j) \otimes (dx^{i_1}) \otimes \ldots \otimes (dx^{1_r}),$$

then we have

$$F^v = F^j_{i_1 \ldots i_r} (\partial/\partial v^j) \otimes (dx^{i_1}) \otimes \ldots \otimes (dx^{i_r}), \tag{2.3}$$

where (x^i, v^i) are local coordinates for TN.

We can easily prove that

$$F^v(X_1^v, \ldots, X_r^v) = 0, \text{ for any } X_1, \ldots, X_r \in \chi(N).$$

Moreover, *if* I *is the identity tensor on* N, *then* $J = I^v$ *is the canonical almost tangent structure on* TN.

Now, let G be a tensor field of type $(0, r), r \geq 1$, on N. The **vertical lift** of G to TN is the tensor field of type $(0, r)$ on TN defined by

$$G^v_y(\bar{X}_1, \ldots, \bar{X}_r) = (G_x((\tau_N)_\star \bar{X}_1, \ldots, (\tau_N)_\star(\bar{X}_r)))^v,$$

where $\bar{X}_1, \ldots, \bar{X}_r \in T_y(TN), y \in T_xN$, $x \in N$, and $f^v = f \circ \tau_N$ denotes the **vertical lift** to TN of the function $f \in C^\infty(N)$. If

$$G = G_{i_1 \ldots i_r} (dx^{i_1}) \otimes \ldots \otimes (dx^{i_r}),$$

then we have

$$G^v = G_{i_1 \ldots i_r} (dx^{i_1}) \otimes \ldots \otimes (dx^{i_r}). \tag{2.4}$$

We easily deduce that

$$G^v(X_1^v, \ldots, X_r^v) = 0, \text{ for any } X_1, \ldots, X_r \in \chi(N).$$

Moreover, if $\alpha = \alpha_i(dx^i)$ is a 1–form on N, then $\alpha^v = \alpha_i(dx^i)$. Thus, α^v is precisely the pullback of α to TN, that is

$$\alpha^v = (\tau_N)^\star \alpha.$$

Next, we define the complete lift of tensor fields on N to TN. Let f be a function on N. Then the **complete lift** of f to TN is the function f^c on TN given by

$$f^c(y) = df(x)(y), \quad y \in T_xN, x \in N.$$

Then we have

$$f^c = y^i(\partial f/\partial x^i). \tag{2.5}$$

Now, let X be a vector field on N. Then X generates a local 1–parameter group of local transformations ϕ_t on N. Let $T\phi_t$ be the (local) 1–parameter group of local transformations of TN determined by ϕ_t. The infinitesimal generator of $T\phi_t$ is called the **complete lift** of X to TN and denoted by X^c. If

$$X = X^i(\partial/\partial x^i),$$

then we have

$$X^c = X^i(\partial/\partial x^i) + v^j(\partial X^i/\partial x^j)(\partial/\partial v^i). \tag{2.6}$$

Thus, we obtain

$$(\partial/\partial x^i)^c = \partial/\partial x^i.$$

From (2.3), (2.4) and (2.6), we easily deduce

$$F^v(X_1^c, \ldots, X_r^c) = (F(X_1, \ldots, X_r))^v,$$

$$(\text{resp.} G^v(X_1^c, \ldots, X_r^c) = (G(X_1, \ldots, X_r))^v)$$

for any $X_1, \ldots, X_r \in \chi(N)$, where F (resp. G) is a tensor field of type $(1, r)$ (resp. $(0, r)$) on N.

From (2.5) and (2.6), we easily deduce the following.

Proposition 2.5.1 *For any $X \in \chi(N)$ and $f \in C^\infty(N)$*

$$X^v f^v = 0, X^v f^c = X^c f^v = (Xf)^v, X^c f^c = (Xf)^c.$$

For the Lie bracket, we have

Proposition 2.5.2 *For any $X, Y \in \chi(N)$*

$$[X^v, Y^v] = 0, [X^v, Y^c] = [X, Y]^v, [X^c, Y^c] = [X, Y]^c.$$

In order to define the complete lifts of tensor fields to TN, we first prove the following.

Proposition 2.5.3 *Let $\bar{F}, \bar{F}'$(resp.$\bar{G}, \bar{G}'$) be tensor fields of type $(1,r)$ (resp.$(0,r)$) on TN such that*

$$\bar{F}(X_1^c, \ldots, X_r^c) = \bar{F}'(X_1^c, \ldots, X_r^c),$$

$$(resp.\bar{G}(X_1^c, \ldots, X_r^c) = \bar{G}'(X_1^c, \ldots, X_r^c))$$

for any $X_1, \ldots, X_r \in \chi(N)$. Then $\bar{F} = \bar{F}'$(resp.$\bar{G} = \bar{G}'$).

Proof: We only prove the case (1,1) (the general cases $(1,r)$ and $(0,r)$ may be proved in a similar way). It is sufficient to prove that if $\bar{F}(X^c) = 0$ for any vector field X on N, then $\bar{F} = 0$. Suppose that

$$\bar{F} = A_i^j(\partial/\partial x^j) \otimes (dx^i) + B_i^j(\partial/\partial x^j) \otimes (dv^i)$$

$$+C_i^j(\partial/\partial v^j) \otimes (dx^i) + D_i^j(\partial/\partial v^j) \otimes (dv^i).$$

If $X = \partial/\partial x^k$, we obtain

$$\bar{F}((\partial/\partial x^k)^c) = \bar{F}(\partial/\partial x^k) = A_i^j(\partial/\partial x^j) + C_i^j(\partial/\partial v^j),$$

which implies $A_i^j = C_i^j = 0$. Then

$$\bar{F} = B_i^j(\partial/\partial x^j) \otimes (dv^j) + D_i^j(\partial/\partial v^j) \otimes (dv^i).$$

Now, let $X = X^k(\partial/\partial x^k)$. Since

$$X^c = X^k(\partial/\partial x^k) + v^s(\partial X^i/\partial x^s)(\partial/\partial v^i),$$

we obtain

$$0 = \bar{F}(X^c) = v^s(\partial X^i/\partial x^s)(B_i^j(\partial/\partial x^j) + D_i^j(\partial/\partial v^j)).$$

Since X^i and v^k are arbitrary, we easily deduce that

$$B_i^j = D_i^j = 0.$$

Hence $\bar{F} = 0$. □

Definition 2.5.4 *Let F (resp. G) be a tensor field of type (1,r) (resp. (0,r)) on N. The* **complete lift** *of F to TN is the tensor field F^c (resp. G^c) of type (1,r) (resp. (0,r)) defined by*

$$F^c(X_1^c, \ldots, X_r^c) = (F(X_1, \ldots, X_r))^c,$$

$$(\text{resp.} G^c(X_1^c, \ldots, X_r^c) = (G(X_1, \ldots, X_r))^c)$$

for any $X_1, \ldots, X_r \in \chi(N)$.

(Obviously, if I denotes the identity tensor on N, then I^c = Identity tensor on TN).

Then, if $F = F_i^j(\partial/\partial x^j) \otimes (dx^i)$ is the local expression of a tensor field of type (1,1) on N, we have

$$F^c = F_i^j(\partial/\partial x^j) \otimes (dx^i) + v^k(\partial F_i^j/\partial x^k)(\partial/\partial v^j) \otimes (dx^i)$$

$$+ F_i^j(\partial/\partial v^j) \otimes (dv^i). \tag{2.7}$$

For a 1–form $\alpha = \alpha_i dx^i$, we obtain

$$\alpha^c = v^k(\partial \alpha^i/\partial x^k)(dx^i) + \alpha_i(dv^i). \tag{2.8}$$

For a tensor field G of type (0,2) we have

$$G^c = v^k(\partial G_{ij}/\partial x^k)(dx^i) \otimes (dx^j) + G_{ij}(dx^i) \otimes (dv^j)$$

$$+ G_{ij}(dv^i) \otimes (dx^j). \tag{2.9}$$

From (2.7), (2.8) and (2.9), we easily deduce the following.

Proposition 2.5.5 *For any $X, Y \in \chi(N)$*

$$F^c(X^v) = (F(X))^v, \alpha^c(X^v) = (\alpha(X))^v,$$

$$G^c(X^v, Y^c) = G^c(X^c, Y^v) = (G(X,Y))^v,$$

$$G^c(X^v, Y^v) = 0.$$

Next, we consider the particular cases of tensor fields type (1,1) and (0,2).

Complete lifts of tensor fields of type (1,1).

Proposition 2.5.6 *(1) If F is a tensor field of type (1,1) on N of rank r, then F^c has rank $2r$; (2) if F, G are tensor fields of type (1,1) on N then we have $(FG)^c = F^cG^c$.*

Proof: (1) is a direct consequence of (2.7). To prove (2), we have

$$(FG)^c(X^c) = ((FG))(X))^c = (F(G(X)))^c = F^c((GX)^c)$$

$$= F^c(G^c(X^c)) = (F^cG^c)(X^c).\square$$

Corollary 2.5.7 *If $P(t)$ is a polynomial in one variable t, then*

$$P(F^c) = (P(F))^c,$$

for any tensor field F of type (1,1) on N.

Complete lifts of tensor fields of type (0,2).

Proposition 2.5.8 *(1) Let G be a tensor field of type (0,2) on N. If G has rank r, then G^c has rank $2r$; (2) if G is symmetric (resp. skew–symmetric), then G^c is symmetric (resp. skew– symmetric).*

Proof: (1) follows from (2.9), (2) is a direct consequence of the definition of G^c.$\square$

Corollary 2.5.9 *If G is a Riemannian metric on N, then G^c is a pseudo Riemannian metric on TN of signature (n,n). (For the definition of signature of pseudo–Riemannian metrics see Exercise 2.11.2).*

Corollary 2.5.10 *If ω is a 2–form on N, then ω^c is a 2–form on TN and we have*

$$d\omega^c = (d\omega)^c.$$

To end this section, we define the complete lifts of distributions on N. Let D be a k–dimensional distribution on N. We define the **complete lift** of D to TN to be the $2k$–dimensional distribution D^c spanned by the vector fields X^v and X^c, X being an arbitrary vector field belonging to D. Obviously, if D is locally spanned by $\{X_1, \ldots, X_k\}$, then D^c is locally spanned by $\{X_1^v, \ldots, X_k^v, X_1^c, \ldots, X_k^c\}$.

2.6 Complete lifts of linear connections to the tangent bundle

Let N be an n–dimensional manifold with a linear connection ∇. Then we define the **complete lift** of ∇ to TN as the unique linear connection ∇^c on TN given by

$$\nabla^c_{X^c} Y^c = (\nabla_X Y)^c,$$

for any $X, Y \in \chi(N)$. This assertion may be verified by an easy computation using Christoffel components. In fact, we obtain

$$\nabla_{\partial/\partial x^i} \partial/\partial x^j = \Gamma^j_{ij}(\partial/\partial x^k) + v^l(\partial \Gamma^k_{ij}/\partial x^l)(\partial/\partial v^k),$$

$$\nabla_{\partial/\partial x^i} \partial/\partial v^j = \Gamma^k_{ij}(\partial/\partial v^k),$$

$$\nabla_{\partial/\partial v^i} \partial/\partial x^j = \Gamma^k_{ij}(\partial/\partial v^k),$$

$$\nabla_{\partial/\partial v^i} \partial/\partial v^j = 0, \tag{2.10}$$

where

$$\nabla_{\partial/\partial x^i} \partial/\partial x^j = \Gamma^k_{ij}(\partial/\partial x^k).$$

If we put

$$\nabla^c_{\partial/\partial x^i} \partial/\partial x^j = \bar{\Gamma}^k_{ij}(\partial/\partial x^k) + \bar{\Gamma}^{n+k}_{ij}(\partial/\partial v^k),$$

$$\nabla^c_{\partial/\partial x^i} \partial/\partial v^j = \bar{\Gamma}^k_{i,n+j}(\partial/\partial x^k) + \bar{\Gamma}^{n+k}_{i,n+j}(\partial/\partial v^k),$$

$$\nabla^c_{\partial/\partial v^i} \partial/\partial x^j = \bar{\Gamma}^k_{n+i,j}(\partial/\partial x^k) + \bar{\Gamma}^{n+k}_{n+i,j}(\partial/\partial v^k),$$

$$\nabla^c_{\partial/\partial v^i} \partial/\partial v^j = \bar{\Gamma}^k_{n+i,n+j}(\partial/\partial x^k) + \bar{\Gamma}^{n+k}_{n+i,n+j}(\partial/\partial v^k),$$

we deduce from (2.10) that

$$\bar{\Gamma}^k_{ij} = \Gamma^k_{ij}, \bar{\Gamma}^k_{i,n+j} = \bar{\Gamma}^k_{n+i,j} = \bar{\Gamma}^k_{n+i,n+j} = \bar{\Gamma}^{n+k}_{n+i,n+j} = 0,$$

$$\bar{\Gamma}^{n+k}_{ij} = v^l(\partial \Gamma^k_{ij}/\partial x^l), \bar{\Gamma}^{n+k}_{i,n+j} = \bar{\Gamma}^{n+k}_{n+i,j} = \Gamma^k_{ij}. \tag{2.11}$$

From (2.11), we easily prove that ∇^c defines a linear connection on TN.

Proposition 2.6.1 *If T and R are respectively the torsion and curvature tensors of ∇, then T^c and R^c are respectively the torsion and curvature tensors of ∇^c.*

Proof: In fact, we have

$$\nabla_{X^c}Y^c - \nabla^c_{Y^c}X^c - [X^c, Y^c] = (\nabla_X Y)^c - (\nabla_Y X)^c - ([X,Y])^c$$

$$= (\nabla_X Y - \nabla_Y X - [X,Y])^c = (T(X,Y))^c = T^c(X^c, Y^c),$$

for any $X, Y \in \chi(N)$. By a similar procedure, we prove that the curvature tensor of ∇^c is, precisely, R^c.□

Corollary 2.6.2 *(1) ∇ is symmetric if and only if ∇^c is symmetric; (2) ∇ is flat if and only if ∇^c is flat.*

Proof: In fact, let S be a tensor field on N of type $(1,r)$ or $(0,r)$. Then $S = 0$ if and only if $S^c = 0$.□

From (2.11), we can easily prove the following

Proposition 2.6.3 *We have*

$$\nabla^c_{X^v}Y^v = (\nabla_X Y)^v, \nabla^c_{X^c}Y^c = (\nabla_X Y)^c,$$

$$\nabla^c_{X^v}Y^c = \nabla^c_{X^c}Y^v = (\nabla_X Y)^v.$$

Proposition 2.6.4 *For a tensor field S of type (1,r) or (0,r) on N we have*

$$\nabla^c S^v = (\nabla S)^v, \nabla^c S^c = (\nabla S)^c.$$

Proof: We only prove the proposition for tensor fields of type (1,1). The general case can be proved by a similar procedure. Let F be a tensor field of type (1,1) on N. Then we have

$$(\nabla^c F^v)(Y^c, X^c) = (\nabla^c_{X^c}F^v)(Y^c) = \nabla^c_{X^c}(F^v Y^c) - F^v(\nabla^c_{X^c}Y^c)$$

$$= \nabla^c_{X^c}(FY)^v - F^v(\nabla_X Y)^c = (\nabla_X(FY))^v - (F(\nabla_X Y))^v$$

$$= (\nabla_X(FY) - F(\nabla_X Y))^v = ((\nabla F)(Y,X))^v$$

$$= (\nabla F)^v(X^c, Y^c).$$

Then $\nabla^c F^v = (\nabla F)^v$. In a similar way, we prove that $\nabla^c F^c = (\nabla F)^c$.□

Corollary 2.6.5 *Let J be the canonical almost tangent structure on TN. Then ∇^c is an almost tangent connection.*

Proof: In fact, we have

$$\nabla^c J = \nabla^c(I^v) = (\nabla I)^v = 0,$$

since I is the identity tensor field on N. □

Corollary 2.6.6 *Let ∇ be the Riemannian connection with respect to a Riemannian metric g on N. Then ∇^c is the pseudo-Riemannian connection with respect to g^c.*

Proof: In fact, from Corollary 2.6.2, we deduce that ∇^c is symmetric. Furthermore, from Proposition 2.6.4, we have $\nabla^c g^c = (\nabla g)^c = 0$. Then the result follows. □

Next, we establish some properties about the geodesics of ∇ and ∇^c. First, let σ be a curve in N and $\dot{\sigma}$ its canonical lift to TN. Then we have

Proposition 2.6.7 *If σ is a geodesic with respect to ∇^c, then $\dot{\sigma}$ is a geodesic with respect to ∇^c.*

Proof: Let σ be a geodesic of ∇. Then σ satisfies the following system of differential equations:

$$(d^2x^k/dt^2) + \Gamma^k_{ij}(dx^i/dt)(dx^j/dt) = 0.$$

Since $\dot{\sigma}(t) = (x^i(t), (dx^i/dt))$, we deduce by a direct computation from (2.11), that $\dot{\sigma}$ is a geodesic of ∇^c. □

Now, let $\bar{\sigma}$ be a curve in TN. In local coordinates, we have

$$\bar{\sigma}(t) = (x^i(t), v^i(t)).$$

We put $\sigma = \tau_N \circ \bar{\sigma}$. Then σ is a curve in N and

$$\bar{\sigma}(t) \in T_{\sigma(t)}N, \text{for any } t.$$

Thus $\bar{\sigma}$ defines a vector field $X(t)$ on N along $\sigma; X(t)$ is locally given by

$$X(t) = v^i(t)(\partial/\partial x^i).$$

Proposition 2.6.8 *Let $\bar{\sigma}$ be a geodesic in TN with respect to ∇^c. Then its projection σ onto N is a geodesic with respect to ∇ and $X(t)$ is a Jacobi field along σ.*

Proof: Let $\bar{\sigma}$ be a geodesic in TN with respect to ∇^c. From (2.11), we deduce

$$(d^2x^k/dt^2) + \Gamma^k_{ij}(dx^i/dt)(dx^j/dt) = 0, \tag{2.12}$$

$$(d^2v^k/dt^2) + (\partial\Gamma^k_{ij}/\partial x^s)v^s(dx^i/dt)(dx^j/dt) \tag{2.13}$$

$$+2\Gamma^k_{ij}(dv^i/dt)(dx^j/dt) = 0.$$

If we set

$$(\delta v^k/dt) = (dv^k/dt) + \Gamma^k_{ij}(dx^i/dt)v^j,$$

then (2.13) becomes

$$(\delta^2 v^k/dt^2) + R^k_{sij}v^s(dx^i/dt)(dx^j/dt) = 0, \tag{2.14}$$

where R^k_{sij} are the components of the curvature tensor of ∇. Now (2.12) implies that σ is a geodesic and (2.14) implies that $X(t)$ is a Jacobi vector field along σ.

Remark 2.6.9 For a discussion of Jacobi fields see Kobayashi and Nomizu [81].

2.7 Horizontal lifts of tensor fields and connections

Let ∇ be a linear connection on N with local components Γ^k_{ij}. We denote by R its curvature tensor, by T its torsion tensor and by R^k_{ijl} and T^k_{ij} their local components. By $\hat{\nabla}$ we denote the opposite connection of ∇ defined by

$$\hat{\nabla}_X Y = \nabla_Y X + [X, Y].$$

Then we have $\hat{\Gamma}^k_{ij} = \Gamma^k_{ji}$. If ∇ is symmetric, then $\hat{\nabla} = \nabla$.

For every local coordinate system (U, x^i) in N, we set

$$D_i = \partial/\partial x^i - v^j \Gamma^k_{ji}(\partial/\partial v^k), 1 \leq i \leq n = dim N,$$

where (x^i, v^i) are the induced coordinates in TU. Then $\{D_i\}$ is a set of linearly independent local vector fields on TN. We define the **horizontal subspace** H_y at $y \in TN$ as the vector subspace of $T_y(TN)$ spanned by $\{D_i(y)\}$. It is a straightforward computation to prove that H_y is independent on the choice of the local coordinate system. Then we obtain an n–dimensional distribution H on TN given by

$$y \in TN \longrightarrow H_y \subset T_y(TN).$$

Since $T_y(TN) = H_y \oplus V_y$, then H is a complementary distribution of the vertical distribution V and we have

$$TTN = H \oplus V \quad \text{(Whitney sum)}.$$

H is called the **horizontal distribution** defined by ∇. Since $V_y = Ker\ d\tau_N(y)$, we deduce that $d\tau_N(y)$ restricted to H_y gives a linear isomorphism

$$d\tau_N(y) : H_y \longrightarrow T_x N, x = \tau_N(y), y \in TN.$$

Consequently, if X is a vector field on N, we can define the **horizontal lift** of X to TN to be the vector field X^H on TN such that

$$d\tau_N(y)(X^H(y)) = X(x), x = p_N(y), y \in TN.$$

Obviously, we have

$$(\partial/\partial x^i)^H = D_i \tag{2.15}$$

and

$$X^H = X^i(\partial/\partial x^i) - v^j X^i \Gamma^k_{ji}(\partial/\partial v^k), \tag{2.16}$$

where $X = X^i(\partial/\partial x^i)$. From (2.15) and (2.16), we deduce that

$$H^H = X^i D_i.$$

Moreover, $X^H f^v = (Xf)^v$, for any function f on N.

The set of local vector fields $\{D_i, V_i = \partial/\partial v^i\}$ is called the **adapted frame** to ∇. The dual coframe $\{\theta^i, \eta^i\}$ is given by

$$\theta^i = dx^i, \eta^i = v^j \Gamma^i_{jk} dx^k + dv^i. \tag{2.17}$$

$\{\theta^i, \eta^i\}$ is called the **adapted coframe** to ∇. A vector field $\bar{X}$ on TN is called **horizontal** if

$$\bar{X}(y) \in H_y, \text{for every } y \in TN.$$

Now, let F be a tensor field of type (1,1) on N. We define a vertical vector field γF on TN by

$$(\gamma F)(y) = (F(y))^v_y, y \in TN.$$

If $F = F^j_i(\partial/\partial x^j) \otimes (dx^i)$, we have

$$\gamma F = v^i F^j_i(\partial/\partial v^j). \tag{2.18}$$

From (2.6) and (2.18), we obtain

$$X^c - X^H = \gamma(\nabla X), \tag{2.19}$$

where ∇X is the tensor field of type (1,1) given by

$$(\nabla X)(Y) = \nabla_Y X.$$

By a straightforward computation in local coordinates, we obtain the following.

Proposition 2.7.1

$$[X^v, Y^H] = [X, Y]^v - (\nabla_X Y)^v = -(\hat{\nabla}_Y X)^v,$$

$$[X^H, Y^H] = [X, Y]^H - \gamma \hat{R}(X, Y),$$

where $\hat{R}$ is the curvature tensor of $\hat{\nabla}$ and $\hat{R}(X, Y)$ is defined by $\hat{R}(X, Y)(Z) = \hat{R}(X, Y, Z)$.

Let ω be a 1–form on N. Then we define the **horizontal lift** ω^H of ω to TN by

$$\omega^H(X^H) = 0, \omega^H(X^v) = (\omega(X))^v.$$

If $\omega = \omega_i dx^i$, we obtain

$$\omega^H = v^j \Gamma^k_{ji} \omega_k + \omega_i dv^i.$$

Hence,

$$(dx^i)^H = v^j \Gamma^k_{ji} dx^k + dv^i.$$

With respect to the adapted coframe $\{\theta^i, \eta^i\}$, we have

$$\omega^H = \omega_i \theta^i. \tag{2.20}$$

Next, we shall define the horizontal lifts of tensor fields of type (1,1) and (0,2) (For a general theory of horizontal lifts see Yano and Ishihara [130]).

Let F be a tensor field on N with local components F^j_i. We define the **horizontal lift** of F to TN with respect to ∇ to be a tensor field F^H of type (1,1) on TN determined by

$$F^H X^H = (FX)^H, F^H X^v = (FX)^v. \tag{2.21}$$

Then we have

$$F^H = F^j_i(\partial/\partial x^j) \otimes (dx^i) + v^s(\Gamma^r_{si} F^k_t - \Gamma^k_{sr} F^r_i)(\partial/\partial v^k) \otimes (dx^i)$$

$$+ F^j_i(\partial/\partial v^j) \otimes (dv^i).$$

With respect to the adapted frame, we have

$$F^H = F^j_i D_j \otimes \theta^i + F^j_i V_j \otimes \eta^i.$$

From (2.21), we easily deduce the following.

Proposition 2.7.2 *Let F, G be tensor fields of type (1,1) on N. Then*
(1) $(FG)^H = F^H G^H$,
(2) $(Id)^H = Id$.

As a direct consequence of Proposition 2.7.2, we have

Proposition 2.7.3 *If $P(t)$ is a polynomial in t, then*

$$(P(F))^H = P(F^H).$$

Moreover, if F has constant rank r, then F^H has constant rank $2r$.

We may extend these definitions for tensor fields of type $(1,r)$, $r \geq 2$ as follows. Let S be a tensor field of type $(1,s)$, $s \geq 2$, on N. Then the **horizontal lift** of S to TN is the tensor field S^H of the same type on TN given by

$$S^H(X_1^v, \ldots, X_s^v) = 0,$$

$$S^H(X_1^H, \ldots, X_{i-1}^H, X_i^v, X_{i+1}^H, \ldots, X_s^H) = (S(X_1, \ldots, X_s))^v,$$

$$S^H(X_1^H, \ldots, X_s^H) = (S(X_1, \ldots, X_s))^H.$$

In a similar way, we define the **horizontal lift** to TN of a tensor field G of type (0,2) on N to be the tensor field G^H of type (0,2) on TN given by

$$G^H(X^v, Y^v)) = 0, \; G^H(X^v, Y^H) = G^H(X^H, Y^v) = (G(X,Y))^v,$$

$$G^H(X^H, Y^H) = 0.$$

If $G = G_{ij}(dx^i) \otimes (dx^j)$, then

$$G^H = G_{ij}\theta^i \otimes \eta^j + G_{ij}\eta^i \otimes \theta^j, \tag{2.22}$$

with respect to the adapted coframe.

We can easily obtain the following.

Proposition 2.7.4 *Let g be a Riemannian metric and ∇ a linear connection on N. Then g^H is a pseudo–Riemannian metric on TN of signature (n,n). Moreover, g^H and g^c coincide if and only if $\nabla g = 0$.*

The proof is left to the reader as an exercise.
(Obviously, we may extend the above definition for tensor fields of type $(0,s), s \geq 3$).

Now, let D be a distribution on N. Then the **horizontal lift** of D to TN is the distribution D^H on TN spanned by X^v and X^H, where X is an arbitrary vector field on N belonging to D. If D is a k–dimensional distribution, then D^H is a $2k$–dimensional distribution. In fact, if D is locally spanned by $\{X_1, \ldots, X_k\}$, then D^H is locally spanned by $\{X_1^v, \ldots, X_k^v, X_1^H, \ldots, X_k^H\}$.

To end this section, we define the horizontal lift of linear connections. Let ∇ be a linear connection on N. The **horizontal lift** of ∇ to TN is the linear connection ∇^H on TN given by

$$\nabla^H_{X^v} Y^v = 0, \ \nabla^H_{X^v} Y^H = 0,$$

$$\nabla^H_{X^H} Y^v = (\nabla_X Y)^v, \ \nabla^H_{X^H} Y^H = (\nabla_X Y)^H. \tag{2.23}$$

From (2.23), we deduce that ∇^H has local components

$$\bar{\Gamma}^k_{ij} = \Gamma^k_{ij},$$

$$\bar{\Gamma}^k_{i,n+j} = \bar{\Gamma}^k_{n+i,j} = \bar{\Gamma}^k_{n+i,n+j} = \bar{\Gamma}^{n+k}_{n+i,n+j} = 0$$

$$\bar{\Gamma}^{n+k}_{ij} = v^l((\partial \Gamma^k_{ij}/\partial x^l) - R^k_{lij})),$$

$$\bar{\Gamma}^{n+k}_{i,n+j} = \Gamma^k_{ij}, \bar{\Gamma}^{n+k}_{n+i,j} = \Gamma^k_{ij}. \tag{2.24}$$

(Here, we use the same notation as in Section 2.6).

From (2.23) or (2.24), we easily obtain

$$\nabla^H_{X^c} Y^c = (\nabla_X Y)^c - \gamma(R(\ , X, Y)), \tag{2.25}$$

where $R(\ , X, Y)$ is the tensor field of type (1,1) on N defined by

$$R(\ , X, Y)Z = R(Z, X, Y).$$

Proposition 2.7.5 *∇^c and ∇^H coincide if and only if ∇ is flat, that is, $R = 0$.*

Proof: Directly from (2.25). □

2.8 Sasaki metric on the tangent bundle

Let g be a Riemannian metric on N with local components g_{ij} and Riemannian connection ∇. In [108], Sasaki defined a Riemannian metric $\bar{g}$ on TN given by

$$\bar{g}_y(X^v, Y^v) = g_x(X, Y),$$

$$\bar{g}_y(X^v, Y^H) = 0,$$

$$\bar{g}_y(X^H, Y^H) = g_x(X, Y), \tag{2.26}$$

where $X, Y \in T_xN, y \in T_xN$ and the horizontal lifts are taken with respect to ∇. It is easy to check that $\bar{g}$ defines a Riemannian metric on TN, called the **Sasaki metric** induced by g.

If H and V denotes the horizontal and vertical distributions, respectively, we deduce that H and V are orthogonal with respect to $\bar{g}$. From (2.26), we see that the local components of $\bar{g}$ are given by the following matrix:

$$\begin{bmatrix} g_{ij} + g_{st}v^r v^k \Gamma^s_{ri}\Gamma^t_{kj} & g_{jk}v^r\Gamma^k_{ri} \\ g_{ik}v^r\Gamma^k_{rj} & g_{ij} \end{bmatrix} \tag{2.27}$$

with respect to the induced coordinates (x^i, v^i).

With respect to the adapted coframe to ∇, we have

$$\bar{g} = \begin{bmatrix} g_{ij} & 0 \\ 0 & g_{ij} \end{bmatrix},$$

or, equivalently,

$$\bar{g} = g_{ij}\theta^i \otimes \theta^j + g_{ij}\eta^i \otimes \eta^j.$$

Let $\bar{\nabla}$ be the Riemannian connection determined by $\bar{g}$. A long but straightforward computation from (2.26) or (2.27), permit us to obtain the local components of $\bar{\nabla}$:

$$\bar{\Gamma}^k_{ij} = \Gamma^k_{ij}, \bar{\Gamma}^k_{i,n+j} = -(1/2)R^k_{isj}v^s,$$

$$\bar{\Gamma}^k_{n+i,j} = -(1/2)R^k_{jsi}v^s, \bar{\Gamma}^k_{n+1,n+j} = 0,$$

$$\bar{\Gamma}_{ij}^{n+k} = -(1/2)R_{ijs}^{k}v^{s}, \bar{\Gamma}_{i,n+j}^{n+k} = \Gamma_{ij}^{k},$$

$$\bar{\Gamma}_{n+i,j}^{n+k} = \bar{\Gamma}_{n+i,n+j}^{n+k} = 0, \tag{2.28}$$

with respect to the adapted frame, where R_{ijk}^{h} are the components of the curvature tensor of ∇ (here, we use the notations introduced in Sections 2.6 and 2.7).

From (2.28), we obtain the following formulas (see Kowalski [84]):

$$\bar{\nabla}_{X^v} Y^v = 0,$$

$$(\bar{\nabla}_{X^H} Y^v)_y = (\nabla_X Y)_y^v + (1/2)(R_x(y, Y_x, X_x))_y^H,$$

$$(\bar{\nabla}_{X^v} Y^H)_y = (1/2)(R_x(y, X_x, Y_x))_y^H,$$

$$(\bar{\nabla}_{X^H} Y^H)_y = (\nabla_X y)_y^H - (1/2)(R_x(X_x, Y_x, y))_y^v, \tag{2.29}$$

for all vector fields X, Y on N, $y \in TN$ and $x = \tau_N(y)$.

Moreover, Kowalski [84] proved that the curvature tensor $\bar{R}$ of $\bar{\nabla}$ is completely determined by the following formulas:

$$\bar{R}(X^v, Y^v, Z^v) = 0,$$

$$\bar{R}(X^v, Y^v, Z^H)_y = (R(X, Y, Z) + (1/4)R(y, X, R(y, Y, Z)) -$$

$$-(1/4)R(y, Y, R(y, X, Z)))_y^H,$$

$$\bar{R}(X^H, Y^v, Z^v)_y = -((1/2)R(Y, Z, X) + (1/4)R(y, Y, R(y, Z, X)))_y^H,$$

$$\bar{R}(X^H, Y^v, Z^H)_y = ((1/4)R(R(y, Y, Z), X, y) + (1/2)R(X, Z, Y))_y^v +$$

$$+((1/2)(\nabla_X R)(y, Y, Z))_y^H,$$

$$\bar{R}(X^H, Y^H, Z^v)_y = (R(X,Y,Z) + (1/4)R(R(y,Z,Y),X,Y)-$$

$$-(1/4)R(R(y,Z,X),Y,y))^v_y+$$

$$+(1/2)(((\nabla_X R)(y,Z,Y) - (\nabla_Y R)(y,Z,X)))^H_y,$$

$$\bar{R}(X^H, Y^H, Z^H) = (1/2)((\nabla_Z R)(X,Y,y))^v_y$$

$$+((R(X,Y,Z) + (1/4)R(y, R(Z,Y,y),X)$$

$$+ (1/4)R(y, R(y, R(X,Z,y),Y) + (1/2)R(y, R(X,Y,y),Z))^H_y, \qquad (2.30)$$

for any $X, Y, Z \in T_xN, y \in TN, x = \tau_N(y)$.

The proof of (2.30) is left to the reader as an exercise. It is sufficient to use (2.29) and the definition of the curvature tensor.

From (2.30), we have

Proposition 2.8.1 $(TN, \bar{g})$ *is flat if and only if* (N, g) *is flat.*

Proof: Obviously, if $R = 0$, then $\bar{R} = 0$. Conversely, suppose that $\bar{R} = 0$. Then

$$\bar{R}(X^H, Y^H, Z^H)_y$$

$$= -((1/2)R(Y,Z,X) + (1/4)R(y,Y,R(y,Z,X)))^H_y = 0$$

If we set $Y = y$, we obtain

$$R(Y,Z,X) = 0,$$

for any X, Y, Z. Hence $R = 0$. □

2.9 Affine bundles

In this section, we introduce the notion of affine bundles. The results obtained here will be used in the next section.

Definition 2.9.1 *An* **affine space** *is a triple (A, V, r), where A is a set, V a finite dimensional real vector space, and $r : A \times V \longrightarrow V$ is a free transitive action of the additive group V on A. We put $r(a, v) = a + v$ and say that the affine space A is* **modeled** *on V.*

Definition 2.9.2 *An* **affine morphism** *of the affine spaces (A, V, r) and (A', V', r') is a pair $(\wedge, \lambda)$ where $\wedge : A \longrightarrow A'$ is a map and $\lambda : V \longrightarrow V'$ is a linear map such that*

$$r'(\wedge(a), \lambda(v)) = \wedge(r(a, v)), a \in A, v \in V, \text{that is },$$

$$\wedge(a) + \lambda(v) = \wedge(a + v).$$

Example.- Let V be a finite dimensional real vector space. Then $(V, V, +)$ is an affine space modeled on V.

Remark 2.9.3 Let A and A' be affine spaces modeled on the same real vector space V. Then A and A' are isomorphic. In fact, A and A' are isomorphic to $(V, V, +)$.

Definition 2.9.4 *An* **affine bundle** *consists of a fibred manifold $\pi : A \longrightarrow N$ and a vector bundle $p : E \longrightarrow N$ of rank m, together with a morphism $r : A \times_N E \longrightarrow A$ of fibred manifolds over id_N, such that for each $x \in N$,*

$$r_x : A_x \times E_x \longrightarrow A_x$$

is a free transitive action of the vector space E_x on the set A_x. Hence, each fibre A_x is an affine space modeled on E_x.

Now, let A be an affine bundle as in Definition 2.9.4. Let U be an open set of N over which A admits a local section s and over which E is locally trivial (let us remark that A always admits local sections since π is a subjective submersion). Then there is a diffeomorphism

$$H : U \times R^m \longrightarrow p^{-1}(U)$$

over the id_U. Let $a \in \pi^{-1}(U)$ with $\pi(a) = x \in N$. There exists a unique element $\phi(a)$ of $\pi^{-1}(x)$ such that

$$a = s(x) + \phi(a).$$

Hence, we can define a diffeomorphism

$$\phi : \pi^{-1}(U) \longrightarrow p^{-1}(U)$$

given by

$$a \longrightarrow \phi(a).$$

Then $\phi^{-1} \circ H$ defines a trivialization of A over U. Thus, A **is a locally trivial fibre bundle over** N **with standard fibre** R^m.

Finally. let $\pi : A \longrightarrow N$ be an affine bundle modeled on a vector bundle $p : E \longrightarrow N$ of rank m. Since N is paracompact, there is a global section s of A. Then we can define an isomorphism of fibred manifolds

$$\phi : A \longrightarrow E$$

given as follows. Let $a \in A_x$. Therefore, there exists a unique $\phi(a) \in E_x$ such that

$$a = s(x) + \phi(a).$$

Thus, **we can define a vector bundle structure on** A **of rank** m **such that** A **becomes a vector bundle isomorphic to** E. Obviously, this vector bundle isomorphism ϕ depends on the choice of s.

2.10 Integrable almost tangent structures which define fibrations

In this section, we prove that an integrable almost tangent manifold which defines a fibration is the tangent bundle of some manifold. This result is due to Crampin and Thompson (see [24]).

Let J be an integrable almost tangent structure on a $2n$–dimensional manifold M. Since the distribution $V = Im\ J = Ker\ J$ is integrable, then it defines a foliation on M. Now, we define an equivalence relation on M as follows: two points of M are equivalent if they lie on the same leaf of the foliation defined by V.

Definition 2.10.1 *We say that J* **defines a fibration** *if the quotient of M by this equivalence relation (that is, the space of leaves) has the structure of a differentiable manifold.*

This will be the case if for every leaf one can find an embedded local submanifold of M of dimension n trough a point of the leaf which intersects each leaf which it does in only one point. In this case, the space of leaves N is an n–dimensional manifold and the canonical projection $p : M \longrightarrow N$ is a subjective submersion (that is, N is a quotient manifold of M). Then $p : M \longrightarrow N$ is a fibred manifold and

$$V_y = T_y(p^{-1}(x)), y \in M, x = p(y),$$

for each $y \in M$.

Bearing in mind the definitions of the Section 2.2, we may define the vertical lift of tangent vectors on N to M as follows. If $u \in T_xN$ and $y \in p^{-1}(x)$ we define $u^v \in T_yM$ by

$$u^v = J_y(\bar{u}),$$

where $\bar{u} \in T_yM$ and $p_*(u) = u$. Since

$$V_y = Ker\{p_* : T_yM \longrightarrow T_xN\}$$

and

$$J_yV_y = 0,$$

then u^v is well–defined. Moreover, we have

$$u^v \in V_y,$$

and the map $u \longrightarrow u^v$ is a linear isomorphism of T_xN with V_y. If X is a vector field on N, then we may define its vertical lift to M given by

$$X^v = J\bar{X},$$

where $\bar{X}$ is any vector field on M which is p–related to X. Obviously, $X^v \in V$.

Proposition 2.10.2 *Let X, Y be two vector fields on N. Then we have*
(1) $[X^v, Y^v] = 0$,
(2) $L_{X^v} J = 0$.

Proof: (1) Let $\bar{X}, \bar{Y}$ be vector fields on M p–related to X, Y. Then

$$[X^v, Y^v] = [J\bar{X}, J\bar{Y}]$$

$$= J[J\bar{X}, \bar{Y}] + J[\bar{X}, J\bar{Y}] \text{ (since } N_J = 0)$$

$$= J[X^v, \bar{Y}] + J[\bar{X}, Y^v].$$

But $(Tp)[X^v, \bar{Y}] = [(Tp)X^v, (Tp)\bar{Y}] = 0$, since $X^v, \bar{Y}$ are p–related to 0, Y, respectively. Similarly, we obtain $(Tp)[\bar{X}, Y^v] = 0$. Hence $[X^v, \bar{Y}], [\bar{X}, Y^v]$ are both vertical and, thus,

$$J[X^v, \bar{Y}] = J[\bar{X}, Y^v] = 0.$$

(2) It is sufficient to prove that

$$(L_{X^v} J)(Z^v) = 0, (L_{X^v} J)(\bar{Z}) = 0,$$

where Z is a vector field on N and $\bar{Z}$ is a vector field on M p–related to Z. But this is a direct consequence of (1).□

Now, let ∇ be a symmetric almost tangent linear connection on M. Then we have

Proposition 2.10.3 *∇ induces by restriction a connection on each leaf of V which is flat.*

Proof: In fact, we have

$$\nabla_{X^v} Y^v = \nabla_{X^v}(JY) = J(\nabla_{X^v}\bar{Y}) \quad \text{(since } \nabla J = 0)$$

$$= J(\nabla_{\bar{Y}} X^v + [X^v, \bar{Y}]) \quad \text{(since } \nabla \text{ is symmetric)}$$

$$= J(\nabla_{\bar{Y}} X^v) + J[X^v, \bar{Y}]$$

$$= \nabla_{\bar{Y}}(JX^v) + J[X^v, \bar{Y}] = 0,$$

where $\bar{Y}$ is any vector field on M p–related to Y.□

Now, we prove the main result of this section.

Theorem 2.10.4 *(Crampin and Thompson [24]) Let (M,J) be an integrable almost tangent structure which defines a fibration $p: M \longrightarrow N$. Let ∇ be any symmetric almost tangent connection on M and suppose that with respect to the flat connection induced on it by ∇, each leaf of the foliation defined by V is geodesically complete. Suppose further that each leaf of this foliation (that is, the fibres of $p: M \longrightarrow N$) is simply connected. Then M is an affine bundle modeled on TN.*

Proof: We define a morphism

$$\rho : M \times_N TN \longrightarrow M$$

as follows. Let $u \in T_xN, x \in N$. Then we define a vertical vector field U on $p^{-1}(x)$ by

$$U_y = u^v \in V_y, y \in p^{-1}(x).$$

Then $\nabla_U U = 0$ by Proposition 2.10.3. Thus U is a geodesic vector field and therefore, complete as a vector field. Let $\phi_u(t,y)$ be the integral curve of U such that

$$\phi_u(0,y) = y,$$

where $\phi_u : R \times p^{-1}(x) \longrightarrow p^{-1}(x)$ is the one–parameter group generated by U. We define ρ by

$$\rho_x(y,u) = \phi_u(1,y).$$

Next, we shall prove that ρ_x defines a transitive, free action of T_xN on $p^{-1}(x)$. In fact, for any $u, v \in T_xN$, the corresponding vector fields U, V satisfy $[U,V] = 0$ (see Proposition 2.10.2). Consequently, their one–parameter groups commute, that is

$$\phi_u(t,\phi_v(s,y)) = \phi_v(s,\phi_u(t,y)).$$

Since U and V commute and are complete, then the composition of their one–parameter groups is a one–parameter group whose infinitesimal generator is $U + V$. Thus we have

$$\phi_u(t,\phi_v(t,y)) = \phi_v(t,\phi_u(t,y)) = \phi_{u+v}(t,y). \tag{2.31}$$

From (2.31), we deduce that

$$\rho_x(\rho_x(y,v),u) = \rho_x(\rho_x(y,u),v) = \rho_x(y,u+v).$$

Hence

$$\rho_x : p^{-1}(x) \times T_xN \longrightarrow p^{-1}(x)$$

defines an action of T_xN on $p^{-1}(x)$.

Next, let $<,>$ be an arbitrary scalar product on T_xN. We define a Riemannian metric g on $p^{-1}(x)$ by $g(U,V) =< u,v >$. Since $\nabla U = \nabla V = 0$, and $g(U,V)$ is constant, we deduce that ∇ is the Riemann connection for g. Hence, $p^{-1}(x)$ is a geodesically complete Riemannian manifold. Then, if y and z are two points of $p^{-1}(x)$, there exists, from the Hopf–Rinow theorem, a geodesic σ such that $\sigma(0) = y$ and $\sigma(1) = z$. Since the tangent vector $\dot{\sigma}(0)$ is vertical, then $\dot{\sigma}(0) = u^v$, where $u \in T_xN$. Therefore, σ is the integral curve of U trough y and

$$z = \phi_u(1,y) = \rho_x(y,u).$$

This proves the transitivity of ρ_x.

Now, let $\Gamma(y)$ be the isotropy group of ρ_x, that is

$$\Gamma(y) = \{u \in T_xN / \rho_x(y,u) = u\}.$$

The map $\beta : T_xN \longrightarrow p^{-1}(x)$ given by $\beta(u) = \rho_x(y,u)$ may be factored as follows:

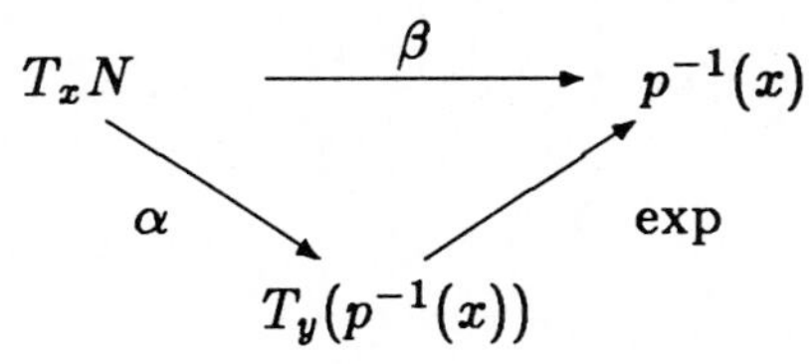

where α is the vertical lift map from T_xN to the tangent space to $p^{-1}(x)$ at y and $exp : T_y(p^{-1}(x)) \longrightarrow p^{-1}(x)$ is the exponential map of ∇ restricted to $p^{-1}(x)$. Since α is a linear isomorphism and exp a local diffeomorphism, then β is a local diffeomorphism. Thus, $\Gamma(y)$ is a discrete subgroup of the additive group T_xN. Therefore, $\Gamma(y)$ must consists of integer linear combinations of some k linearly independent vector $v_1, \ldots, v_k, 0 \leq k \leq n$. Moreover, since T_xN acts transitively on $p^{-1}(x)$, then this space is diffeomorphic to the coset space $T_xN/\Gamma(y)$. Then we deduce that $p^{-1}(x)$ is the product of a k–torus T^k and R^{n-k}. Thus, $\Gamma(y)$ must be trivial, since $p^{-1}(x)$ is simply connected. Therefore, the action ρ_x is free. This ends the proof. □

Corollary 2.10.5 *If (M,J) verifies all the hypotheses of the theorem and in addition $p : M \longrightarrow N$ admits a global section, then M is isomorphic (as a vector bundle) to TN. (This isomorphism depends on the choice of the section).*

Corollary 2.10.6 *If (M,J) verifies all the hypotheses of the theorem except the hypotheses that the leaves of the foliation defined by V are simply connected, then TN is a covering space of M and the leaves of V are of the form*

$$T^k \times R^{n-k},$$

where T^k is the k–dimensional torus, $0 \leq k \leq n$. Moreover, if it is assumed that the leaves of V are compact, then TN is a covering space of M and the fibres are diffeomorphic to T^n.

Remark 2.10.7 In de León, Méndez and Salgado [33], [34], we introduce the concept of a p–almost tangent structure and prove similar results for integrable p–almost tangent structures which define fibrations.

2.11 Exercises

2.11.1 Prove Proposition 2.1.4.
2.11.2 (1) Let g be a symmetric tensor of type (0,2) on an n–dimensional vector space E. Prove that, if g has rank r, then there exists a basis $\{e_1, \ldots, e_n\}$ of E with dual basis $\{e^1, \ldots, e^n\}$ such that

$$g = \sum_{i=1}^{r} a_i\, e^i \otimes e^i,$$

where $a_i = \pm 1$, or, equivalently, the matrix of g is

$$\left[\begin{array}{ccc|c} a_1 & & 0 & \\ & \ddots & & 0 \\ 0 & & a_r & \\ \hline & 0 & & 0 \end{array}\right]$$

(2) If g satisfies

$$g(v, w) = 0 \text{ for all } w \text{ implies } v = 0,$$

then g has rank n and we have

$$g = \sum_{i=1}^{n} a_i e^i \otimes e^i,$$

where $a_i = \pm 1$, $1 \le i \le n$. We say that g **has signature** (p, q) with $p + q = n$ if $a_1 = \ldots = a_p = 1$ and $a_{p+1} = \ldots = a_n = -1$.
(3) Prove that if g is a pseudo–Riemannian metric on an n–dimensional manifold M then g_x has the same signature (p, q), $p + q = n$, for all $x \in M$. We say that g has signature (p, q).
(4) Prove Corollary 2.5.9.

2.11.3 Prove that if M is complete with respect to a linear connection ∇ then TM is complete with respect to ∇^c, and conversely.

2.11.4 Prove Proposition 2.7.4.

2.11.5 Prove that ∇^H is of zero curvature if and only if ∇ is of zero curvature.

2.11.6 Prove (2.30).

Chapter 3

Structures on manifolds

3.1 Almost product structures

In this section, we introduce some definitions and basic facts about almost product structures. For more details, we remit to Fujimoto [57], Walker [121], [122], Willmore [128].

Definition 3.1.1 *Let M be a differentiable m–dimensional manifold. An* **almost product structure** *on M is a tensor field F of type (1,1) on M such that $F^2 = Id$. M, endowed with an almost product structure F is said to be an* **almost product manifold.**

We set

$$P = (1/2)(Id + F),\ Q = (1/2)(Id - F).$$

Then we have

$$P^2 = P,\ PQ = QP = 0,\ Q^2 = Q. \tag{3.1}$$

Conversely, if (P, Q) is a pair of tensor fields of type (1,1) on M satisfying (3.1), then we put

$$F = P - Q,$$

and F is an almost product structure on M.

We set

$$\mathbf{P} = ImP,\ \mathbf{Q} = ImQ.$$

Then $\mathbf{P}$ and $\mathbf{Q}$ are complementary distributions on M, i.e.,

$$T_xM = \mathbf{P}_x \oplus \mathbf{Q}_x, x \in M.$$

If P has constant rank p and Q has constant rank q, respectively then $\mathbf{P}$ is a p–dimensional distribution and $\mathbf{Q}$ a q–dimensional distribution on M, respectively, and $p + q = m$. Conversely, if there exist on M two complementary distributions $\mathbf{P}$ and $\mathbf{Q}$, then P and Q are defined to be the corresponding projectors

$$P_x : T_xM \longrightarrow \mathbf{P}_x, Q_x : T_xM \longrightarrow \mathbf{Q}_x, x \in M.$$

Let $e_1, \ldots, e_p$ be a basis of $\mathbf{P}_x$ and $e_{p+1}, \ldots, e_m$ a basis of $\mathbf{Q}_x, x \in M$. Hence $\{e_1, \ldots, e_p, e_{p+1}, \ldots, e_m\}$ is a basis of T_xM which is called an **adapted frame** at x. Let $\{e_1, \ldots, e_m\}, \{e'_1, \ldots, e'_m\}$ be two adapted frames at x. Therefore, we have

$$e'_i = A_i^j e_j, 1 \leq i, j \leq p,$$

$$e'_i = B_i^j e_j, p + 1 \leq i, j \leq m,$$

where $A \in Gl(p, R), B \in Gl(q, R)$. Then $\{e_i; 1 \leq i \leq m\}$ and $\{e'_i; 1 \leq i \leq m\}$ are related by the $m \times m$ matrix

$$\alpha = \begin{bmatrix} A & 0 \\ 0 & B \end{bmatrix} \in Gl(m, R).$$

Clearly, $\alpha \in Gl(p, R) \times Gl(q, R)$, where $Gl(p, R) \times Gl(q, R)$ is identified to the Lie subgroup G of $Gl(m, R)$ given by

$$G = \left\{ \begin{bmatrix} A & 0 \\ 0 & B \end{bmatrix} / A \in Gl(p, R), B \in Gl(q, R) \right\}$$

We note that, with respect to an adapted frame, P, Q and F have the following matricial representations

$$P_0 = \begin{bmatrix} I_p & 0 \\ 0 & 0 \end{bmatrix}, Q_0 = \begin{bmatrix} 0 & 0 \\ 0 & I_q \end{bmatrix}, F_0 = \begin{bmatrix} I_p & 0 \\ 0 & I_q \end{bmatrix}.$$

We set

$$B = \{\text{adapted frames at all points of } M\}.$$

One can easily proves that B defines a $(Gl(p,R) \times Gl(q,R))$–structure on M. Conversely, if B is a $(Gl(p,R) \times Gl(q,R))$–structure on M, then we define P and Q to be the tensor fields of type (1,1) on M which have matricial representations P_0 and Q_0 with respect to any frame of B at x, for each $x \in M$.

Summing up, we have proved the following.

Proposition 3.1.2 *Giving an almost product structure is the same as giving a $(Gl(p,R) \times Gl(q,R))$–structure.*

We say that an almost product structure F is **integrable** if there exists a local coordinate neighborhood $(U, x^1, \ldots, x^m)$ at each point of M such that the local frame field

$$\sigma : x \in U \longrightarrow \sigma(x) = ((\partial/\partial x^1)_x, \ldots, (\partial/\partial x^m)_x)$$

is a section of B, that is, $\sigma(x)$ is an adapted frame at x, for each $x \in U$. Therefore we have

Proposition 3.1.3 *F is integrable if and only if* **P** *and* **Q** *are integrable.*

Next, we shall give a characterization of the integrability of an almost product structure F in terms of the Nijenhuis tensors N_P, N_Q and N_F.

Proposition 3.1.4 *The following four assertions are equivalent:*
(1) The almost product structure F is integrable.
(2) $N_F = 0$.
(3) $N_P = 0$.
(4) $N_Q = 0$.

Proof: First, we note that

$$N_P = (1/2)N_F, N_Q = -(1/2)N_F.$$

Hence (2), (3) and (4) are equivalent. Next, we shall prove that (1) and (2) are equivalent. Let us recall that N_P and N_Q are given by

$$N_P(X,Y) = [PX, PY] - P[PX, Y] - P[X, PY] + P[X,Y],$$

$$N_Q(X,Y) = [QX, QY] - Q[QX, Y] - Q[X, QY] + Q[X,Y].$$

since $P^2 = P$ and $Q^2 = Q$. If F is integrable, then **P** and **Q** are integrable. Hence

$$N_P(X,Y) = [PX,PY] - P[PX,Y] - P[X,PY] + P[X,Y]$$

$$= [PX,PY] - P[PX,PY+QY] - P[PX+QX,PY] + P[PX+QX,PY+QY]$$

(since $Z = PZ + QZ$, for any vector field Z on M)

$$= Q[PX,PY] + P[QX,QY].$$

But $[PX,PY] \in \mathbf{P}$ and $[QX,QY] \in \mathbf{Q}$, since **P** and **Q** are integrable. Thus $N_P = 0$ and, therefore, $N_F = 0$. Conversely, suppose that $N_F = 0$. Then $N_P = N_Q = 0$, and thus

$$[PX,PY] \in P, [QX,QY] \in Q.$$

Consequently, **P** and **Q** are integrable and, so, F is integrable, by Proposition 3.1.3. □

Let ∇ be a linear connection on M. Since

$$\nabla F = 2(\nabla P) = 2(\nabla Q),$$

we have

Proposition 3.1.5 *The following three assertions are equivalent:* *(1)* $\nabla F = 0$; *(2)* $\nabla P = 0$; *(3)* $\nabla Q = 0$.

Definition 3.1.6 *A linear connection* ∇ *on* M *such that* $\nabla F = 0$ *is said to be an* **almost product connection**.

Proposition 3.1.7 *There exists an almost product connection on every almost product manifold.*

Proof: Let $\bar{\nabla}$ be an arbitrary linear connection on M. We define a tensor field of type (1,2) on M by

$$S(X,Y) = (1/2)\{(\bar{\nabla}_{FY}F)X + F(\bar{\nabla}_Y F)X - F((\bar{\nabla}_X F)Y)\}.$$

Since

$$(\bar{\nabla}_X F)F = -F(\bar{\nabla}_X F),$$

we can easily prove, from a straightforward computation, that $\nabla = \bar{\nabla} - S$ is an almost product connection on M.□

Now, let ∇ be a symmetric linear connection on M. Then we obtain

$$N_P(X,Y) = (\nabla_{PX}P)Y - (\nabla_{PY}P)X - (P(\nabla_X P))Y + (P(\nabla_Y P))X,$$

$$N_Q(X,Y) = (\nabla_{QX}Q)Y - (\nabla_{QY}Q)X - (Q(\nabla_X Q))Y + (Q(\nabla_Y Q))X, \quad (3.2)$$

$$N_F(X,Y) = (\nabla_{FX}F)Y - (\nabla_{FY}F)X - (F(\nabla_X F))Y + (F(\nabla_Y F))X.$$

From (3.2), we easily deduce the following.

Theorem 3.1.8 *If there exists a symmetric almost product connection on M then the almost product structure F is integrable. (See Exercise 3.8.3).*

The converse is also true (see Fujimoto [57]).

3.2 Almost complex manifolds

Definition 3.2.1 *An* **almost complex structure** *on a differentiable manifold M is a tensor field J of type (1,1) such that $J^2 = -Id$. A manifold M with an almost complex structure J is called an* **almost complex manifold**.

Let J be an almost complex structure on M. Then, for each point x of M, J_x is an endomorphism of the tangent space T_xM such that $J_x^2 = -Id$. Hence T_xM may be turned into a complex vector space by defining scalar multiplication by complex numbers as follows:

$$(a + \sqrt{-1}b)X = aX + b(JX), X \in T_xM, a, b \in R.$$

Therefore the real dimension of T_xM must be even, namely $2n$. We deduce that every almost complex manifold M has even dimension $2n$. In fact, let $\{X_1, \ldots, X_n\}$ be a basis for T_xM as a complex vector space. Then $\{X_1, \ldots, X_n, JX_1, \ldots, JX_n\}$ is a basis for T_xM as a real vector space. In fact, $\{X_1, \ldots, X_n, JX_1, \ldots, JX_n\}$ is a set of linearly independent vectors, since, if

$$\Sigma(a^i X_i + b^i(JX_i)) = 0,$$

then we have

$$0 = \Sigma(a^i X_i + b^i(JX_i)) = \Sigma(a^i + \sqrt{-1}b^i)X_i,$$

which implies $a^i + \sqrt{-1}b^i = 0, 1 \leq i \leq n$. Thus, $a^i = b^i = 0, 1 \leq i \leq n$. Moreover, if $X \in T_x M$, then

$$X = \Sigma(a^i + \sqrt{-1}b^i)X_i = \Sigma a^i X_i + \Sigma b^i(JX_i).$$

Thus, $\{X_i, JX_i\}$ span $T_x M$. This basis is called an **adapted** (or **complex) frame** at x.

Let now $\{X_i, JX_i\}$, $\{X_i', JX_i'\}$ be two bases as above. Therefore, we have

$$X_i' = (A_i^j + \sqrt{-1}B_i^j)X_j = A_i^j X_j + B_i^j(JX_j),$$

and, consequently,

$$JX_i' = A_i^j(JX_j) + B_i^j(J^2 X_j) = -B_i^j X_j + A_i^j(JX_j),$$

where A, B are $n \times n$ matrices. Then the two complex frames are related by the $2n \times 2n$ matrix

$$\alpha = \begin{bmatrix} A & -B \\ B & A \end{bmatrix}.$$

Clearly, $\alpha \in Gl(2n, R)$.

Now, let G be the set of such matrices; G is a closed subgroup of $Gl(2n, R)$ and therefore a Lie subgroup of $Gl(2n, R)$. If $Gl(n, C)$ is the complex linear general group, we have a real representation of $Gl(n, C)$ into $Gl(2n, R)$ given by

$$\rho : Gl(n, C) \longrightarrow Gl(2n, R)$$

$$\alpha = A + \sqrt{-1}B \longrightarrow \begin{bmatrix} A & -B \\ B & A \end{bmatrix}$$

In fact, ρ is a Lie group monomorphism. Hence $Gl(n,C)$ may be identified with $\rho(Gl(n,C)) = G$.

We note that, with respect to a complex frame, J is represented by the matrix

$$J_0 = \begin{bmatrix} 0 & -I_n \\ I_n & 0 \end{bmatrix},$$

where I_n is the $n \times n$ identity matrix. It is easy to prove that $Gl(n,C)$ can be described as the invariance group of the matrix J_0, that is,

$$Gl(n,C) = \{\alpha \in Gl(2n,R)/\alpha J_0 = J_0 \alpha\}.$$

Now, we set

$$CM = \{\text{complex frames at all points of } M\}.$$

We shall prove that CM defines a $Gl(n,C)$–structure on M. In order to do this, we note that the tangent bundle TM becomes a complex vector bundle of rank n. Then, for each $x \in M$, there are n local vector fields $X_1, \dots, X_n$ on a neighborhood U of x such that $\{X_1(y), \dots, X_n(y)\}$ is a basis for T_yM as a complex vector space for any $y \in U$. If we define

$$\sigma(y) = (X_1(y), \dots, X_n(y), JX_1(y), \dots, JX_n(y)),$$

then σ is a local section of FM over U such that $\sigma(U) \subset CM$. Thus, CM is a $Gl(n,C)$–structure on M.

Conversely, let B be a $Gl(n,C)$–structure on M. We define a tensor field of type (1,1) on M as follows. We set

$$J_x(X) = p(J_0(p^{-1}(X))),$$

where $X \in T_xM$, $x \in M$ and $p \in CM$ is a linear frame at x. Obviously, $J_x(X)$ is independent of the choice of p and $J_x^2 = -Id$. Then J defines an almost complex structure on M.

Summing up, we have proved the following.

Proposition 3.2.2 *Giving an almost complex structure is the same as giving a $Gl(n,C)$–structure on M.*

Definition 3.2.3 *Let M be a topological space such that each point has a neighborhood U homeomorphic to an subset of C^n. Each pair (U, ϕ), where U is an open set of M and ϕ is a homeomorphism of U to a open subset $\phi(U)$ of C^n is called a* **coordinate neighborhood**; *to $x \in U$ we assign the n* **complex coordinates** *$z^1(x), \ldots, z^n(x)$ of $\phi(x) \in C^n$. Two coordinate neighborhoods $(U, \phi), (V, \psi)$ are said to be* **compatible** *if the mappings $\phi \circ \psi^{-1}$ and $\psi \circ \phi^{-1}$ are holomorphic. A* **complex structure** *on M is a family $U = \{(U_\alpha, \phi_\alpha)\}$ of coordinate neighborhoods such that*
(1) the U_α cover M;
(2) for any α, β the neighborhoods (U_α, ϕ_α) and (U_β, ϕ_β) are compatible;
(3) U is maximal (in the obvious sense).
M, endowed with a complex structure, is said to be a **complex manifold** *of complex dimension n.*

Let M be a complex manifold of complex dimension n. Then M becomes a C^∞–manifold of real dimension $2n$. In fact, each coordinate neighborhood U with complex coordinates $z^1, \ldots, z^n$ gives real coordinates $x^1, \ldots, x^n, y^1, \ldots, y^n$ by setting

$$z^i = x^i + \sqrt{-1}y^i, 1 \leq i \leq n.$$

We shall prove that every complex manifold carries a natural almost complex structure. Let $(z^1, \ldots, z^n)$ be a complex local coordinate system on a neighborhood U. We define an endomorphism

$$J_x : T_xM \longrightarrow T_xM, x \in U,$$

by

$$J_x(\partial/\partial x^i) = \partial/\partial y^i, J_x(\partial/\partial y^i) = -(\partial/\partial x^i), 1 \leq i \leq n.$$

We prove that the definition of J does not depend on the choice of the complex local coordinate system. If $(w^1, \ldots, w^n)$ is another complex local coordinate system on a neihgborhood $V, U \cap V \neq \emptyset$ and

$$w^i = u^i + \sqrt{-1}v^i, 1 \leq i \leq n,$$

then the change of coordinates $w^i = w^i(z^j)$ is a holomorphic function. Hence the following Cauchy–Riemann conditions hold:

$$(\partial u^k/\partial x^i) = (\partial v^k/\partial y^i), (\partial u^k/\partial y^i) = -(\partial v^k/\partial x^i). \tag{3.3}$$

On the other hand, we have

$$\begin{aligned} \partial/\partial x^i &= (\partial u^k/\partial x^i)(\partial/\partial u^k) + (\partial v^k/\partial x^i)(\partial/\partial v^k) \\ \partial/\partial y^i &= (\partial u^k/\partial y^i)(\partial/\partial u^k) + (\partial v^k/\partial y^i)(\partial/\partial v^k). \end{aligned} \tag{3.4}$$

Let $J'_x : T_xM \longrightarrow T_xM, x \in U \cap V$, defined by

$$J'_x(\partial/\partial u^i) = \partial/\partial v^i, J'_x(\partial/\partial v^i) = -(\partial/\partial u^i).$$

From (3.4), we have

$$J'_x(\partial/\partial x^i) = (\partial u^k/\partial x^i)J'_x(\partial/\partial u^k) + (\partial v^k/\partial x^i)J'_x(\partial/\partial v^k)$$

$$= (\partial u^k/\partial x^i)(\partial/\partial v^k) - (\partial v^k/\partial x^i)(\partial/\partial u^k)$$

$$= (\partial u^k/\partial y^i)(\partial/\partial v^k) + (\partial v^k/\partial y^i)(\partial/\partial v^k) \text{ (since (3.3))}$$

$$= \partial/\partial y^i$$

Similarly, we deduce that

$$J'_x(\partial/\partial y^i) = -(\partial/\partial x^i).$$

Hence $J'_x = J_x$ and, therefore, J is well–defined.

To end this section, we shall give a characterization of the integrability of almost complex structures.

Definition 3.2.4 *An almost complex structure J on a $2n$–dimensional manifold M is said to be* **integrable** *if it is integrable as a $Gl(n, C)$–structure.*

Therefore, if J is integrable, for each point $x \in M$, there exists a local coordinate system $(x^1, \ldots, x^n, y^1, \ldots, y^n)$ such that

$$J(\partial/\partial x^i) = \partial/\partial y^i, J(\partial/\partial y^i) = -(\partial/\partial x^i), 1 \leq i \leq n.$$

In fact, the local section

$$\sigma : y \longrightarrow ((\partial/\partial x^i)_y, (\partial/\partial y^i)_y)$$

of FM takes values in CM. Hence, if J is integrable, M becomes a complex manifold; it is sufficient to set

$$z^i = x^i + \sqrt{-1}y^i$$

as complex local coordinates (details are left to the reader as an exercise). Hence, an integrable almost complex structure J is called a **complex structure**. If we denote by J_0 the canonical complex structure on $C^n = R^{2n}$, then an almost complex structure is integrable if and only if the corresponding $Gl(n, C)$–structure is locally isomorphic to J_0.

Definition 3.2.5 *Let J be an almost complex structure on M. The* **Nijenhuis tensor** *N_J of J is a tensor field of type (1,2) on M given by*

$$N_J(X, Y) = [JX, JY] - J[JX, Y] - J[X, JY] - [X, Y],$$

$X, Y \in \chi(M)$.

Obviously, if J is integrable, then the Nijenhuis tensor N_J vanishes. The converse is true; it is the theorem of Newlander and Niremberg [100]. It is beyond the scope of this book to give a proof of this theorem.

Theorem 3.2.6 *(Newlander–Niremberg) An almost complex structure J is integrable if and only if its Nijenhuis tensor N_J vanishes.*

3.3 Almost complex connections

Let M be an almost complex manifold of dimension $2n$ with almost complex structure J.

Definition 3.3.1 *A linear connection ∇ on M is said to be an* **almost complex connection** *if $\nabla J = 0$.*

We shall prove the existence of an almost complex connection on M. We need the following lemma.

Lemma 3.3.2 *Let ∇ be a symmetric linear connection on M. Then*

$$N_J(X, Y) = (\nabla_{JX} J)Y - (\nabla_{JY} J)X + J((\nabla_Y J)X - (\nabla_X J)Y).$$

Proof: Since ∇ is symmetric, we have

$$[X,Y] = \nabla_X Y - \nabla_Y X.$$

Then we obtain

$$N_J(X,Y) = [JX,JY] - J[JX,Y] - J[X,JY] - [X,Y]$$

$$= \nabla_{JX}(JY) - \nabla_{JY}(JX) - J(\nabla_{JX}Y - \nabla_Y(JX))$$

$$-J(\nabla_X(JY) - \nabla_{JY}X) - (\nabla_X Y - \nabla_Y X)$$

$$= \nabla_{JX}(JY) - J(\nabla_{JX}Y) - (\nabla_{JY}(JX) - J(\nabla_{JY}X))$$

$$+J(\nabla_Y(JX)) + \nabla_Y X - J(\nabla_X(JY)) - \nabla_X Y$$

$$= (\nabla_{JX}J)Y - (\nabla_{JY}J)X + J((\nabla_Y J)X) - J((\nabla_X J)Y). \square$$

Proposition 3.3.3 *There exists an almost complex connection $\bar{\nabla}$ on $\bar{M}$ such that its torsion tensor $\bar{T}$ is given by*

$$\bar{T} = (1/4)N_J,$$

where N_J is the Nijenhuis tensor of J.

Proof: Let ∇ be an arbitrary symmetric linear connection on M. We define a tensor field Q of type (1,2) by

$$Q(X,Y) = (1/4)\{(\nabla_{JY}J)X + J((\nabla_Y J)X) + 2J((\nabla_X J)Y)\},$$

for any vector fields X, Y on M. Consider the linear connection $\bar{\nabla}$ given by

$$\bar{\nabla}_X Y = \nabla_X Y - Q(X,Y).$$

First, we prove that $\bar{\nabla}$ is, in fact, an almost complex connection. We have

$$Q(X,JY) = (1/4)\{-(\nabla_Y J)X + J((\nabla_{JY}J)X) + 2J((\nabla_X J)(JY)\},$$

$$J(Q(X,Y)) = (1/4)\{J((\nabla_{JY}J)X - (\nabla_Y J)X - 2(\nabla_X J)Y\}.$$

On the other hand, since

$$(\nabla_X J)J = -J(\nabla_X J),$$

we obtain

$$J((\nabla_X J)(JY)) = -(\nabla_X J)(J^2 Y) = (\nabla_X J)Y.$$

Hence, we deduce

$$Q(X,JY) - JQ(X,Y) = (1/2)J((\nabla_X J)(JY)) + (1/2)(\nabla_X J)Y = (\nabla_X J)Y.$$

Consequently, we have

$$\bar{\nabla}_X(JY) = \nabla_X(JY) - Q(X,JY)$$

$$= (\nabla_X J)Y + J(\nabla_X Y) - Q(X,JY)$$

$$= (\nabla_X J)Y + J(\nabla_X Y) - JQ(X,Y) - (\nabla_X J)Y$$

$$= J(\nabla_X Y) - JQ(X,Y)$$

$$= J(\nabla_X Y) - Q(X,Y))$$

$$= J(\bar{\nabla}_X Y),$$

and, then, $\bar{\nabla} J = 0$.

The torsion $\bar{T}$ of $\bar{\nabla}$ is given by

$$\bar{T}(X,Y) = \bar{\nabla}_X Y - \bar{\nabla}_Y X - [X,Y]$$

$$= \nabla_X Y - \nabla_Y X - [X,Y] - Q(X,Y) + Q(Y.X)$$

$$= -Q(X,Y) + Q(Y,X),$$

since

$$T(X,Y) = \nabla_X Y - \nabla_Y X - [X,Y] = 0.$$

Hence

$$\bar{T}(X,Y) = -Q(X,Y) + Q(Y,X)$$

$$= (1/4)\{-(\nabla_{JY}J)X - J((\nabla_Y J)X) - 2J((\nabla_X J)Y)$$

$$+(\nabla_{JX}J)Y + J((\nabla_X J)Y) + 2J((\nabla_Y J)X)\}$$

$$= (1/4)\{(\nabla_{JX}J)Y - (\nabla_{JY}J)X + J((\nabla_Y J)X - (\nabla_X J)Y)\}$$

$$= 4N_J(X,Y) \quad \text{(by Lemma 3.3.2).} \square$$

Corollary 3.3.4 *An almost complex structure J on M is integrable if and only if M admits a symmetric almost complex linear connection.*

Proof: If J is integrable, then the torsion $\bar{T}$ of the connection $\bar{\nabla}$ constructed in Proposition 3.3.3 vanishes. Conversely, suppose that there exists a symmetric almost complex connection ∇ on M. From $\nabla J = 0$, we deduce that $Q = 0$. Hence $\bar{\nabla} = \nabla$ and $N_J = 4\bar{T} = 0$. □

The following result gives some properties of the torsion and curvature tensors of an almost complex connection.

Proposition 3.3.5 *Let M be an almost complex manifold with almost complex structure J and ∇ an almost complex connection on M. Then the torsion tensor T and the curvature tensor R satisfy the following identities:*
(1) $T(JX,JY) - J(T,JX,Y)) - J(T(X,JY)) - T(X,Y) = -N_J(X,Y)$;
(2) $R(X,Y)J = JR(X,Y)$.

Proof: (1) We have

$$T(JX,JY) = \nabla_{JX}(JY) - \nabla_{JY}(JX) - [JX,JY],$$

$$T(JX,Y) = \nabla_{JX}Y - \nabla_Y(JX) - [JX,Y],$$

$$T(X,JY) = \nabla_X(JY) - \nabla_{JY}X - [X,JY],$$

$$T(X,Y) = \nabla_X Y - \nabla_Y X - [X,Y].$$

Hence,

$$T(JX,JY) - J(T(JX,Y)) - J(T(X,JY)) - T(X,Y)$$

$$= \nabla_{JX}(JY) - J(\nabla_{JX}Y) - \nabla_{JY}(JX) + J(\nabla_{JY}X)$$

$$+J((\nabla_Y(JX)) + \nabla_Y X - J(\nabla_X(JY)) - \nabla_X Y$$

$$-[JX,JY] + J[JX,Y] + J[X,JY] + [X,Y]$$

$$= (\nabla_{JX}J)Y - (\nabla_{JY}J)X - (\nabla_Y J)(JX)$$

$$-(\nabla_X J)(JY) - N_J(X,Y)$$

$$= -N_J(X,Y),$$

since $\nabla J = 0$.
(2) is proved by a similar device. □

3.4 Kähler manifolds

In this section, we introduce an important class of almost complex manifolds.

Definition 3.4.1 *A* **Hermitian metric** *on an almost complex manifold with almost complex structure J is a Riemannian metric g on M such that*

$$g(JX, JY) = g(X, Y),$$

for any vector fields X, Y on M.

Hence, a Hermitian metric g defines a Hermitian inner product g_x on T_xM for each $x \in M$ with respect to its structure of complex vector space given by J_x, that is,

$$g_x(J_xX, J_xY) = g_x(X, Y), X, Y \in T_xM.$$

Then $J_x : T_xM \longrightarrow T_xM$ is an isometry.

An almost complex manifold M with a Hermitian metric is called an **almost Hermitian manifold**. If M is a complex manifold, then M is called a **Hermitian manifold**.

Proposition 3.4.2 *Every almost complex manifold M admits a Hermitian metric.*

Proof: Let h be an arbitrary Riemannian metric on M. We set

$$g(X, Y) = h(X, Y) + h(JX, JY).$$

Then g is a Hermitian metric. □

Now, let M be a $2n$–dimensional almost Hermitian manifold with almost complex structure J and Hermitian metric g. The triple (M, J, g) is called an **almost Hermitian structure**.

Before proceeding further, we prove the following lemma.

Lemma 3.4.3 *Let V be a $2n$–dimensional real vector space with complex structure J (that is, J is a linear endomorphism of V satisfying $J^2 = -Id$) and a Hermitian inner product $<,>$ (i.e., $< JX, JY > = < X, Y >, X, Y \in V$). Then there exists an orthonormal basis $\{X_1, \ldots, X_n, JX_1, \ldots, JX_n\}$.*

Proof: We use induction in dim V. Let X_1 be a unit vector. Then $\{X_1, JX_1\}$ is orthonormal, since

$$< X_1 JX_1 >=< JX_1, J^2 X_1 >= - < JX_1, X_1 >= - < X_1, JX_1 >,$$

$$< JX_1, JX_1 >=< X_1, X_1 >= 1.$$

Now, if W is the subspace spanned by $\{X_1, JX_1\}$, we denote by $W^\perp$ the orthonormal complement so that $V = W \oplus W^\perp$. The subspace $W^\perp$ is invariant by J. In fact, if $X \in W^\perp$, we have

$$< X_1, JX >= - < JX_1, X >= 0,$$

$$< JX_1, JX >=< X_1, X >= 0,$$

and, hence, $JX \in W^\perp$. By the induction assumption, $W^\perp$ has an orthonormal basis of the form $\{X_2, \ldots, X_n, JX_2, \ldots, JX_n\}$. Therefore $\{X_1, \ldots, X_n, JX_1, \ldots, JX_n\}$ is the required basis. □

Let (M, J, g) be an almost Hermitian structure. For each point $x \in M$, by the lemma, there exists an orthonormal basis $\{X_1, \ldots, X_n, JX_1, \ldots, JX_n\}$ of $T_x M$. This basis is called an **adapted** (or **unitary**) **frame** at x. Let now $\{X_i, JX_i\}, \{X'_i, JX'_i\}$ be two unitary frames at x. Therefore we have

$$X'_i = A^j_i X_j + B^j_i (JX_j),$$

$$JX'_i = -B^j_i X_j + A^j_i (JX_j),$$

where $A, B \in Gl(n, R)$. Then the matrix

$$\begin{bmatrix} A & -B \\ B & A \end{bmatrix}$$

belongs to $Gl(n, C) \cap O(2n)$. It is easy to see that $Gl(n, C) \cap O(2n) = U(n)$, where $Gl(n, C)$ and $O(2n)$ are considered as subgroups of $Gl(2n, R)$. In fact, $U(n)$ consists of elements of $Gl(n, C)$ whose real representation (by ρ) are in $O(2n)$. If we set

$$UM = \{\text{unitary frames at all points of } M\},$$

then we can easily prove that UM is a $U(n)$–structure on M. Proceeding as in Section 3.2, we deduce that ***giving an almost Hermitian structure is the same as giving a $U(n)$–structure.***

Let (M, J, g) be an almost Hermitian structure. We define on M a 2–form Ω by

$$\Omega(X,Y) = g(X, JY),$$

for any vector fields X and Y on M. Ω is called the **fundamental** or **Kähler form** of (M, J, g).

Proposition 3.4.4 *Ω is invariant by J, that is,*

$$\Omega(JX, JY) = \Omega(X,Y).$$

Proof: In fact

$$\Omega(JX, JY) = g(JX, J^2Y) = -g(JX,Y) = g(X, JY) = \Omega(X,Y). \square$$

In general, J is not parallel with respect to the Riemannian connection ∇ defined by g.

Proposition 3.4.5 *We have*

$$2g((\nabla_X J)Y, Z) = 3d\Omega(X, JY, JZ) - 3d\Omega(X,Y,Z) + g(N_J(Y,Z), JX),$$

for any vector fields X, Y and Z on M.

Proof: We have

$$g((\nabla_X J)Y, Z) = g(\nabla_X(JY), Z) - g(J(\nabla_X Y), Z)$$

$$= g(\nabla_X(JY), Z) + g(\nabla_X Y, JZ).$$

Since

$$2g(\nabla_X Y, JZ) = Xg(Y, JZ) + Yg(X, JZ) - (JZ)g(X,Y)$$

$$+g([X,Y], JZ) + g([JZ, X], Y) + g(X, [JZ, Y]),$$

$$2g(\nabla_X(JY),Z) = Xg(JY,Z) + (JY)g(X,Z) - Zg(X,JY)$$

$$+g([X,JY],Z) + g([Z,X],JY) + g(X,[Z,JY]),$$

$$3d\Omega(X,Y,Z) = X\Omega(Y,Z) + Y\Omega(Z,X) + Z\Omega(X,Y)$$

$$-\Omega([X,Y],Z) - \Omega([Z,X],Y) - \Omega([Y,Z],X),$$

$$3d\Omega(X,JY,JZ) = X\Omega(JY,JZ) + (JY)\Omega(JZ,X) + (JZ)\Omega(X,JY)$$

$$-\Omega([X,JY],JZ) - \Omega([JZ,X],JY) - \Omega([JY,JZ],X),$$

and

$$N_J(Y,Z) = [JY,JZ] - J[JY,Z] - J[Y,JZ] - [Y,Z],$$

we deduce our proposition by a direct computation. □

Corollary 3.4.6 *Let (M,J,g) be an almost Hermitian structure. Then the following conditions are equivalent:*
(1) The Riemannian connection ∇ defined by g is an almost complex connection;
(2) $N_J = 0$ and the Kähler form Ω is closed, i.e., $d\Omega = 0$.

Proof: If $N_J = 0$ and $d\Omega = 0$, then $\nabla J = 0$ by Proposition 3.4.5. Conversely, suppose that $\nabla J = 0$. Since ∇ is symmetric, we deduce that J is integrable (by Corollary 3.3.4). Moreover, since $\nabla J = 0$ and $\nabla g = 0$, we easily deduce that $\nabla\Omega = 0$. Then $d\Omega = 0$. □

Corollary 3.4.7 *If M is a Hermitian manifold, then the following conditions are equivalent:*
(1) ∇ is an almost complex connection;
(2) Ω is closed.

Proof: It is a direct consequence of Corollary 3.4.6, since $N_J = 0$. □

Next, we introduce two important classes of almost Hermitian manifolds.

Definition 3.4.8 *An almost Hermitian manifold is called* **almost Kähler** *if its Kähler form Ω is closed. If, moreover, M is Hermitian, then M is called a* **Kähler manifold**.

From Corollary 3.4.7, we deduce that a Hermitian manifold is Kähler if and only if ∇ is an almost complex connection (for an exhaustive classification of almost Hermitian structures, see Gray and Hervella [70]).

Remark 3.4.9 It is easy to prove that the Kähler form of an almost Hermitian manifold satisfy

$$\Omega^n = \Omega \wedge \ldots \wedge \Omega \neq 0 \ \ (n \text{ times, where } \dim M = 2n).$$

Then we deduce:
(1) Ω^n is a volume form and, hence, every almost complex manifold is orientable;
(2) if M is an almost Kähler manifold, then Ω defines a symplectic structure on M (see Chapter 5).

3.5 Almost complex structures on tangent bundles (I)

In this section, we shall prove that the tangent bundle TM of a given manifold M carries interesting examples of almost complex structures.

3.5.1 Complete lifts

Let M be an almost complex manifold of dimension $2n$ and almost complex structure F. Let TM be its tangent bundle and F^c the complete lift of F to TM defined by $F^cX^c = (FX)^c$. From Proposition 2.5.6 and Corollary 2.5.7, we obtain

$$(F^c)^2 = -Id.$$

Hence F^c *defines an almost complex structure on* TM.

Now, if N is the Nijenhuis tensor of F^c, we have

$$N(X^c, Y^c) = [F^cX^c, F^cY^c] - F^c[F^cX^c, Y^c] - F^c[X^c, F^cY^c] + [X^c, Y^c]$$

$$= ([FX, FY] - F[FX, Y] - F[X, FY] + [X, Y])^c$$

$$= (N_F(X,Y))^c,$$

where N_F is the Nijenhuis tensor of F. Hence,

$$N = (N_F)^c.$$

Therefore, we have

Proposition 3.5.1 *F^c is integrable if and only if F is integrable.*

3.5.2 Horizontal lifts

Let F be an almost complex structure on M. Consider the horizontal lift F^H of F to TM with respect to a linear connection ∇ on M. Let $\hat{\nabla}$ be the opposite connection with curvature tensor $\hat{R}$. From Proposition 2.7.3, we have

$$(F^H)^2 = -Id,$$

and so, F^H in an **almost complex structure on** TM. Let N be the Nijenhuis tensor of F^H. From a straightforward computation, we obtain

$$N(X^v, Y^v)) = 0,$$

$$N(X^v, Y^H) = (N_F(X,Y) - ((\nabla_{FX}F)Y + (\nabla_X F)(FY)))^v,$$

$$N(X^H, Y^H) = (N_F(X,Y))^H$$

$$-\gamma(\hat{R}(FX,FY) - F\hat{R}(FX,Y) - F\hat{R}(X,FY) - \hat{R}(X,Y)).$$

Hence we have

Proposition 3.5.2 *If F^H is integrable, then F is integrable. Conversely, suppose that ∇ is an almost complex connection (i.e. $\nabla F = 0$); then, if F is integrable and $\hat{\nabla}$ has zero curvature, F^H is integrable. Particularly, let F be a complex structure on M and ∇ is a symmetric almost complex connection, then F^H is integrable if ∇ has zero curvature.*

3.5.3 Almost complex structure on the tangent bundle of a Riemannian manifold

Let M be a differentiable manifold with a linear connection ∇. Let T and R be the torsion and curvature tensors of ∇. We denote by $\hat{\nabla}$ the opposite connection with curvature tensor $\hat{R}$. We define a tensor field of type (1,1) on TM by

$$FX^H = -X^v, FX^v = X^H, \tag{3.5}$$

for any vector field X on M. From (3.5), we deduce that $F^2 = -Id$, and, so *F is an almost complex structure on TM*. With respect to the adapted frame we have

$$FD_i = -V_i, FV_i = D_i.$$

Next, we study the integrability of F. Let N be the Nijenhuis tensor of F. We obtain

$$N(X^v, Y^v) = (T(X,Y))^H - \gamma\hat{R}(X,Y),$$

$$N(X^v, Y^H) = (T(X,Y))^v + F\gamma\hat{R}(X,Y),$$

$$N(X^H, Y^H) = (T(X,Y))^H - \gamma\hat{R}(X,Y), \tag{3.6}$$

for any vector fields X, Y on M.

Proposition 3.5.3 *F is integrable if and only if $T = 0$ and $R = 0$.*

Proof: Suppose that F is integrable. From (3.6), we deduce that $T = 0$ and $\hat{R} = 0$. Since $T = 0$, ∇ is symmetric and, hence $\hat{\nabla} = \nabla$. Then $\hat{R} = R = 0$. Conversely, suppose that $T = 0$ and $R = 0$. Therefore $\hat{\nabla} = \nabla$ and, then $\hat{R} = R = 0$. Consequently, $N = 0$, and, thus, F is integrable. □

Now, suppose that (M, g) is a Riemannian manifold and ∇ the Riemannian connection defined by g. Since ∇ has zero torsion, (3.6) becomes

$$\begin{aligned} N_F(X^v, Y^v) &= -\gamma R(X,Y), \\ N_F(X^v, Y^H) &= F\gamma R(X,Y), \\ N_F(X^H, Y^H) &= \gamma R(X,Y), \end{aligned} \tag{3.7}$$

where F is defined by ∇ according to (3.5). From (3.7), we easily deduce the following

Proposition 3.5.4 *F is integrable if and only if (M,g) is flat, i.e., $R = 0$.*

Consider the Sasaki metric $\bar{g}$ on TM determined by g. Then we have

Proposition 3.5.5 *$\bar{g}$ is an Hermitian metric for F.*

Proof: We must check that

$$\bar{g}(F\bar{X}, F\bar{Y} = \bar{g}(\bar{X}, \bar{Y}), \tag{3.8}$$

for any vector fields $\bar{X}, \bar{Y}$ on TM. It is sufficient to prove (3.8) when $\bar{X}, \bar{Y}$ are horizontal and vertical lifts of vector fields on M. Thus we have

$$\bar{g}(FX^v, FY^v) = \bar{g}(X^H, Y^H) = (g(X,Y))^v = \bar{g}(X^v, Y^v),$$

$$\bar{g}(FX^v, FY^H) = -\bar{g}(X^H, Y^v) = 0 = \bar{g}(X^v, Y^H),$$

$$\bar{g}(FX^H, FY^H) = -\bar{g}(X^v, Y^v) = (g(X,Y))^v = \bar{g}(X^H, Y^H),$$

for any vector fields X, Y on M. □

Therefore $(TM, F, \bar{g})$ is an almost Hermitian structure. Let us consider the Kähler form Ω associated to $(TM, F, \bar{g})$. We recall that Ω is given by

$$\Omega(\bar{X}, \bar{Y}) = \bar{g}(\bar{X}, F\bar{Y}),$$

for any vector fields, $\bar{X}, \bar{Y}$ on TM. Hence we have

$$\Omega(X^v, Y^v) = \Omega(X^H, Y^H) = 0, \Omega(X^v, Y^H) = -(g(X,Y))^v, \tag{3.9}$$

for any vector fields X, Y on M. If we compute $d\Omega$ acting on horizontal and vertical lifts, we deduce, by a straightforward computation from (3.9) that $d\Omega = 0$. Therefore, by using Propositions 3.4.5 and 3.5.5, we obtain

Theorem 3.5.6 *$(TM, F, \bar{g})$ is an almost Kähler structure. Furthermore, $(TM, F\bar{g})$ is a Kähler structure if and only if (M,g) is flat, i.e., $R = 0$.*

Remark 3.5.7 Since $(TM, F, \bar{g})$ is an almost Kähler structure, then Ω is always a symplectic form (see Chapter 5).

3.6 Almost contact structures

In this section we shall give alternative definitions of almost contact structures. Roughly speaking, almost contact structures are the odd–dimensional counterpart to almost complex structures. We remit to Blair [8] for an extensive study of such a type of structures.

Definition 3.6.1 *Let M be a $(2n+1)$–dimensional manifold. If M carries a 1–form η such that*

$$\eta \bigwedge (d\eta)^n \neq 0,$$

then M is said to be a **contact manifold** *or to have a* **contact structure**. *We call η a* **contact form**.

Example.- We set

$$\eta = dz + \sum_i y^i dx^i,$$

where $(x^i, y^i, z; 1 \leq i \leq n)$ are the canonical coordinates in R^{2n+1}. Then η is a contact form on R^{2n+1}. Moreover, a contact form on a $(2n+1)$–dimensional manifold M can be locally expressed in this way (see Chapter 6).

If η is a contact form on M, then there exists a unique vector field ξ on M such that

$$\eta(\xi) = 1, i_\xi(d\eta) = 0.$$

(See Chapter 6 for a proof). We call ξ the **Reeb vector field**.

Next, we generalize the notion of contact structure.

Definition 3.6.2 *(Blair [8]).- An* **almost contact structure** *on a $(2n+1)$–dimensional manifold M is a triple (ϕ, ξ, η) where ϕ is a tensor field of type (1,1), ξ a vector field and η a 1–form on M such that*

$$\phi^2 = -Id + \xi \otimes \eta, \eta(\xi) = 1. \tag{3.10}$$

From (3.10), it follows that

$$\phi(\xi) = 0.\eta\phi = 0, rank\ (\phi) = 2n.$$

If there exists a Riemannian metric g on M such that

$$g(\phi X, \phi Y) = g(X,Y) - \eta(X)\eta(Y), \tag{3.11}$$

for any vector fields X, Y on M, then g is said to be a **compatible** or **adapted metric** and (ϕ, ξ, η, g) is called an **almost contact metric structure**. From (3.10) and (3.11), we deduce that

$$\eta(X) = g(\xi, X), g(\xi, \xi) = 1.$$

Proposition 3.6.3 *Let (ϕ, ξ, η) be an almost contact structure on M. Then M admits a compatible metric.*

Proof: Let g' be an arbitrary Riemannian metric on M. We set

$$g''(X,Y) = g'(\phi^2 X, \phi^2 Y) + \eta(X)\eta(Y).$$

Then g'' is a Riemannian metric satisfying

$$g''(\xi, X) = \eta(X).$$

Now, we define a Riemannian metric g on M by

$$g(X,Y) = (1/2)(g''(X,Y) + g''(\phi X, \phi Y) + \eta(X)\eta(Y)).$$

We have

$$g(\phi X, \phi Y) = (1/2)(g''(\phi X, \phi Y) + g''(-X + \eta(X), -Y + \eta(Y)\xi)$$

$$= (1/2)(g''(\phi X, \phi Y) + g''(X,Y) - \eta(X0\eta(Y))$$

$$= (1/2)(g''(\phi X, \phi Y) + g''(X,Y) + \eta(X)\eta(Y)) - \eta(X)\eta(Y)$$

$$= g(X,Y) - \eta(X)\eta(Y). \square$$

Let now (ϕ, ξ, η, g) be an almost contact metric structure on M. Let $x \in M$. We choose a unit tangent vector $X_1 \in T_xM$ orthogonal to ξ_x. Then ϕX_1 is also a unit tangent vector orthogonal to both ξ_x and X_1; in fact,

(1) $g(\phi X_1, \xi_x) = \eta\phi(X_1) = 0$;
(2) $g(\phi X_1, X_1) = g(\phi^2 X_1, \phi X_1) + \eta\phi(X_1)\eta(X_1)$

$$= g(-X_1 + \eta(X_1)\xi_x, \phi X_1) = -g(X_1, \phi X_1) = -g(\phi X_1, X_1),$$

which implies $g(\phi X_1, X_1) = 0$;
(3) $g(\phi X_1, \phi X_1) = g(X_1, X_1) - \eta(\phi X_1)\eta(\phi X_1) = g(X_1, X_1) = 1$.

Now, take $X_2 \in T_xM$ to be a unit tangent vector orthogonal to ξ_x, X_1 and ϕX_1; then ϕX_2 is a unit tangent vector orthogonal to $\xi_x, X_1, \phi X_1$ and X_2. Proceeding in this way we obtain an orthonormal basis $\{X_i, \phi X_i, \xi_x\}$ on T_xM, that is, a frame at x, which is called a ϕ**–basis** or **adapted frame**. With respect to an adapted frame ϕ, ξ, η and g are represented by the matrices

$$\phi_0 = \begin{bmatrix} 0 & -I_n & 0 \\ I_n & 0 & 0 \\ 0 & 0 & 0 \end{bmatrix}, \xi_0 = \begin{bmatrix} 0 \\ \vdots \\ 0 \\ 1 \end{bmatrix}, \eta_0 = (0, \ldots, 0, 1),$$

$$g_0 = \begin{bmatrix} I_n & 0 & 0 \\ 0 & I_n & 0 \\ 0 & 0 & 1 \end{bmatrix},$$

respectively. Let $\{X_i, \phi X_i, \xi_x\}, \{X_i', \phi X_i', \xi_x\}$ be two adapted frames at x. Then we have

$$X_i' = A_i^j X_j + B_i^j (\phi X_j),$$

$$\phi X_i' = -B_i^j X_j + A_i^j (\phi X_j),$$

where $A, B \in Gl(n, R)$. Hence the two frames are related by the $(2n+1) \times (2n+1)$ matrix

$$\alpha = \begin{bmatrix} A & -B & 0 \\ B & A & 0 \\ 0 & 0 & 1 \end{bmatrix},$$

Obviously $\alpha \in U(n) \times 1$. We set

$$B = \{\text{adapted frames at all points of } M\}.$$

One can easily proves that B is a $(U(n) \times 1)$–structure on M (it is sufficient to repeat the above construction to obtain a local frame field $\{X_i, \phi X_i, \xi\}$ on a neighborhood of each point of M). Conversely, suppose that B is a $(U(n) \times 1)$–structure on M. Then we define an almost contact metric structure (ϕ, ξ, η, g) on M as follows. With respect to a frame of B at $x, \phi_x \xi_x, \eta_x$ and g_x are given by the matrices ϕ_0, ξ_0, η_0 and g_0, respectively, for each $x \in M$.

Summing up, we have proved the following.

Proposition 3.6.4 *Giving an almost contact metric structure is the same as giving a $(U(n) \times 1)$–structure.*

Let (ϕ, ξ, η, g) be an almost contact structure on M. We define the **fundamental 2–form** Ω on M by

$$\Omega(X, Y) = g(X, \phi Y).$$

Then we deduce that $\eta \wedge \Omega^n \neq 0$, that is, $\eta \wedge \Omega^n$ is a volume form on M.

Definition 3.6.5 *An almost contact metric structure (ϕ, ξ, η, g) is said to be a* **contact metric structure** *if $\Omega = d\eta$.*

There exists in the literature an alternative definition of almost contact structure which generalizes Definition 3.6.1.

Definition 3.6.6 *(Libermann [87]). An* **almost contact structure** *or* **almost cosymplectic structure** *on a $(2n+1)$–dimensional manifold M is a pair (η, Ω), where η is a 1–form and Ω a 2–form on M such that $\eta \wedge \Omega^n \neq 0$.*

The following result relate these definitions.

Proposition 3.6.7 *Let M be a $(2n+1)$–dimensional manifold. We have: (1) If M admits an almost cosymplectic structure (η, Ω) then M admits an almost contact metric structure. (2) If M admits a contact form η, then there is an almost contact metric structure (ϕ, ξ, η, g) such that the fundamental form Ω is, precisely, $d\eta$.*

Proof: (1) Since $\eta \wedge \Omega^n \neq 0$, then M is orientable and, then, there exists a non–vanishing vector field ξ' on M such that $i_{\xi'}\Omega = 0$ (see Blair [8]). Let g' be a Riemannian metric on M and define a vector field ξ by

$$\xi = \xi' / \parallel \xi' \parallel .$$

Thus ξ is a unit vector field on M. We now define a 1–form η' by

$$\eta'(X) = g'(X, \xi).$$

Let D be the orthogonal complement of ξ, i.e., $TM = D \oplus < \xi >$. Then Ω is a symplectic form on the vector bundle D (we also denote by D the corresponding distribution of sections of D). We consider a metric g'' on D and an endomorphism ϕ' of D such that

$$g''(X, \phi' Y) = \Omega(X, Y), \phi'^2 = -Id, \tag{3.12}$$

for any vector field $X \in D$ (see Exercise 3.8.1). Next, we define a Riemannian metric g by

$$g(X, Y) = g''(X, Y), g(X, \xi) = 0, g(\xi, \xi) = 1, \tag{3.13}$$

for any vector fields $X, Y \in D$, and a tensor field ϕ of type (1,1) by

$$\phi(X) = \phi'(X), \phi(\xi) = 0,$$

for any vector field $X \in D$. Thus (ϕ, ξ, η') is an almost contact structure on M.

(2) Let ξ be the Reeb vector field, i.e., $\eta(\xi) = 1$ and $i_\xi(d\eta) = 0$. Let h be a Riemannian metric on M and define a Riemannian metric g' by

$$g'(X, Y) = h(-X + \eta(X)\xi, -Y + \eta(Y)\xi) + \eta(X)\eta(Y).$$

Hence

$$g'(X, \xi) = \eta(X).$$

Let g'' be a metric on D and ϕ an endomorphism of D such that (3.12) holds and define g by (3.13). Then (ϕ, ξ, η, g) is an almost contact metric structure whose fundamental form is $d\eta$. □

Next, we study the integrability of almost contact metric structures. Let (ϕ, ξ, η) be an almost contact structure on a $(2n+1)$–dimensional manifold M. Consider the product manifold $M \times R$ and define there a tensor field J of type (1,1) by

$$J(X, f(d/dt)) = ((\phi X - f\xi, \eta(X)(d/dt)), \tag{3.14}$$

for any vector field X on M and any C^∞ function f on $M \times R$, t being the canonical coordinate in R. From (3.14) we deduce that $J^2 = -Id$, and, thus, J is an *almost complex structure on* $M \times R$.

Definition 3.6.8 *(ϕ, ξ, η) is said to be* **normal** *if J is integrable.*

Proposition 3.6.9 *(ϕ, ξ, η) is normal if and only if*

$$N_\phi + 2\xi \otimes (d\eta) = 0.$$

Proof: Let J be the almost complex structure defined by (3.14) and N its Nijenhuis tensor. A straightforward computation shows that

$$N((X,0),(Y,0)) = (N_\phi(X,Y) + 2d\eta(X,Y)\xi, ((L_{\phi X}\phi)Y - (L_{\phi Y}\phi)X)(d/dt)),$$

$$N((X,0),(0,d/dt)) = ((L_\xi\phi)X, ((L_\xi\eta)X)(d/dt)),$$

$$N((0,d/dt),(0,d/dt)) = 0. \tag{3.15}$$

From (3.15), it follows that the vanishing of N implies that

$$N_\phi + 2\xi \otimes (d\eta) = 0.$$

Conversely, suppose that $N_\phi + 2\xi \otimes (d\eta) = 0$. Then we have

$$0 = N_\phi(X,\xi) + 2(d\eta)(X,\xi)\xi$$

$$= \eta[\xi, X] - \xi(\eta(X)) = -(L_\xi\eta)X,$$

and thus

$$0 = \phi([\xi, X] + \phi[\xi, \phi X] - (\xi(\eta(X))\xi)$$

$$= \phi[\xi, X] - [\xi, \phi X] + \eta([\xi, \phi X])\xi = -(L_\xi\phi)X,$$

since $\eta([\xi, X]) = \xi(\eta(\phi X)) = 0$.
Finally,

$$0 = N_\phi(\phi X, Y) + 2(d\eta)(\phi X, Y)\xi$$

$$= -[\phi X, Y] - [X, \phi Y] - (\phi Y)(\eta(X))\xi - \eta(X)[\phi Y, \xi]$$

$$-\phi[-X + \eta(X)\xi, Y] - \phi[\phi X, \phi Y] + (\phi X)(\eta(Y)\xi.$$

Applying now η, we obtain

$$0 = (\phi X)(\eta(Y)) - \eta([\phi X, Y]) - (\phi Y)(\eta(X)) + \eta([\phi Y, X])$$

$$= (L_{\phi X}\eta)Y - (L_{\phi Y}\eta)X.$$

Thus, $N = 0$. □

To end this section, we introduce an important class of almost contact metric structures.

Let η be a contact form on M. By Proposition 3.6.7, there exists a contact metric structure (ϕ, ξ, η, g) on M.

Definition 3.6.10 *A normal contact metric structure* (ϕ, ξ, η, g) *is said to be* **Sasakian**.

A Sasakian structure is the odd–dimensional counterpart to Kähler manifold. In fact, we may prove that, if (ϕ, ξ, η, g) is a Sasakian structure and ∇ is the Riemannian connection defined by g, then we have

$$(\nabla_X \phi)Y = g(X, Y) - \eta(Y)X$$

(see Blair [8]).

Remark 3.6.11 *For a classification of almost contact metric structures we remit to Oubiña [102].*

3.7 f–structures

In this section we study f–structures on manifolds and give integrability conditions of an f–structure.

Definition 3.7.1 *(Yano [129]) A non–null tensor field f of constant rank, say r, on an m–dimensional manifold M satisfying*

$$f^3 + f = 0$$

is called an f–**structure** *(or $f(3,1)$*–**structure***).*

If $m = r$, then an f–structure gives an almost complex structure on M and $m = r$ is even. If M is orientable and $m - 1 = r$, then an f– structure gives an almost contact structure on M and m is odd (see Yano [129]).

We set

$$\ell = -f^2, \quad \mathbf{m} = f^2 + Id,$$

where Id is the identity tensor field on M. Then we have

$$\ell + \mathbf{m} = Id, \ \ell^2 = \ell, \ \mathbf{m}^2 = \mathbf{m}, \ \ell\mathbf{m} = \mathbf{m}\ell = 0,$$

$$f\ell = \ell f = f, \ \mathbf{m}f = f\mathbf{m} = 0.$$

Therefore we obtain two complementary distributions $L = Im\,\ell$ and $\mathcal{M} = Im\,\mathbf{m}$ corresponding to the projection tensor ℓ and $\mathbf{m}$, respectively. If rank $f = r$, then L is r–dimensional and $\mathcal{M}$ is $(m - r)$–dimensional.

For each point $x \in M$ we have

$$T_xM = L_x \oplus \mathcal{M}_x.$$

Since $f_x^2(\ell X) = -\ell X$ for all $X \in T_xM$, then L_x is a vector space with complex structure f_x/L_x. Hence r must be even, say $r = 2n$.

Now let $\{e_1, \ldots, e_n, fe_1, \ldots, fe_n\}$ be a basis of L_x and $\{e_{2n+1}, \ldots, e_m\}$ a basis of M_x. Hence $\{e_1, \ldots, e_m, fe_1, \ldots, fe_n, e_{2n+1}, \ldots, e_m\}$ is a basis of T_xM which is called an **adapted frame** (or f–**basis**) at x. If $\{e'_1, \ldots, e'_n, fe'_1, \ldots, fe'_n, e'_{2n+1}, \ldots, e'_m\}$ is another f–basis at x we have

$$e'_i = A_i^j e_j + B_i^j (fe_j),$$

$$fe'_i = -B_i^j e_j + A_i^j (fe_j),$$

$$e'_a = C_a^b e_b,$$

$1 \leq i,j \leq n$, $2n+1 \leq a,b \leq m$, where A, B are $n \times n$ matrices and $C \in Gl(m-r, R)$. Then the two f–basis at x are related by the $m \times m$ matrix

$$\alpha = \begin{bmatrix} A & -B & 0 \\ B & A & 0 \\ 0 & 0 & C \end{bmatrix}$$

Clearly $\alpha \in Gl(m, R)$.

Now let G be the set of such matrices; G is a closed subgroup of $Gl(m, R)$ and therefore a Lie subgroup of $Gl(m, R)$ which may be canonically identified to $Gl(m, C) \times Gl(m-2n, R)$.

If we set

$$B = \{f\text{–basis at all points of } M\}$$

it is not hard to prove that B is a $(Gl(n, C) \times Gl(m-2n, R))$–structure on M. The converse is also true, i.e., if B is a $(Gl(n, C) \times Gl(m-2n, R))$–structure on M then it determines an f–structure on M (details are left to the reader as an exercise).

Thus we have:

Proposition 3.7.2 *Giving an f–structure is the same as giving a $(Gl(n, C) \times Gl(m-2n, R))$–structure on M.*

Remark 3.7.3 With respect to an f–basis at x f, ℓ and $\mathbf{m}$ are represented by the matrices

$$f = \begin{bmatrix} 0 & -I_n & 0 \\ I_n & 0 & 0 \\ 0 & 0 & 0 \end{bmatrix}, \ell = \begin{bmatrix} I_n & 0 & 0 \\ 0 & I_n & 0 \\ 0 & 0 & 0 \end{bmatrix},$$

$$\mathbf{m} = \begin{bmatrix} 0 & 0 & 0 \\ 0 & 0 & 0 \\ 0 & 0 & I_{m-2n} \end{bmatrix},$$

respectively.

Definition 3.7.4 *A Riemannian metric g on M is said to be* **adapted** *to an f–structure f if*
(1) $g(\ell X, \mathrm{m}Y) = 0$, i.e., L and $\mathcal{M}$ are orthogonal with respect to g;
(2) $g(f\ell X, f\ell Y) = g(\ell X, \ell Y)$.
In such a case (M, f, g) is called a **metric f–structure**.

Proposition 3.7.5 *There always exists an adapted metric.*

Proof: Let h be an arbitrary Riemannian metric on M. We define h' by

$$h'(X,Y) = h(\ell X, \ell Y) + h(\mathrm{m}X, \mathrm{m}Y)$$

Then h' is a Riemannian metric on M such that L and $\mathcal{M}$ are orthogonal. Hence an adapted metric g to f is given by

$$g(X,Y) = h'(fX, fY) + h'(X,Y) + h'(\mathrm{m}X, \mathrm{m}Y). \square$$

Now, proceeding as in Section 3.4, it is not hard to prove the following.

Proposition 3.7.6 *Giving a metric f–structure is the same as giving a $(U(n) \times O(m-2n))$–structure on M.*

Now, suppose that L is integrable. Then f operates as an almost complex structure on each integral manifold of L.

Definition 3.7.7 *When L is integrable and the induced almost complex structure is integrable on each integral manifold of L, we say that the f–structure f is* **partially integrable**.

Proposition 3.7.8 *f is partially integrable if and only if $N_f(\ell X, \ell Y) = 0$, where N_f is the Nijenhuis torsion of f.*

The proof is a direct consequence of Theorem 3.2.6.

Definition 3.7.9 *An f–structure f on M is* **integrable** *if it is integrable as a $(Gl(n,C) \times Gl(m-2n,R))$–structure, i.e., for each $x \in M$ there exists a coordinate neighborhood U with local coordinates $(x^1, \ldots, x^n, x^{n+1}, \ldots, x^{2n}, x^{2n+1}, \ldots, x^m)$ such that f is locally given in U by*

$$f = \begin{bmatrix} 0 & I_n & 0 \\ -I_n & 0 & 0 \\ 0 & 0 & 0 \end{bmatrix}.$$

Theorem 3.7.10 *f is integrable if and only if $N_f = 0$.*

We remit to Yano and Kon [131] for a proof.

In a similar way, we can consider $f(3,-1)$–structures on manifolds.

Definition 3.7.11 *An $f(3,-1)$–**structure** on an m–dimensional manifold M is given by a non–null tensor field f of type (1,1) on M of constant rank r satisfying*

$$f^3 - f = 0.$$

If we set

$$\ell = f^2, \ \mathbf{m} = -f^2 + Id$$

we have

$$\ell + \mathbf{m} = Id, \ \ell^2 = \ell, \ \mathbf{m}^2 = \mathbf{m}, \ \ell\mathbf{m} = \mathbf{m}\ell = 0.$$

$$f\ell = \ell f = f, \ f\mathbf{m} = \mathbf{m}f = 0.$$

Then $L = Im\,\ell$ and $\mathcal{M} = Im\,\mathbf{m}$ are complementary distributions on M of rank r and $m-r$, respectively. When $m = r$ then f is an almost product structure on M.

Since $f_x^2(\ell X) = \ell X$, for all $X \in T_xM$, then f_x acts on L_x as an almost product structure operator. If we set

$$P = \frac{1}{2}(I+f)\ell, \ Q = \frac{1}{2}(I-f)\ell$$

then $P^2 = P$, $Q^2 = Q$ and $PQ = QP = 0$. Hence $\mathbf{P} = Im\,P$ and $\mathbf{Q} = Im\,Q$ determines two distributions on M of dimension p and q, respectively, such that $p+q = r$ and

$$T_xM = \mathbf{P}_x \oplus \mathbf{Q}_x \oplus \mathbf{M}_x,$$

since $\mathbf{L}_x = \mathbf{P}_x \oplus \mathbf{Q}_x$.

Now, proceeding as above, we have

Proposition 3.7.12 *The following three assertions are equivalent:*
(1) M possesses an $f(3,-1)$–structure of rank r;
(2) M possesses a $(Gl(p,R) \times Gl(q,R) \times Gl(m-r,R))$–structure;
(3) M possesses a $(O(p) \times O(q) \times O(m-r))$–structure.

Proposition 3.7.13 *An $f(3,-1)$–structure is integrable if and only if $N_f = 0$.*

3.8 Exercises

3.8.1 (1) Let (E, ω) be a symplectic vector space (see Chapter 5). Prove that there exists a complex structure J and a Hermitian inner product g on E such that the 2–form Ω given by $\Omega(x, y) = g(x, Jy)$ is precisely ω.

(*Hint*: Choose any inner product $<,>$ on E. Then we define a linear isomorphism $k : E \longrightarrow E$ by $\omega(x, y) =< x, ky >$. If $k^2 = -Id$, we are done. Otherwise we consider the polar decomposition $k = RJ$, where R is positive definite symmetric, J is orthogonal and $JR = RJ$. Since $k^t = -k$, we deduce $J^t = -J$ and then $J^2 = -Id$. Also, $\omega(Jx, Jy) = \omega(x, y)$. Now, we define a Hermitian inner product g by $g(x, y) = \omega(x, Jy)$).

(2) Let (S, ω) be a symplectic manifold. Prove that there exists an almost Hermitian structure (J, g) on S such that its Kähler form is precisely ω.

3.8.2 Prove that, if V is an involutive distribution on a manifold M, then there exists a symmetric linear connection ∇ on M such that $\nabla_X Y \in V$ for all $Y \in V$ (see Walker (1958)).

3.8.3 If F is an integrable almost product structure on a manifold M, then prove that any linear connection ∇ on M is an almost product connection. Hence there exists a symmetric almost product linear connection on M.

3.8.4 (1) Let f be an f–structure on M of rank r. Prove that f^c is an f–structure of rank $2r$ on TM.

(2) Prove that f^c is partially integrable (resp. integrable) if and only if f is partially integrable (resp. integrable).

(3) Let ∇ be a linear connection on M. Prove that the horizontal lift f^H of f to TM with respect to ∇ is also an f–structure of rank $2r$ on TM.

Chapter 4

Connections in tangent bundles

4.1 Differential calculus on TM: Vertical derivation and vertical differentiation

In this section we develop a differential calculus on tangent bundles determined by the canonical almost tangent structure and the Liouville vector field.

Let M be a differentiable m–dimensional manifold and TM its tangent bundle. We define a canonical vector field C on TM as follows:

$$C(y) = (y)^v_y, \; y \in TM;$$

C is called the **Liouville vector field** on TM (sometimes we use the notation C_M).

We locally have

$$C = v^i(\partial/\partial v^i) \tag{4.1}$$

Let J be the canonical almost tangent structure on TM. From (4.1), we easily deduce that $JC = 0$.

Remark 4.1.1 In Section 4.2, we shall give an alternative definition of C.

We now consider the **adjoint operator** J^* of J; J^* is defined by

$$J^*(f) = f, \; f \in C^\infty(TM),$$

$$(J^\star\omega)(X_1,\ldots,X_p) = \omega(JX_1,\ldots,JX_p),$$

$$X_1,\ldots,X_p \in \chi(TM),\ \omega \in \wedge^p(TM). \tag{4.2}$$

From (4.2), we deduce that $J^\star$ is locally characterized by

$$J^\star(f) = f,\ f \in C^\infty(TM),$$

$$J^\star(dx^i) = 0,\ J^\star(dv^i) = dx^i,$$

where (x^i, v^i) are the induced coordinates in TM. Then $J^\star$ does not commute with the exterior derivative d on TM.

Proposition 4.1.2 *We have*

$$i_X J^\star = J^\star \circ i_{JX}.$$

Proof: In fact,

$$(i_X J^\star)(f) = i_X(J^\star f) = i_X f = 0.$$

On the other hand,

$$(J^\star \circ i_{JX})(f) = J^\star(i_{JX} f) = J^\star(0) = 0.$$

Moreover, if $\omega \in \wedge^p(TM)$, we have

$$((i_X J^\star)(\omega))(X_1,\ldots,X_{p-1}) = (i_X(J^\star\omega))(X_1,\ldots,X_{p-1})$$

$$= (J^\star\omega)(X,X_1,\ldots,X_{p-1}) = \omega(JX,JX_1,\ldots,JX_{p-1})$$

$$= (J^\star\omega)(X,X_1,\ldots,X_{p-1}) = (i_X(J^\star\omega))(X_1,\ldots,X_{p-1})$$

$$= ((i_X J^\star)(\omega))(X_1,\ldots,X_{p-1}),$$

$X_1,\ldots,X_{p-1} \in \chi(TM)$.□

Corollary 4.1.3 *We have*

$$i_C J^\star = 0.$$

Proof: In fact, $JC = 0$.□

4.1.1 Vertical derivation

We define the **vertical derivation** i_J as follows:

$$i_J f = 0, f \in C^\infty(TM),$$

$$(i_J\omega)(X_1, \ldots, X_p) = \sum_{i=1}^{p} \omega(X_1, \ldots, JX_i, \ldots, X_p), \tag{4.3}$$

$\omega \in \wedge^p(TM), X_1, \ldots, X_p \in \chi(TM)$.

Then i_J is a derivation of degree 0 of $\wedge(TM)$ (see Section 1.13) and a derivation of degree 0 and type i_* in the sense of Frölicher and Nijenhuis.

From (4.3), we deduce

$$i_J(df) = J^\star(df), \; f \in C^\infty(TM).$$

Then we have

$$i_J(dx^i) = 0, i_J(dv^i) = dx^i. \tag{4.4}$$

From (4.2) and (4.4), we easily deduce the following.

Proposition 4.1.4 *We have*
(1) $i_J \circ J^\star = J^\star \circ i_J = 0.$
(2) $[i_X, i_J] = i_X i_J - i_J i_X = i_{JX},$
(3) $[i_J, L_C] = i_J L_C - L_C i_J = i_J.$

Proof: We only prove (2); (1) and (3) are left to the reader as an exercise. If $f \in C^\infty(TM)$, we have

$$[i_X, i_J](f) = (i_X(i_J(f))) - (i_J(i_X(f))) = 0,$$

and

$$(i_{JX})(f) = 0.$$

If $\omega \in \wedge^p(TM)$, we obtain

$$([i_X, i_J]\omega)(X_1, \ldots, X_{p-1}) = (i_X(i_J\omega)(X_1, \ldots, X_{p-1})$$

$$-(i_J(i_X\omega))(X_1, \ldots, X_{p-1})$$

$$= (i_J\omega)(X, X_1, \ldots, X_{p-1}) - \sum_{i=1}^{p-1}(i_X\omega)(X_1, \ldots, JX_i, \ldots, X_{p-1})$$

$$= \omega(JX, X_1, \ldots, X_{p-1}) + \sum_{i=1}^{p-1}\omega(X, X_1, \ldots, JX_i, \ldots, X_{p-1})$$

$$- \sum_{i=1}^{p-1}\omega(X, X_1, \ldots, JX_i, \ldots, X_{p-1})$$

$$= \omega(JX, X_1, \ldots, X_{p-1})$$

$$= (i_{JX}\omega)(X_1, \ldots, X_{p-1}),$$

$X_1, \ldots, X_{p-1} \in \chi(TM)$.□
From (4.3), we deduce by a straightforward computation, the following:

Proposition 4.1.5 *We have*

$$i_J(\omega \wedge \tau) = (i_J\omega) \wedge \tau + \omega \wedge (i_J\tau),$$

$\omega, \tau \in \wedge(TM)$.

4.1.2 Vertical differentiation

We define the **vertical differentiation** d_J on TM by

$$d_J = [i_J, d] = i_J d - d i_J. \tag{4.5}$$

Then

$$d_J : \wedge^p(TM) \longrightarrow \wedge^{p+1}(TM).$$

and d_J is a skew–derivation of degree 1 of $\wedge(TM)$ and a derivation of degree 1 and type d_* in the sense of Frölicher and Nijenhuis.

From (4.5) we easily deduce that

$$d_J f = J^*(df),\ d_J(df) = -d(J^*(df)),\ f \in C^\infty(TM).$$

Then in local coordinates we have

$$d_J f = \sum_{i=1}^{m} (\partial f/\partial v^i) dx^i,$$

$$d_J(dx^i) = d_J(dv^i) = 0. \tag{4.6}$$

Proposition 4.1.6 *We have*

$$d_J d = -d d_J.$$

Proof: In fact,

$$d_J d = -d i_J d = -d d_J. \square$$

Proposition 4.1.7 *We have*
(1) $d_J^2 = 0$,
(2) $d_J(\omega \wedge \tau) = (d_J\omega) \wedge \tau + (-1)^p \omega \wedge (d_J \tau)$, *if* $\omega \in \wedge^p(TM)$.

Proof: (2) is a direct consequence of Theorem 1.12.1, and Proposition 4.1.5. To prove (1), it is sufficient to check that $d_J^2 f = 0$ and $d_J^2(df) = 0$, for any function f on TM. Let $f \in C^\infty(TM)$. Hence

$$d_J^2 f = d_J(d_J f) = d_J(\sum_i (\partial f/\partial v^i) dx^i)$$

$$= \sum_i [d_J(\partial f/\partial v^i) \wedge dx^i + (\partial f/\partial v^i) d_J(dx^i)] \quad \text{(by (1))}$$

$$= \sum_i d_J(\partial f/\partial v^i) \wedge dx^i = \sum_{i,j} (\partial^2 f/\partial v^j \partial v^i)\, dx^j \wedge dx^i$$

$$= \sum_{i<j} (\partial^2 f/\partial v^i \partial v^j)\, dx^j \wedge dx^i + \sum_{i>j} (\partial^2 f/\partial v^i \partial v^j)\, dx^j \wedge dx^i$$

$$= \sum_{i<j}(\partial^2 f/\partial v^i \partial v^j)\, dx^j \wedge dx^i + \sum_{i<j}(\partial^2 f/\partial v^i \partial v^j)\, dx^i \wedge dx^j$$

$$= \sum_{i<j}(\partial^2 f/\partial v^i \partial v^j))(dx^j \wedge dx^i + dx^i \wedge dx^j) = 0,$$

since $dx^j \wedge dx^i = -dx^i \wedge dx^j$.
Moreover, if $\omega = df, f \in C^\infty(TM)$, we obtain

$$d_J^2(df) = d_J(d_J df) = -d_J(dd_J f) = d(d_J^2 f) = 0. \square$$

Proposition 4.1.8 *We have*
(1) $[i_J, d_J] = i_J d_J - d_J i_J = 0$,
(2) $[i_C, d_J] = i_C d_J + d_J i_C = i_J$,
(3) $[d_J, L_C] = d_J L_C - L_C d_J = d_J$,
(4) $J^\star d_J = 0$,
(5) $d_J J^\star = J^\star d$.

The proof is left to the reader as an exercise.

4.2 Homogeneous and semibasic forms

In this section, we shall introduce two important classes of differential forms on tangent bundles.

4.2.1 Homogeneous forms

Let $h_t : TM \longrightarrow TM$ be the homothetia of ratio e^t, that is,

$$h_t(y) = e^t y, \; y \in T_x M, \; x \in M.$$

Then h_t is a vector bundle isomorphism (in fact, its inverse is h_{-t}). An easy computation shows that h_t is a (global) 1–parameter group on TM. Moreover, we have the following.

Proposition 4.2.1 *The Liouville vector field C on TM generates the 1–parameter group h_t.*

Proof: In fact, h_t is locally given by

$$h_t(x^i, v^i) = (x^i, e^t v^i).$$

Then, if X is the infinitesimal generator of h_t, we have

$$X(x^i) = (d/dt)_{/t=0}(x^i \circ h_t) = 0, \text{ and}$$

$$X(v^i) = (d/dt)_{/t=0}(v^i \circ h_t) = (d/dt)_{/t=0}(e^t v^i) = v^i.$$

Thus, we obtain

$$X = v^i(\partial/\partial v^i).$$

Hence $X = C$. □

Definition 4.2.2 *A differentiable function f on TM is said to be* **homogeneous of degree r** *if*

$$L_C f = rf. \tag{4.7}$$

Hence f is homogeneous of degree r if and only if

$$rf(x^i, v^i) = \sum_i v^i(\partial f/\partial v^i).$$

Proposition 4.2.3 *A differentiable function f on TM is homogeneous of degree r if and only if*

$$h_t^* f = e^{rt} f, \ t \in R.$$

Proof: Suppose that f is homogeneous of degree r. Hence $L_C f = rf$. Then, for each point $y \in TM, (h_t^* f)(y) = (f \circ h_t)(y)$ is a solution of the differential equation $(du/dt) = ru$ with initial condition $u(0) = f(y)$. From the uniqueness of solutions of differential equations, we deduce that $h_t^* f = e^{rt} f$. Conversely, if $h_t^* f = e^{rt} f$, we have

$$L_C f = \lim_{t \longrightarrow 0} [(h_t^* f - f)/t]$$

$$= \lim_{t \longrightarrow o} [(e^{rt} f - f)/t] \quad \text{(by Proposition 4.2.1)}$$

$$= (\lim_{t \longrightarrow 0} [(e^{rt} - 1)/t]) f = rf. \square$$

Next, we extend the notion of homogeneity to vector fields and forms.

Definition 4.2.4 *A vector field X on TM is said to be* **homogeneous of degree** r *if* $[C,X] = (r\text{-}1)X$.

The following result can be proved in a similar way as the Proposition 4.2.3.

Proposition 4.2.5 *A vector field X on TM is homogeneous of degree r if and only if the following diagram*

$$\begin{array}{ccc} TTM & \xrightarrow{e^{(r-1)t}Th_t} & TTM \\ X \big\downarrow & & \big\downarrow X \\ TM & \xrightarrow{h_t} & TM \end{array}$$

is commutative, i.e.,

$$X \circ h_t = e^{(r-1)t}(Th_t) \circ X.$$

Suppose that

$$X = X^i(\partial/\partial x^i) + Y^i(\partial/\partial v^i).$$

Then, from (4.7), we easily deduce that X is homogeneous of degree r if and only if

$$(r-1)X^i = \sum_j v^j(\partial X^i/\partial v^j), \ rY^i = \sum_j v^j(\partial Y^i/\partial v^j),$$

$$1 \leq i \leq m = dim\ M.$$

Hence X is homogeneous of degree r if and only if X^i (resp. Y^i) are homogeneous of degree r–1 (resp. r).

Proposition 4.2.6 *Let X and Y be homogeneous vector fields on TM of degree r and s, respectively. Then $[X,Y]$ is homogeneous of degree $r+s\text{-}1$.*

Proof: We have

$$L_C[X,Y] = [[C,X],Y] + [X,[C,Y]] \text{ (by Jacobi identity)}$$

$$= (r-1)[X,Y] + (s-1)[X,Y]$$

$$= (r + s - 2)[X,Y]. \square$$

Now, let $\bar{\chi}(TM)$ be the set of all homogeneous vector fields of degree 1 on TM. From Proposition 4.2.6, we deduce that $\bar{\chi}(TM)$ is a Lie subalgebra of $\chi(TM)$.

Definition 4.2.7 *Let ω be a differential form on TM; ω is said to be* **homogeneous of degree** r *if $L_C\omega = r\omega$.*

As above, we deduce that ω is homogeneous of degree r if and only if

$$h_t^*\omega = e^{rt}\omega, \; t \in R.$$

Moreover, if ω is a Pfaff form locally expressed by

$$\omega = a_i dx^i + b_i dv^i,$$

then ω is homogeneous of degree r if and only if a_i (resp. b_i) is homogeneous of degree r (resp. (r-1)), $1 \leq i \leq m$.

Some properties of homogeneous vector fields and forms follow.

Proposition 4.2.8 *(1) If ω and τ are homogeneous forms of degree r and s, respectively, then $\omega \wedge \tau$ is homogeneous of degree r+s.*
(2) Let X be an homogeneous vector field of degree r and f an homogeneous function of degree s. Then Xf is an homogeneous function of degree r+s-1.
(3) Let ω be an homogeneous p-form of degree r and $X_1, \dots, X_p$ p homogeneous vector fields of degree s. Then $\omega(X_1, \dots, X_p)$ is an homogeneous function of degree r+p(s-1).

Next, we study the behaviour of d, i_C, i_J and d_J acting on homogeneous forms.

Proposition 4.2.9 *Let ω be an homogeneous p–form of degree r on TM. Then $d\omega, i_C\omega, i_J\omega$ and $d_J\omega$ are homogeneous forms of degree r, r, r-1 and r-1, respectively.*

Proof: In fact, we have

$$L_C(d\omega) = d(L_C\omega) = r(d\omega),$$

$$L_C(i_J\omega) = i_C(L_C\omega)$$

$$(\text{since } [L_C, i_C] = L_C i_C - i_C L_C = 0)$$

$$= r(i_C\omega),$$

$$L_C(i_J\omega) = i_J(L_C\omega) - i_J\omega \text{ (by Proposition (4.1.4))}$$

$$= r(i_J\omega) - i_J\omega = (r-1)(i_J\omega),$$

$$L_C(d_J\omega) = d_J(L_C\omega) - d_J\omega \text{ (by Proposition (4.1.8))}$$

$$= r(d_J\omega) - d_J\omega = (r-1)(d_J\omega). \square$$

Finally, we introduce the notion of homogeneity of p–**vector forms** (i.e., tensor fields of type $(1, p)$).

Definition 4.2.10 *A p–vector form L on TM is said to be* **homogeneous of degree** r *if*

$$L_C L = (r-1)L.$$

Remark 4.2.11 Sometimes, the functions, vector fields and forms are supposed to be defined only on $\mathcal{T}M = TM$ - {zero section}. All the definitions and results in this section hold in such a case.

4.2.2 Semibasic forms

Definition 4.2.12 *Let α be a differentiable form on TM; α is said to be* **semibasic** *if $\alpha \in Im J^*$.*

Let us denote by S^pB the set of all semibasic p–forms on TM, $p \geq 0$. (Obviously, $S^0B = C^\infty(TM)$). Since $J^*(dx^i) = 0, J^*(dv^i) = dx^i$, then S^pB is locally spanned by

$$\{dx^{i_1} \wedge \ldots \wedge dx^{i_p}; 1 \leq i_1 < i_2 < \ldots \leq i_p \leq m\}$$

(hence the name).
If we set

$$SB = \bigoplus_{p=0}^{m} S^pB, m = \dim M,$$

then SB is called the **(graduated) algebra of semibasic forms** on TM.

Next, we study the behaviour of d, i_C, i_J and d_J acting on semibasic forms.

Proposition 4.2.13 *If $\alpha \in S^pB$, then (1) $i_C\alpha = i_J\alpha = 0$; (2) $d_J\alpha \in S^{p+1}B$.*

Proof: (1) follows from $i_CJ^* = i_JJ^* = 0$ (by Corollary 4.1.3 and Proposition 4.1.4). (2) follows from $d_JJ^* = J^*d$ (by Proposition 4.1.8).□

Corollary 4.2.14 *If f is a differentiable function on TM, then d_Jf is a semibasic Pfaff form.*

Now, let f be a differentiable function on TM. Since

$$df = \sum_i (\partial f/\partial x^i)dx^i + (\partial f/\partial v^i)dv^i,$$

we deduce that df is not, in general, a semibasic form on TM.

Next, we stablish some important properties of semibasic Pfaff forms.

Proposition 4.2.15 *Let α be a Pfaff form on TM. Then α is semibasic if and only if $\alpha(X) = 0$, for any vertical vector field X on TM.*

Proof: In fact, if $\alpha = J^*\beta$, then $\alpha(JX) = 0$, since $J^2 = 0$. Conversely, if $\alpha(X) = 0$, for any vertical vector field X on TM, then α is locally given by $\alpha = a_i(x,v)dx^i$. Hence α is semibasic.□

Let α be a semibasic Pfaff form on TM. Then we can define a mapping

$$D : TM \longrightarrow T^*M$$

as follows. Let $y \in T_xM, x \in M$; then $D(y) \in T_x^*M$ is given by

$$D(y)(X) = \alpha_y(\bar{X}),$$

where $X \in T_xM$ and $\bar{X} \in T_y(TM)$ is an arbitrary tangent vector such that $(\tau_M)_*(\bar{X}) = X$. In fact, if $\bar{Y} \in T_y(TM)$ is any other tangent vector such that $(\tau_M)_*(\bar{Y}) = X$, then $(\tau_M)_*(\bar{X} - \bar{Y}) = 0$ and so $\bar{X} - \bar{Y}$ is vertical. Hence $\alpha_y(\bar{X} - \bar{Y}) = 0$ by Proposition 4.2.15. We note that the diagram

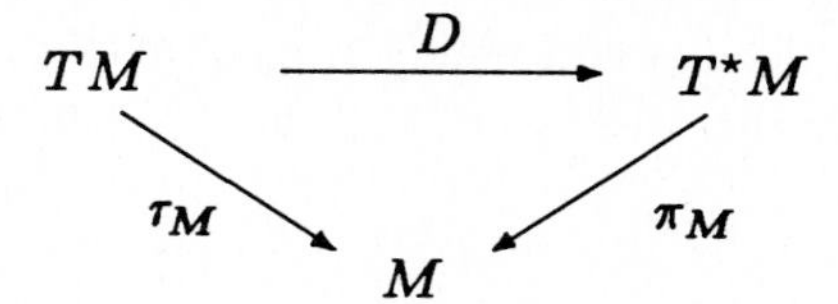

is commutative, where $\tau_M : TM \longrightarrow M$ and $\pi_M : TM \longrightarrow M$ are the canonical projections. Hence D is a bundle morphism, but it is not, in general, a vector bundle homomorphism.

Conversely, suppose that $D : TM \longrightarrow T^*M$ is a mapping such that $\pi_M \circ D = \tau_M$. Then D determines a semibasic Pfaff form α on TM as follows:

$$\alpha_y(X) =< (\tau_M)_*X, D(y) >,$$

where $y \in T_xM, X \in T_y(TM)$. Clearly α is semibasic, since, if X is vertical, then $(\tau_M)_*X = 0$.

Summing up, we have proved the following.

Theorem 4.2.16 *The correspondence*

$$\alpha(X) =< (\tau_M)_*X, D(\tau_M(X)) >, \; X \in TTM,$$

determines a bijection between the semibasic Pfaff forms on TM and the mappings $D : TM \longrightarrow T^*M$ such that $\pi_M \circ D = \tau_M$.

Let us remark that if the semibasic Pfaff form α is locally expressed by

$$\alpha = \alpha_i(x, v)dx^i,$$

then we easily see that corresponding mapping D is locally given by

$$D(x^i, v^i) = (x^i, \alpha_i). \tag{4.8}$$

Let λ be the Liouville form on T^*M. We recall that λ is locally given by

$$\lambda = p_i dx^i,$$

where (x^i, p_i) are the induced coordinates in T^*M. Then we have the following.

Corollary 4.2.17 $D^*\lambda = \alpha$.

Proof: Using (4.8), we obtain

$$D^*\lambda = D^*(p_i dx^i) = (p_i \circ D)d(x^i \circ D) = \alpha_i dx^i = \alpha. \Box$$

To end this section, we introduce the notion of semibasic vector forms.

Definition 4.2.18 *A vector p–form L on TM is said to be* **semibasic** *if*
(1) $L(X_1, \ldots, X_p)$ is a vertical vector field, for any vector fields $X_1, \ldots, X_p$;
(2) $L(X_1, \ldots, X_p) = 0$, when some X_i is a vertical vector field.
If L is skew-symmetric, then (1) and (2) become:
(1)' $JL = 0$;
(2)' $i_{JX}L = 0$, for any vector field X on TM.

4.3 Semisprays. Sprays. Potentials

We know that a vector field on a manifold M is the geometrical interpretation of a system of ordinary differential equations (see Section 1.7). The aim of this section is to introduce a class of vector fields on the tangent bundle TM which interprets geometrically a system of second order differential equations.

Definition 4.3.1 *A* **semispray** *(or* **second order differential equation***) on M is a vector field on TM (that is, a section of the tangent bundle of $TM, \tau_{TM} : TTM \longrightarrow TM$), C^∞ on TM, which is also a section of the vector bundle $T\tau_M : TTM \longrightarrow TM$.*

Let ξ be a semispray on M. Then ξ is locally given by

$$\xi = \bar{\xi}^i(\partial/\partial x^i) + \xi^i(\partial/\partial v^i).$$

Since τ_M is locally given by $\tau_M(x^i, v^i) = (x^i)$, we deduce that

$$T\tau_M(a^i(\partial/\partial x^i) + b^i(\partial/\partial v^i)) = a^i(\partial/\partial x^i).$$

Hence we have

$$T\tau_M(\xi(x^i, v^i)) = T\tau_M(x^i, v^i, \bar{\xi}^i, \xi^i) = (x^i, \bar{\xi}^i) = (x^i, v^i),$$

because ξ is a section of $T\tau_M$. Thus, we obtain $\bar{\xi}^i = v^i$. Therefore, a semispray is locally given by

$$\xi = v^i(\partial/\partial x^i) + \xi^i(\partial/\partial v^i), \tag{4.9}$$

where $\xi^i = \xi^i(x, v)$ are C^∞ on $\mathcal{T}M$.

From (4.9), we easily deduce the following.

Proposition 4.3.2 *A vector field ξ on TM, C^∞ on $\mathcal{T}M$, is a semispray if and only if $J\xi = C$.*

Definition 4.3.3 *Let ξ be a semispray on M. A curve σ on M is called a* **path** *(or* **solution***) of ξ if σ is an integral curve of ξ, that is,*

$$\ddot{\sigma}(t) = \xi(\dot{\sigma}(t)).$$

In local coordinates, if $\sigma(t) = (x^i(t))$, then we deduce that σ is a path of ξ if σ satisfies the following system of second order differential equations:

$$(d^2x^i/dt^2) = \xi^i(x, (dx/dt), \; 1 \leq i \leq m = \dim M.$$

We shall express the non-homogeneity of a semispray.

Definition 4.3.4 *Let ξ be a semispray on M. We call* **deviation** *of ξ the vector field $\xi^\star = [C, \xi] - \xi$.*

A simple computation shows that $J[C, \xi] = C$. Then $\xi^\star$ is a vertical vector field.

Definition 4.3.5 *A semispray ξ is called a* **spray** *if $\xi^\star = 0$ and ξ is C^1 on the zero section. If, moreover, ξ is C^2 on the zero section, then ξ is called a* **quadratic spray**.

Remark 4.3.6 Let ξ be a semispray. Since $\xi^\star = 0$ if and only if $[C, \xi] = \xi$, we deduce that ξ is a spray if the functions ξ^i are homogeneous of degree 2 and C^1 on the zero section (if, moreover, the functions ξ^i are quadratic on v^i, then ξ is a quadratic spray). Thus, a spray is the geometrical interpretation of a system of second order differential equations homogeneous of degree 2 with respect to the first derivatives.

Proposition 4.3.7 *Let ξ be a semispray on M. Then we have*

$$J[\xi, JX] = -(JX),\ X \in \chi(TM).$$

Proof: In fact, since $N_J = 0$, we deduce

$$0 = N_J(\xi, X) = [C, JX] - J[C, X] - J[\xi, JX].$$

On the other hand,

$$-(JX) = (L_C J)(X) = [C, JX] - J[C, X].$$

Then we obtain

$$J[\xi, JX] = -(JX).\square$$

Next, we introduce the potential of a semibasic form.

Definition 4.3.8 *Let α (resp. L) be a semibasic scalar p-form (resp. a vector l-form) on TM. Then the* **potential** *α^0 of α (resp. L^0 of L) is the scalar (p-1)-form (resp. vector (l-1)-form) given by*

$$\alpha^0 = i_\xi \alpha \ \text{(resp. } L^0 = i_\xi L),$$

where ξ is an arbitrary semispray on M.

Obviously, α^0 (resp. L^0) is independent of the choice of ξ. In fact, if ξ' is another semispray, then $\xi - \xi'$ is vertical.

Proposition 4.3.9 *α^0 and L^0 are semibasic.*

(Let us remark that the scalar p–form α is not necessarily skew-symmetric).

Proposition 4.3.10 *Let α be a semibasic scalar p-form on TM which is homogeneous of degree r, with $p+r \neq 0$. Then we have*

$$\alpha = (1/(p+r))((d_J\alpha)^0 + d_J\alpha^0).$$

Hence if α is d_J-closed, we deduce

$$\alpha = (1/(p+r))d_J\alpha^0,$$

that is, α is expressed as the derivative of its potential.

Proof: From Exercise 4.12.4, we have

$$(d_J\alpha)^0 + d_J\alpha^0 = i_\xi d_J\alpha + d_J i_\xi \alpha$$

$$= L_C\alpha - i_{[\xi,J]}\alpha.$$

But

$$(i_{[\xi,J]}\alpha)(X_1,\ldots,X_p) = \sum_{i=1}^{p} \alpha(X_1,\ldots,[\xi,J]X_i,\ldots,X_p).$$

Now,

$$[\xi,J]X_i = [\xi,JX_i] - J[\xi,X_i]$$

implies that

$$J([\xi,J]X_i) = J[\xi,JX_i] = -JX_i,$$

by Proposition 4.3.7. Hence

$$J([\xi,J]X_i + X_i) = 0$$

and then $[\xi,J]X_i + X_i$ is a vertical vector field. Since α is semibasic we have

$$(i_{[\xi,J]}\alpha)(X_1,\ldots,X_p) = \sum_{i=1}^{p} \alpha(X_1,\ldots,-X_i,\ldots,X_p)$$

$$= -p\alpha(X_1,\ldots,X_p)$$

Thus,

$$L_C\alpha - i_{[\xi,J]}\alpha = r\alpha + p\alpha = (p+r)\alpha,$$

which proves the Proposition. □

Remark 4.3.11 A similar result holds for vector forms (see Klein and Voutier [80] and Grifone [73]).

4.4 Connections in fibred manifolds

Let $p : E \longrightarrow M$ be a fibred manifold such that dim $M = m$ and dim $E = m + n$. We denote by $V(E)$ the vertical bundle of E, that is,

$$V(E) = \bigcup_{e \in E} V_e(E),$$

where $V_e(E) = Ker\{p_* : T_e E \longrightarrow T_{p(e)}\}$. Then $V(E)$ is a vector bundle over E of rank n. In fact, $V(E)$ is vector subbundle of TE. Furthermore, we have

$$V_e(E) = T_e(E_{p(e)}),$$

where $E_{p(e)}$ is the fibre over $p(e)$.

Definition 4.4.1 *A* **connection** *in E is a vector bundle H over E (or a distribution H on E) such that*

$$TE = V(E) \oplus H,$$

that is,

$$T_e E = V_e(E) \oplus H_e, \; e \in E.$$

H is called the **horizontal bundle** *(or* **horizontal distribution***) and H_e (resp. $V_e(E)$) is called the* **horizontal** *(resp.* **vertical***)* **subspace at** *e. The tangent vectors to E which belong to H are called the* **horizontal tangent vectors**. *If a vector field X verifies that $X(e) \in H_e$, for any $e \in E$, then X is called a* **horizontal vector field**.

If H is a connection in E, then a tangent vector X of E at $e \in E$ may be decomposed as follows:

$$X = vX + hX;$$

vX (resp. hX) is called the **vertical** (resp. **horizontal**) **component** of X.

Then we have two canonical projectors v and h which will be called the **vertical** and **horizontal projectors**, respectively.

Now, we consider the induced bundle $E \times_M TM$. Let us recall that

$$E \times_M TM = \{(e, y) \in E \times TM / p(e) = \tau_M(y)\}$$

is a vector bundle over E such that the following diagram

$$\begin{array}{ccc} E \times_M TM & \xrightarrow{p_2} & TM \\ {\scriptstyle p_1}\downarrow & & \downarrow{\scriptstyle \tau_M} \\ E & \xrightarrow{p} & M \end{array}$$

is commutative, where p_1 and p_2 are the canonical projections given by $p_1(e, y) = e$ and $p_2(e, y) = y$. Consider the following commutative diagram:

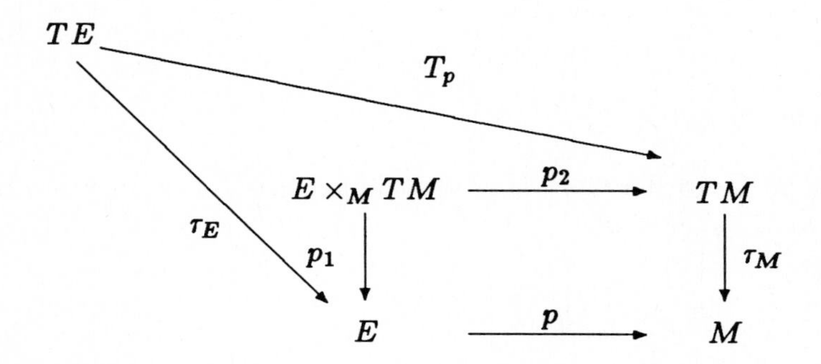

Then there exists a unique vector bundle homorphism

$$s : TE \longrightarrow E \times_M TM$$

such that

$$p_2 \circ s = Tp, \; p_1 \circ s = \tau_E;$$

s is defined by

$$s(X) = (\tau_E(X), Tp(X)), \; X \in TE.$$

Hence we obtain a sequence of vector bundle homomorphisms

$$0 \longrightarrow VE \xrightarrow{i} TE \xrightarrow{s} E \times_M TM \longrightarrow 0, \tag{4.10}$$

where $i : VE \longrightarrow TE$ is the canonical injection. One can easily prove that (4.10) is, in fact, exact.

The following result is left to the reader as an exercise.

Proposition 4.4.2 *(see Vilms [119]).- A connection in E determines a left splitting of (4.10) and, conversely, each left splitting of (4.10) determines a connection in E.*

4.5 Connections in tangent bundles

In this section, we characterize the connections in tangent bundles by means of the canonical almost tangent structure. This theory is due to J. Grifone [73].

Let TM the tangent bundle of a manifold M and denote by $\tau_M : TM \longrightarrow M$ the canonical projection. Suppose that a connection H in TM is given with horizontal and vertical projectors h and v, respectively. Since $V = V(TM) = Ker\ J = Im\ J$, where J is the canonical almost tangent structure on TM, we can easily prove that

$$Jh = J,\ hJ = 0,\ Jv = 0,\ vJ = J.$$

We now put $\Gamma = 2h - Id_{TM}$. Then Γ is a tensor field of type (1,1) on TM (or a vector 1–form on TM) such that

$$J\Gamma = J,\ \Gamma J = -J. \tag{4.11}$$

Conversely, if Γ is a tensor field of type (1,1) on TM satisfying (4.11), then we easily deduce that

$$\Gamma^2 = Id_{TM}.$$

In fact,

$$\Gamma(\partial/\partial v^i) = \Gamma(J(\partial/\partial x^i)) = -\partial/\partial v^i, \tag{4.12}$$

and

$$J(\Gamma(\partial/\partial x^i)) = J(A_i^j(\partial/\partial x^j) + B_i^j(\partial/\partial v^j)) = A_i^j(\partial/\partial v^i) = \partial/\partial v^i.$$

Hence, we have $A_i^j = \delta_i^j$, that is,

$$\Gamma(\partial/\partial x^i) = \partial/\partial x^i + B_i^j(\partial/\partial v^j). \tag{4.13}$$

From (4.12) and (4.13), we easily deduce that $\Gamma^2 = Id_{TM}$. Now, if we put

$$h = (1/2)(Id_{TM} + \Gamma),\ v = (1/2)(Id_{TM} - \Gamma),$$

then Im $v = V$, and $H =$ Im h defines a connection in TM.

Thus, we give the following definition.

Definition 4.5.1 *A* **connection** Γ *in TM is a tensor field of type (1,1) on TM, C^∞ on $\mathcal{T}M$, such that*

$$J\Gamma = J,\ \Gamma J = -J.$$

Proposition 4.5.2 *A connection Γ in TM defines an almost product structure on TM, C^∞ on $\mathcal{T}M$, such that, for each $y \in TM$, the eigenspace corresponding to the eigenvalue -1 is V_y and conversely.*

Proof: In fact, $\Gamma^2 = Id_{TM}$. Now, let $\Gamma X = -X$. Then we have

$$J\Gamma(X) = -(JX) = JX.$$

Then $JX = 0$, and, hence, X is vertical. On the other hand, if X is vertical, then $X = JY$, for some tangent vector Y. Then $\Gamma X = -X$. Conversely, let Γ be an almost product structure on TM such that the eigenspace corresponding to the eigenvalue -1 is V_y, for any $y \in TM$. Then $\Gamma J = -J$. Furthermore, let $X \in T_y(TM), y \in TM$. Then we have

$$\Gamma(\Gamma X - X) = \Gamma^2 X - \Gamma X = -(\Gamma X - X),$$

since $\Gamma^2 = Id_{TM}$. This implies that

$$\Gamma X - X \in V_y(TM) = ImJ_y.$$

Thus $\Gamma X - X = JY$, for some $Y \in T_y(TM)$. Hence, we have

$$J(\Gamma X - X) = J\Gamma(X) - JX = 0,$$

and, consequently, we deduce that $J\Gamma = J$. $\square$

Let

$$h = (1/2)(Id_{TM} + \Gamma),\ v = (1/2)(Id_{TM} - \Gamma).$$

Then $V = Im\ v$ and $H = Im\ h$ defines a connection in TM (in the sense of the previous section). We easily deduce that

$$h^2 = h,\ v^2 = v,\ Jh = J,\ hJ = 0,\ Jv = 0,\ vJ = J,$$

$$T(TM) = V \oplus H.$$

From (4.12) and (4.13), we deduce that Γ is locally given by

$$\Gamma(\partial/\partial x^i) = \partial/\partial x^i - 2\Gamma_i^j(\partial/\partial v^j),$$

$$\Gamma(\partial/\partial v^i) = \partial/\partial v^i. \tag{4.14}$$

Since

$$d\tau_M(y) : T_y(TM) \longrightarrow T_xM,\ x = \tau_M(y), y \in TM$$

defines a linear isomorphism

$$d\tau_M(y) : H_y \longrightarrow T_xM,$$

we can define the **horizontal lift** of a vector field X on M with respect to Γ to be the vector field X^H on TM such that

$$X_y^H \in H_y,\ d\tau_M(y)(X_y^H) = X_x,\ x = \tau_M(y), y \in TM.$$

We locally have

$$X^H = X^i(\partial/\partial x^i) - X^i\Gamma_i^j(\partial/\partial v^j), \tag{4.15}$$

where $X = X^i(\partial/\partial x^i)$.

The functions Γ_i^j are C^∞ on τM and will be called the **Christoffel components** of Γ. From (4.14), we deduce that h is locally given by

$$h(\partial/\partial x^i) = \partial/\partial x^i - \Gamma_i^j(\partial/\partial v^j),$$

$$h(\partial/\partial v^i) = 0. \tag{4.16}$$

Next, we express the nonhomogeneity of a connection.

Definition 4.5.3 *The* **tension** *of a connection Γ in TM is the tensor field* $\mathbf{H}$ *of type (1,1) on TM given by*

$$\mathbf{H} = (1/2)(L_C\Gamma),$$

that is,

$$\mathbf{H}(X) = (1/2)([C, \Gamma X] - \Gamma[C, X]),$$

for any vector field X on TM.

Obviously,

$$\mathbf{H}(X) = (L_C h)(X) = [C, hX] - h[C, X],$$

where $h = (1/2)(I + \Gamma)$ is the horizontal projector of Γ.

Proposition 4.5.4 $\mathbf{H}$ *is semibasic.*

Proof: In fact, we have

$$\mathbf{H}(\partial/\partial x^i) = (\Gamma_i^j - v^k(\partial\Gamma_i^j/\partial v^k))(\partial/\partial v^j),$$

$$\mathbf{H}(\partial/\partial v^i) = 0. \tag{4.17}$$

From (4.17), we easily deduce that $\mathbf{H}$ is semibasic. □

Definition 4.5.5 *A connection Γ in TM is said to be* **homogeneous** *if Γ is homogeneous of degree 1, that is, its tension* $\mathbf{H}$ *vanishes.*

Obviously, Γ is homogeneous if the functions Γ_i^j are homogeneous of degree 1.

Definition 4.5.6 *An homogeneous connection Γ is said to be a* **linear connection** *if Γ is C^1 on the zero section.*

If Γ is a linear connection in TM, then the functions Γ_i^j are linear on v^k. Then, we may writte

$$\Gamma_i^j(x, v) = v^k \Gamma_{ik}^j,$$

where

$$\Gamma^j_{ik} = (\partial \Gamma^j_i / \partial v^k).$$

Next, we define the covariant derivative determined by a connection in TM.

Before proceeding further, we consider the linear isomorphism

$$T_x M \longrightarrow V_y,\ y \in T_x M,$$

given by the vertical lift operator

$$z \longrightarrow (z^v)_y.$$

We denote by

$$\phi_y : V_y \longrightarrow T_x M$$

the inverse linear isomorphism. With respect to the induced coordinates, we have

$$\phi_y(\sum_i A^i(\partial/\partial v^i)) \longrightarrow \sum_i A^i(\partial/\partial x^i). \tag{4.18}$$

Let Γ be a connection in TM. We define the **covariant derivative of a vector field** Y **on** M **with respect to a vector field** X on M to be the vector field $\nabla_X Y$ on M given by

$$(\nabla_X Y)_x = \phi_{Y(x)}(v(dY(x)(X_x)),\ x \in M. \tag{4.19}$$

Suppose that $X = X^i(\partial/\partial x^i)$ and $Y = Y^i(\partial/\partial x^i)$. From (4.18) and (4.19), we obtain

$$\nabla_X Y = X^i((\partial Y^j/\partial x^i) + \Gamma^j_i(x,v))\partial/\partial x^j. \tag{4.20}$$

From (4.20), one easily deduces that

$$\nabla_{X_1+X_2} Y = \nabla_{X_1} Y + \nabla_{X_2} Y,$$

$$\nabla_{fX} Y = f(\nabla_X Y), \tag{4.21}$$

for any $f \in C^\infty(M), X_1, X_2, X, Y \in \chi(M)$. Moreover, (4.20) shows that $(\nabla_X Y)_x$ depends only on the value of X at x. Thus we can define the

covariant derivative of Y with respect to a tangent vector $z \in T_x M$ as follows:

$$\nabla_z Y = (\nabla_X Y)_x,$$

where X is an arbitrary vector field on M such that $X_x = z$.

If $\sigma(t)$ is a curve in M and $\tau(t)$ is a vector field along $\sigma(t)$, that is,

$$\tau(t) \in T_{\sigma(t)}M, \ t \in R,$$

then the **covariant derivative of** $\tau(t)$ **along** $\sigma(t)$ is defined by

$$(\nabla_\sigma \tau)(t) = \nabla_{\dot{\sigma}(t)} Y,$$

where Y is an arbitrary vector field on M which extends $\tau(t)$. Then $(\nabla_\sigma \tau)(t)$ is a new vector field along $\sigma(t)$. We say that $\tau(t)$ is **parallel along** $\sigma(t)$ if

$$\nabla_\sigma \tau = 0.$$

Definition 4.5.7 *Let Γ be a connection in TM. A curve σ in M is said to be a* **path** *of Γ if the vector field $\dot{\sigma}(t)$ is parallel along $\sigma(t)$, i.e.,*

$$\nabla_\sigma \dot{\sigma} = 0.$$

If Γ is an homogeneous or linear connection, then the paths of Γ are called **geodesics**.

Hence a curve σ in M is a path of Γ if and only if its canonical lift $\dot{\sigma}$ to TM is an horizontal curve.

Suppose that $\sigma(t) = (x^i(t))$. From (4.20), we obtain

$$\nabla_\sigma \dot{\sigma} = ((d^2 x^i/dt^2) + \Gamma^i_j(x, dx/dt)(dx^j/dt))\partial/\partial x^i.$$

Hence σ is a path of Γ if and only if it satisfies the following system of second order differential equations:

$$d^2 x^i/dt^2 + \Gamma^i_j(x, dx/dt)(dx^j/dt) = 0, \ 1 \leq i \leq m = \dim M. \tag{4.22}$$

If Γ is a linear connection, then (4.22) becomes

$$d^2 x^i/dt^2 + \Gamma^i_{jk}(x)(dx^i/dt)(dx^j/dt) = 0,$$

which are the usual differential equations for geodesics of linear connections ∇ on M (see below and Section 1.18).

Now, let Γ be a linear connection in TM. Then (4.20) becomes

$$\nabla_X Y = X^i((\partial Y^j/\partial x^i) + v^k \Gamma^j_{ik})\partial/\partial x^j. \quad (4.23)$$

From (4.23), we see that

$$\nabla_X(Y_1 + Y_2) = \nabla_X Y_1 + \nabla_X Y_2,$$

$$\nabla_X(fY) = (Xf)Y + f(\nabla_X Y), \quad (4.24)$$

for any $f \in C^\infty(M), X, Y_1, Y_2, Y \in \chi(M)$. From (4.21) and (4.24), we deduce that ∇ is a linear connection on M. Then **a linear connection in** TM **defines a linear connection on** M (in the sense of Section 1.18). Moreover, the horizontal lifts of vector fields on M to TM with respect to a linear connection Γ in TM coincide with the corresponding ones defined in Chapter 2.

Let Γ be a connection in TM with Christoffel components Γ^j_i. We set

$$D_i = h(\partial/\partial x^i) = \partial/\partial x^i - \Gamma^j_i(\partial/\partial v^j),\ V_i = \partial/\partial v^i. \quad (4.25)$$

Then $\{D_i, V_i\}$ is a local frame field which is called the adapted frame with respect to Γ. The dual coframe field is given by $\{\theta^i, \eta^i\}$, where

$$\theta^i = dx^i,\ \eta^i = \Gamma^i_j dx^j + dv^i, \quad (4.26)$$

We have

$$hD_i = D_i,\ hV_i = 0,\ vD_i = 0,\ vV_i = 0,\ JD_i = V_i,\ JV_i = 0.$$

From (4.25), we obtain

$$[D_i, D_j] = ((\partial\Gamma^k_i/\partial x^k) - (\partial\Gamma^k_j/\partial x^i) + \Gamma^l_i(\partial\Gamma^k_j/\partial v^l)$$

$$-\Gamma^l_j(\partial\Gamma^k_i/\partial v^l))V_k,$$

$$[D_i, V_j] = (\partial\Gamma^k_i/\partial v^j)V_k,$$

$$[V_i, V_j] = 0. \quad (4.27)$$

If Γ is linear, then (4.25) and (4.26) are, precisely, the adapted frame and coframe defined by ∇ (see Chapter 2).

Remark 4.5.8 In Definition 4.5.1 we restrict ourselves to the tangent bundle without the zero section. So we can distinguish between linear and strictly homogeneous connections. In fact, if Γ is homogeneous and C^∞ on all TM then Γ is a linear connection (see Grifone [73] and Vilms [119]).

Remark 4.5.9 From Section 4.4, we obtain an exact sequence of vector bundle homomorphisms

$$0 \longrightarrow V(TM) \longrightarrow TTM \longrightarrow TM \times_M TM \longrightarrow 0, \tag{4.28}$$

and, by Proposition 4.4.2, we deduce that each connection in TM determines a left splitting of (4.28) and, conversely, each left splitting of (4.28) defines a connection in TM.

Remark 4.5.10 If $p : E \longrightarrow M$ is a vector bundle, we may consider homogeneous and linear connections in E. In this situation, we can define a covariant derivative of a section of E with respect to a vector field on M (see Vilms [119]). If the connection is linear, then we obtain a connection in E in the same sense of Koszul [83]).

Remark 4.5.11 The extension of the theory for connections in tangent bundles of higher order may be founded in de Andrés et al. (see [27]).

4.6 Semisprays and connections

In this section, we study the relations between semisprays on TM and connections in TM.

Let Γ be a connection in TM, ξ' an arbitrary semispray on TM and $h = (1/2)(I + \Gamma)$ the horizontal projector of Γ. We set

$$\xi = h\xi'. \tag{4.29}$$

If ξ'' is any other semispray on TM, we have

$$h(\xi' - \xi'') = h\xi' - h\xi'' = 0,$$

since $\xi' - \xi''$ is a vertical vector field. Hence the vector field ξ defined by (4.29) does not depends on the choice of ξ'. Moreover, ξ is a semispray; in fact,

$$J\xi = Jh(\xi') = J\xi' = C,$$

since $Jh = J$. The semispray ξ is called the **associated semispray to** Γ.

Proposition 4.6.1 *If* $\mathbf{H}$ *is the tension of* Γ*, then we have*

$$\mathbf{H}^0 = \xi^\star,$$

where ξ *is the associated semispray to* Γ.

Proof: In fact,

$$\mathbf{H}^0 = \mathbf{H}(\xi) = [C, h\xi] - h[C, \xi].$$

But $h(\xi) = \xi$, since ξ is an horizontal vector field, and

$$h(\xi^\star) = h[C, \xi] - h\xi = h[C, \xi] - \xi,$$

since $\xi^\star$ is vertical. Hence

$$\mathbf{H}^0 = [C, \xi] - \xi = \xi^\star. \square$$

A simple computation from (4.16) shows that ξ is locally given by

$$\xi = v^i(\partial/\partial x^i) - v^j \Gamma^i_j(\partial/\partial v^i). \tag{4.30}$$

If Γ is a linear connection, then (4.30) becomes

$$\xi = v^i(\partial/\partial x^i) - v^j v^k \Gamma^i_{jk}(\partial/\partial v^i).$$

Corollary 4.6.2 *If* Γ *is homogeneous (resp. linear) then* ξ *is a spray (resp. a quadratic spray).*

Remark 4.6.3 If Γ is a linear connection defined by a Riemannian metric g on M, then ξ is the geodesic spray defined by g (see Boothby [9]).

Now, let ξ be an arbitrary semispray on TM. We set

$$\Gamma = -L_\xi J,$$

that is,

$$\Gamma(X) = (-L_\xi J)(X) = -[\xi, JX] + J[\xi, X],$$

for any vector field X on TM.

Proposition 4.6.4 *Γ is a connection in TM whose associated semispray is $\xi + (1/2)\xi^*$.*

Proof: We have

$$J\Gamma(X) = -J[\xi, JX] = JX,$$

$$\Gamma J(X) = J[\xi, JX] = -JX,$$

by Proposition 4.3.7. Hence Γ is a connection in TM. On the other hand, if ξ' is the associated semispray to Γ, we have

$$\xi' = h\xi = (1/2)(I + \Gamma)\xi = (1/2)\xi + (1/2)\Gamma\xi$$

$$= (1/2)\xi + (1/2)(-[\xi, J\xi]) = (1/2)\xi + (1/2)[C, \xi]$$

$$= (1/2)\xi + (1/2)(\xi^* + \xi)$$

$$= \xi + (1/2)\xi^*. \square$$

If $\xi = v^i(\partial/\partial x^i) + \xi^i(\partial/\partial v^i)$, then $\Gamma = -L_\xi J$ is given by

$$\Gamma(\partial/\partial x^i) = \partial/\partial x^i + (\partial\xi^j/\partial v^i),\ \Gamma(\partial/\partial v^i) = -(\partial/\partial v^i). \tag{4.31}$$

Then $\Gamma_i^j = -(1/2)(\partial\xi^j/\partial v^i)$.

From (4.9), (4.22) and (4.31), we obtain the following:

Proposition 4.6.5 *A connection Γ and its associated semispray have the same paths.*

Corollary 4.6.6 *(1) If ξ is a spray, then $\Gamma = -L_\xi J$ is an homogeneous connection whose associated semispray is, precisely, ξ. (2) If ξ is a quadratic spray, then $\Gamma = -L_\xi J$ is a linear connection.*

Proof: Clearly, if ξ is a spray, then $\xi^* = 0$. So the associated semispray to Γ is, precisely, ξ, by Proposition 4.6.4. Moreover, if $\mathbf{H}$ is the tension of Γ, we have

$$\mathbf{H}(\partial/\partial x^i) = (\Gamma_i^j - v^k(\partial\Gamma_i^j/\partial v^k))(\partial/\partial v^j)$$

$$= -(1/2)((\partial\xi^j/\partial v^i) - v^k(\partial^2\xi^j/\partial v^i\partial v^k)).$$

Since ξ is a spray, we know that the functions ξ^j are homogeneous of degree 2, and so,

$$2\xi^j = v^k(\partial\xi^j/\partial v^k).$$

Hence

$$2(\partial\xi^j/\partial v^i) = (\partial\xi^j/\partial v^i) + v^k(\partial^2\xi^j/\partial v^i\partial v^k),$$

which implies

$$(\partial\xi^j/\partial v^i) = v^k(\partial^2\xi^j/\partial v^i\partial v^k).$$

Therefore,

$$\mathbf{H}(\partial/\partial x^i) = 0,$$

and then $\mathbf{H} = 0$, since $\mathbf{H}$ is semibasic.
(2) follows by a similar devide in local coordinates. □

Example

Let us consider a semispray ξ on $R^4 = TR^2$ given by

$$\xi = u\partial/\partial x + v\partial/\partial y + f(x,y)\partial/\partial u + g(x,y)\partial/\partial v,$$

where (x, y) (resp. (x, y, u, v)) are the canonical coordinates on R^2 (resp. R^4) and f, g are C^∞ functions on R^4.

The Liouville vector field on TR^2 is given by

$$C = u\partial/\partial u + v\partial/\partial v.$$

Then we have

$$[C, \xi] = u\partial/\partial x + v\partial/\partial y - f(x,y)\partial/\partial u - g(x,y)\partial/\partial v,$$

and so

$$\xi^* = [C, \xi] - \xi = -2f(x,y)\partial/\partial u - 2g(x,y)\partial/\partial v.$$

We set $\Gamma = -L_\xi J$. Then we obtain

$$\Gamma(\partial/\partial x) = \partial/\partial x, \ \ \Gamma(\partial/\partial u) = -\partial/\partial u,$$

$$\Gamma(\partial/\partial y) = \partial/\partial y, \ \ \Gamma(\partial/\partial v) = -\partial/\partial v.$$

Hence

$$h(\partial/\partial x) = \partial/\partial x, \ \ h(\partial/\partial u) = 0,$$

$$h(\partial/\partial y) = \partial/\partial y, \ \ h(\partial/\partial v) = 0.$$

Then the associated semispray to Γ is

$$\bar{\xi} = h(\xi) = u\partial/\partial x + v\partial/\partial y.$$

Since the Christoffel components of Γ are all zero, we deduce that the tension of Γ vanishes and so, Γ is a linear connection in TR^2.

If we compute the covariant derivative associated to Γ we obtain

$$\nabla_X Y = [X^1(\partial Y^1/\partial x) + X^2(\partial Y^1/\partial y)]\partial/\partial x$$

$$+[X^1(\partial Y^2/\partial x) + X^2(\partial Y^2/\partial y)]\partial/\partial y,$$

where $X = X^1\partial/\partial x + X^2\partial/\partial y$, $Y = Y^1\partial/\partial x + Y^2\partial/\partial y$ are vector fields on R^2.

Hence ∇ is the canonical flat linear connection on R^2. In fact, ∇ is the Riemannian connection determined by the standard Riemannian metric g on R^2 given by

$$g(\partial/\partial x, \partial/\partial x) = g(\partial/\partial y, \ \partial/\partial y) = 1,$$

$$g(\partial/\partial x, \ \partial/\partial y) = 0.$$

4.7 Weak and strong torsion

In this section, the torsion forms of a connection in TM is defined and related with the torsion tensor for linear connection.

Definition 4.7.1 *Let Γ be a connection in TM with horizontal projector h. The* **weak torsion** *of Γ is the vector 2–form (or tensor field of type (1,2)) on TM given by*

$$t = [J, h],$$

that is,

$$[J, h](X, Y) = [JX, hY] + [hX, JY] + J[X, Y] - J[X, hY]$$

$$-J[hX, Y] - h[X, JY] - h[JX, Y],$$

for any vector fields X, Y on TM.

Let $X = X^i(\partial/\partial x^i) + \bar{X}^i(\partial/\partial v^i)$, $Y = Y^i(\partial/\partial x^i) + \bar{Y}^i(\partial/\partial v^i)$. Then a simple computation shows that

$$t(X, Y) = X^i Y^j((\partial\Gamma_i^k/\partial v^j) - (\partial\Gamma_j^k/\partial v^i))(\partial/\partial v^k). \tag{4.32}$$

From (4.32), we see that t is a skew–symmetric semibasic form. Then we define the **strong torsion** T of Γ to be the vector 1–form (or tensor field of type (1,1)) on TM given by

$$T = t^0 - \mathbf{H},$$

where $\mathbf{H}$ is the tension of Γ. We locally have

$$T = (v^i(\partial\Gamma_i^k/\partial v^j) - \Gamma_j^k)(\partial/\partial v^k) \otimes (dv^j), \tag{4.33}$$

which shows that T is also semibasic.

Proposition 4.7.2 *We have*

$$\xi^\star + T^0 = 0,$$

where ξ is the associated semispray to Γ.

Proof: In fact,

$$T^0 = (t^0 - \mathbf{H})^0 = (t^0)^0 - \mathbf{H}^0 = -\mathbf{H}^0 = -\xi^\star,$$

by Proposition 4.6.1. □

With respect to the adapted frame, we obtain

$$t(D_i, D_j) = ((\partial\Gamma^k_i/\partial v^j) - (\partial\Gamma^k_j/\partial v^i))V_k,$$

$$t(D_i, V_j) = t(V_i, V_j) = 0, \tag{4.34}$$

and

$$T(D_i) = (v^j(\partial\Gamma^k_j/\partial v^i) - \Gamma^k_i)V_k,\ T(V_i) = 0. \tag{4.35}$$

Proposition 4.7.3 *If $T = 0$, then $t = 0$.*

Proof: If $T = 0$, then from (4.35), we have

$$\Gamma^k_i = v^r(\partial\Gamma^k_r/\partial v^i).$$

Hence we obtain

$$t(D_i, D_j) = ((\partial\Gamma^k_i/\partial v^j) - (\partial\Gamma^k_j/\partial v^i))V_k$$

$$= ((\partial\Gamma^k_j/\partial v^i) + v^r(\partial^2\Gamma^k_r/\partial v^i\partial v^j)$$

$$-(\partial\Gamma^k_i/\partial v^j) - v^r(\partial^2\Gamma^k_r/\partial v^i\partial v^j))V_k$$

$$= ((\partial\Gamma^k_j/\partial v^i) - (\partial\Gamma^k_i/\partial v^j))V_k$$

$$= t(D_j, D_i).$$

Since $t(D_j, D_i) = -t(D_i, D_j)$, we deduce that $t(D_i, D_j) = 0$. Hence $t = 0$, since it is semibasic. □

Let Γ be a linear connection. Then (4.32) and (4.33) become

$$t(X,Y) = X^i Y^j(\Gamma^k_{ij} - \Gamma^k_{ji})(\partial/\partial v^k), \tag{4.36}$$

$$T(X) = v^i X^j(\Gamma^k_{ij} - \Gamma^k_{ji})(\partial/\partial v^k). \tag{4.37}$$

Now, let U, V be vector fields on M. We define the **torsion tensor T** of Γ by

$$\mathbf{T}(U,V)_x = \phi_y(t(U^H, V^H)_y), \ y \in T_x M, \ x \in M. \tag{4.38}$$

We note that (4.38) does not depends on the choice of y. In fact,

$$t(U^H, V^H) = U^i V^j(\Gamma^k_{ij} - \Gamma^k_{ji})(\partial/\partial v^k), \tag{4.39}$$

where $U = U^i(\partial/\partial x^i), V = V^i(\partial/\partial x^i)$, and the right–side of (4.39) depends only on x. If we develop (4.38) in local coordinates, we obtain

$$\mathbf{T}(U,V) = U^i V^j(\Gamma^k_{ij} - \Gamma^k_{ji})(\partial/\partial x^k),$$

$$= \nabla_U V - \nabla_V U - [U,V],$$

where ∇ is the covariant derivative defined by Γ. Then **T** is, precisely, the torsion tensor defined by ∇ (see Section 1.18).

4.8 Decomposition theorem

In this section, we shall prove that the strong torsion and the associated semispray characterize a connection in TM.

We shall need the following result.

Proposition 4.8.1 *Let Γ and Γ' be two connections in TM with the same strong torsion and the same associated semispray. Then $\Gamma = \Gamma'$.*

Proof: Let ξ, t, T and **H** (resp. ξ', t', T' and **H**$'$) be the associated semispray, weak torsion, strong torsion and tension of Γ (resp. Γ'). If we put $B = \Gamma' - \Gamma$, we have $JB = 0$ and $BJ = 0$. Therefore B is a semibasic vector 1–form on TM. Moreover

$$B\xi = \Gamma'\xi - \Gamma\xi = \xi' - \xi = 0,$$

since $\xi' = \xi$. Now, let h (resp. h') be horizontal projector corresponding to Γ (resp. Γ'). Then

$$t' = [J, h'] = [J, h] + (1/2)[J, B] = t + (1/2)[J, B],$$

$$\mathbf{H}' = (1/2)[C, \Gamma'] = \mathbf{H} + (1/2)[C, B],$$

$$T' = (t')^0 - \mathbf{H}' = T + (1/2)([J, B]^0 - [C, B]),$$

since $h' = h + (1/2)B$. But $T' = T$. Hence

$$[J, B]^0 = [C, B].$$

Therefore we deduce

$$BJ[\xi, X] - J[\xi, BX] - B[\xi, JX] = 0, \tag{4.40}$$

for every vector field X on TM. Since $BX \in Im\, J$ and B is semibasic, we obtain

$$J[\xi, BX] = -BX$$

by Proposition 4.3.7. On the other hand, we have

$$BJ[\xi, X] - B[\xi, JX] = B(J[\xi, X] - [\xi, JX])$$

$$= -B(L_\xi J)X = -B(L_\xi J)(\bar{h}X),$$

since B is semibasic (here $\bar{h}$ is the horizontal projector of $\bar{\Gamma} = -L_\xi J$). But

$$(L_\xi J)\bar{h} = -\bar{h}$$

Then we deduce

$$BJ[\xi, X] - B[\xi, JX] = B(\bar{h}X) = BX.$$

Thus (4.40) becomes

$$BX + BX = 2(BX) = 0.$$

So $BX = 0$, and then B vanishes. This ends the proof. □

Theorem 4.8.2 (Decomposition theorem). *Let ξ be a semispray and T a semibasic 1–vector form on TM such that $T^0 + \xi^\star = 0$. Then there exists a unique connection Γ in TM such that its associated semispray is ξ and its strong torsion is T. The connection Γ is given by*

$$\Gamma = -L_\xi J + T.$$

Proof: **EXISTENCE**. Let $\Gamma = -L_\xi J + T$. Then $J\Gamma = J$ and $\Gamma J = -J$, since T is semibasic. So Γ is a connection in TM. Now, if h is the horizontal projector of Γ, we have

$$\begin{aligned} h\xi &= (1/2)(I+\Gamma)(\xi) = (1/2)(I - L_\xi J)(\xi) + (1/2)T(\xi) \\ &= \xi + (1/2)\xi^\star + (1/2)T^0 \\ &= \xi, \end{aligned}$$

by Proposition 4.6.4.

Furthermore, the weak torsion of Γ is

$$t = [J,h] = \bar{t} + (1/2)[J,T]$$

where $\bar{t}$ is the weak torsion of $\bar{\Gamma} = -L_\xi J$. Then

$$t = 1/2[J,T]$$

(see Exercise 4.12.5). If we compute the tension $\mathbf{H}$ of Γ, we obtain

$$\mathbf{H} = (1/2)(L_C\Gamma) = \bar{\mathbf{H}} + (1/2)(L_C T)$$

$$= (1/2)(L_C T - L_{\xi^\star} T),$$

where $\bar{\mathbf{H}}$ is the tension of $\bar{\Gamma} = -L_\xi J$ (see Exercise 4.12.5).

Therefore the strong torsion of Γ is

$$T' = t^0 - \mathbf{H} = 1/2([J,T]^0 + [C,T] - [\xi^\star, J]).$$

But an easy computation shows that

$$([J,T]^0 + [C,T] - [\xi^\star, J])(X) = -T([\xi,J](X)) - J[\xi,TX].$$

Consequently, we have

$$T'X = T'(\bar{h}X) = -(1/2)\{T([\xi, J](\bar{h}X) + J[\xi, T(\bar{h}X)]\}$$

$$= -(1/2)\{-T(\bar{h}X) - T(\bar{h}X)\} = T(\bar{h}X) = TX,$$

since T and T' are semibasic and where $\bar{h}$ is the horizontal projector of $\bar{\Gamma}$. Then $T' = T$.

UNIQUENESS. It is a direct consequence of Proposition 4.8.1.□

Corollary 4.8.3 *Let Γ be a connection in TM. Then the strong torsion of Γ vanishes if and only if its weak torsion and tension vanish.*

Proof: If $T = 0$ then $\Gamma = -L_\xi J$ and therefore $t = 0$ (see Exercise 4.12.5). On the other hand, since

$$T = t^0 - \mathbf{H},$$

we obtain $\mathbf{H} = 0$.

The converse is trivial.□

Remark 4.8.4 From Corollary 4.8.3 we deduce that there are no non–homogeneous connections with zero strong torsion.

4.9 Curvature

In this section, we introduce the curvature form of a connection in TM. We shall show that the curvature is the obstruction to the integrability of the horizontal distribution determined by the connection.

Definition 4.9.1 *Let Γ be a connection in TM. The* **curvature form** *of Γ is the vector 2-form (or tensor field of type (1,2)) on TM given by*

$$R = -N_h,$$

where h is the horizontal projector of Γ.

Since $h^2 = h$, we have

$$R(X,Y) = -[hX, hY] + h[hX, Y] + h[X, hY] - h[X, Y]. \tag{4.41}$$

From (4.41), we obtain

$$R(X,Y) = X^i Y^j((\partial\Gamma^k_j/\partial x^i) - (\partial\Gamma^k_i/\partial x^j) + \Gamma^l_j(\partial\Gamma^k_i/\partial v^l)$$

$$- \Gamma^l_i(\partial\Gamma^k_j/\partial v^l))(\partial/\partial v^k), \tag{4.42}$$

where $X = X^i(\partial/\partial x^i) + \bar{X}^i(\partial/\partial v^i), Y = Y^i(\partial/\partial x^i) + \bar{Y}^i(\partial/\partial v^i)$.

From (4.42), we see that R is a skew–symmetric semibasic form. If Γ is a linear connection, then (4.42) becomes

$$R(X,Y) = X^i Y^j v^k((\partial\Gamma^k_{jr}/\partial x^i) - (\partial\Gamma^k_{ir}) + \Gamma^l_{jr}\Gamma^k_{il}$$

$$- \Gamma^l_{ir}\Gamma^k_{jl})(\partial/\partial v^k). \tag{4.43}$$

With respect to the adapted frame, we have

$$R(D_i, D_j) = ((\partial\Gamma^k_j/\partial x^i) - (\partial\Gamma^k_l/\partial x^j) + \Gamma^l_j(\partial\Gamma^k_i/\partial v^l))$$

$$- \Gamma^l_i(\partial\Gamma^k_j/\partial v^l))V_k, \tag{4.44}$$

$$R(D_i, V_j) = R(V_i, V_j) = 0.$$

Proposition 4.9.2 *Let Γ be a connection in TM. Then the horizontal distribution H is integrable if and only if $R = 0$.*

Proof: Suppose that $R = 0$. Then, from (4.41), we obtain

$$[hX, hY] = h[hX, Y] + h[X, hY] - h[X, Y]$$

$$= h([hX, Y] + [X, hY] - [X, Y]).$$

Hence H is integrable. Conversely, suppose that H is integrable. We have

$$R(hX, hY) = -[hX, hY] + h[hX, hY] \text{ (since } h^2 = h)$$

$$= -v[hX, hY]$$

$$= 0 \text{ (since } H \text{ is integrable)}$$

But, since R is semimbasic, then $R = 0$.□

Now, let Γ be a linear connection in TM. We define the **curvature tensor R** of Γ to be the tensor field of type (1,3) on M given by

$$\mathbf{R}_x(U,V,W) = \phi_U(R_U(V^H,W^H)), \tag{4.45}$$

where $U,V,W \in T_xM, x \in M$. If we develop (4.45) in local coordinates we easily obtain

$$\mathbf{R}(U,V,W) = \nabla_U\nabla_V W - \nabla_V\nabla_U W - \nabla_{[U,V]}W,$$

for any vector fields $U,V,W \in \xi(M)$. Then $\mathbf{R}$ is the curvature tensor of the linear connection ∇ on M defined by Γ.

4.10 Almost complex structures on tangent bundles (II)

In this section, we generalize for connections in TM the results obtained in Section 3.5, for linear connections.

Let Γ be a connection in TM with horizontal projector h. Then we define a tensor field F of type (1,1) on TM by

$$F(hX) = -JX,\ F(JX) = hX, \tag{4.46}$$

for any vector field X on TM. It follows from (4.46) that F **is an almost complex structure on** TM.

With respect to the adapted frame, we have

$$FD_i = -V_i,\ FV_i = D_i. \tag{4.47}$$

From a straightforward computation from (4.47), we easily see that

$$N_F(D_i, D_j) = ((\partial\Gamma_i^k/\partial v^j) - (\partial\Gamma_j^k/\partial v^i))D_k$$

$$-((\partial\Gamma_i^k/\partial x^j) - (\partial\Gamma_j^k/\partial x^i) + \Gamma_i^l(\partial\Gamma_j^k/\partial v^l) - \Gamma_j^l(\partial\Gamma_i^k/\partial v^l))V_k,$$

$$N_F(D_i, V_j) = -((\partial\Gamma_i^k/\partial v^j) - (\partial\Gamma_j^k/\partial v^i))V_k$$

$$-((\partial\Gamma_i^k/\partial x^j) - (\partial\Gamma_j^k/\partial x^i) + \Gamma_i^l(\partial\Gamma_j^k/\partial v^l) - \Gamma_j^l(\partial\Gamma_i^k/\partial v^l))D_k,$$

$$N_F(V_i,V_j) = ((\partial\Gamma_i^k/\partial x^j) - (\partial\Gamma_j^k/\partial x^i) + \Gamma_i^l(\partial\Gamma_j^k/\partial v^l) - \Gamma_j^l(\partial\Gamma_i^k/\partial v^l))V_k$$

$$-((\partial\Gamma_i^j/\partial v^j) - (\partial\Gamma_j^k/\partial v^i))D_k. \quad (4.48)$$

From (4.48), we obtain the following.

Theorem 4.10.1 *F is integrable if and only if $R = 0$ and $t = 0$.*

Corollary 4.10.2 *Let Γ be a linear connection in TM with covariant derivative ∇. Then the almost complex structure F associated to Γ coincide with the corresponding one defined by ∇. Therefore, F is integrable if and only if Γ is flat, that is, $\mathbf{R} = 0$ and $\mathbf{T} = 0$.*

Now, let $\bar{g}$ be a metric in the vertical bundle V and Γ a connection in TM.

Definition 4.10.3 *The* **Riemannian prolongation** *of $\bar{g}$ along Γ is the Riemannian metric g_Γ on TM defined by*

$$g_\Gamma(X,Y) = \bar{g}(JX,JY) + \bar{g}(vX,vY),$$

where v is the vertical projector of Γ.

It is easy to see that, in fact, g_Γ defines a Riemannian metric on TM.

The next proposition (which proof is omitted) characterizes the Riemannian prolongations of metrics in V along connections in TM.

Proposition 4.10.4 *A Riemannian metric g on TM is the Riemannian prolongation of a metric $\bar{g}$ in V along Γ if and only if*
(1) $g(hX,vY) = 0$,
(2) $g(hX,hY) = g(JX,JY) = \bar{g}(JX,JY)$,
for any vector fields X and Y on TM.

Next, let F be the almost complex structure determined by Γ. We have

Proposition 4.10.5 *(1) g_Γ is a Hermitian metric. (2) The Kähler form K_Γ of the almost Hermitian structure (TM,F,g_Γ) is given by*

$$K_\Gamma(X,Y) = g_\Gamma(JX,Y) - g_\Gamma(X,JY).$$

Proof: (1) In fact, we easily see that

$$g_\Gamma(FX, FY) = g_\Gamma(X, Y).$$

(2) The Kähler form K_Γ of (TM, F, g_Γ) is defined by

$$K_\Gamma(X, Y) = g_\Gamma(X, FY)$$

But

$$g_\Gamma(X, FY) = g_\Gamma(hX + vX, FhY + FvY)$$

$$= g_\Gamma(hX, FhY) + g_\Gamma(hX, FvY) + g_\Gamma(vX, FhY) + g_\Gamma(vX, FvY)$$

$$= g_\Gamma(hX, FvY) + g_\Gamma(vX, FhY)$$

$$= g_\Gamma(hX, hFY) - g_\Gamma(vX, JY),$$

since $Fv = hF$ and $Fh = -J$.

On the other hand

$$g_\Gamma(hX, hFY) = g_\Gamma(JX, JFY) \quad \text{(by Proposition 4.10.4)}$$

$$= g_\Gamma(JX, vY) \quad (\text{since } JF = v).$$

Hence

$$g_\Gamma(X, FY) = g_\Gamma(JX, vY) - g_\Gamma(vX, JY)$$

$$= g_\Gamma(JX, Y) - g_\Gamma(X, JY). \square$$

Now, let us suppose that (M, g) is a Riemannian manifold. We can define a metric $\bar{g}$ in V as follows:

$$\bar{g}_z(X^v, Y^v) = g_x(X, Y),$$

where $X, Y \in T_xM$, $z \in T_xM$. Let Γ be the Riemannian connection defined by g and $\tilde{g}$ the Sasaki metric on TM determined by g. It is easy to see that $\tilde{g}$ is the Riemannian prolongation of $\bar{g}$ along Γ. Then we reobtain the results of section 3.5.3.

Remark 4.10.6 The results of 4.10 has been extended for connections in tangent bundles of higher order in de León [29], de León et al. [32].

4.11 Connection in principal bundles

Let $P(M,G)$ be a principal bundle over M with structure group G and projection $\pi : P \longrightarrow M$.

Definition 4.11.1 *A* **connection** Γ *in* P *is a connection in the fibred manifold* $\pi : P \longrightarrow M$ *such that the horizontal distribution* H *is invariant by* G*, i.e.,*

$$H_{pa} = dR_a(p)H_p,$$

for all $p \in P$, $a \in G$.

Then, given a connection Γ in P, we have a decomposition

$$T_pP = H_p \oplus V_p, \text{ for all } p \in P,$$

where H_p (resp. V_p) is the horizontal (resp. vertical) subspace at p.

Let g be the Lie algebra of G. Then a connection Γ in P determines a 1–form ω on P with values in g as follows. For each $X \in T_pP$, we define $\omega(X)$ to be the unique $A \in g$ such that

$$(\lambda A)_p = vX,$$

where vX is the vertical component of X. The form ω is called the **connection form** of Γ. Obviously, $\omega(X) = 0$ if and only if X is horizontal.

Proposition 4.11.2 *The connection form* ω *satisfies the following conditions:*
(1) $\omega(\lambda A) = A$, *for all* $A \in g$.
(2) $R_a^{\star}\omega = Ad(a^{-1})\omega$, *i.e.,*

$$\omega((TR_a)X) = Ad(a^{-1})(\omega(X)), \text{ for all } X \in TP \text{ and } a \in G.$$

Conversely, given a g*–valued 1–form* ω *on* P *satisfying (1) and (2), there is a unique connection* Γ *in* P *whose connection form is* ω.

Proof: If ω is the connection form of a connection Γ, then (1) and (2) follows directly from the definition of ω and Proposition 1.20.6. Conversely, given a 1–form ω satisfying (1) and (2), we define

$$H_p = \{X \in T_pP / \omega(X) = 0\}. \square$$

Now, let $\{\psi_{\alpha\beta}\}$ be the transition functions corresponding to an open covering $\{U_\alpha\}$ of M (see Section 1.20). Let $\sigma_\alpha : U_\alpha \longrightarrow P$ be the section over U_α defined by $\alpha_\alpha(x) = \psi_\alpha^{-1}(x, e)$, $x \in U_\alpha$, where $\psi_\alpha : \pi^{-1}(U_\alpha) \longrightarrow U_\alpha \times G$ is the local trivialization of P over U_α. If θ is the canonical g–valued 1–form on G (see Exercise 1.22.21), then we define on $U_\alpha \cap U_\beta$ a g–valued 1–form $\theta_{\alpha\beta}$ by

$$\theta_{\alpha\beta} = \psi^\star_{\alpha\beta}\theta.$$

For each α, we define a g–valued 1–form ω_α on U_α by

$$\omega_\alpha = \sigma^\star_\alpha \omega.$$

It is not hard to prove that

$$\omega_\beta = Ad(\psi_{\alpha\beta}^{-1})\omega_\alpha + \theta_{\alpha\beta} \text{ on } U_\alpha \cap U_\beta. \tag{4.49}$$

(see Kobayashi and Nomizu [81] for a detailed proof).

Next we shall define the curvature form of a connection Γ in P.

Definition 4.11.3 *Let ω be the connection form of Γ. Then the* **curvature form** *of Γ is defined to be the g–valued 2–form Ω on P given by*

$$\Omega(X, Y) = d\omega(hX, hY), \text{ for all } X, Y \in TP.$$

It is easy to prove that:

- Ω is **horizontal**, i.e., $\Omega(X, Y) = 0$ whenever at least one of the tangent vectors X, Y are vertical.

- $R^\star_a\Omega = Ad(a^{-1})\Omega$, for all $a \in G$.

Proposition 4.11.4 (Structure equation) *We have*

$$d\omega(X, Y) = -\frac{1}{2}[\omega(X), \omega(Y)] + \Omega(X, Y),$$

for all $X, Y \in TP$.

The proof is left to the reader as an exercise.

The structure equation is sometimes written, for the sake of simplicity, as

$$d\omega = -\frac{1}{2}[\omega, \omega] + \Omega.$$

Now, let $\pi : E \longrightarrow M$ be a vector bundle of rank n associated with P (see Exercise 1.22.27). Then a connection Γ in P induces a connection $\tilde{\Gamma}$ in E as follows. For each $e \in E$, choose $(p, \xi) \in P \times R^n$ such that $e = [p, \xi]$. Consider the map

$$P \xrightarrow{F} E$$

defined by $F(q) = [q, \xi]$. Then the horizontal subspace $\tilde{H}_e$ at e is defined to be the image of H_p by F, i.e.,

$$\tilde{H}_e = dF(p)H_p.$$

One can easily prove that $\tilde{H}$ defines a connection in E.

Next, we consider the frame bundle FM of an m–dimensional manifold M.

Let Γ be a connection in FM with connection form ω. With respect to the canonical basis $\{E^i_j\}$ of $gl(m, R)$, $m = dim\, M$, we set

$$\omega = \omega^i_j E^j_i,$$

where ω^i_j are 1–forms on FM.

Let (U, x^i) be a coordinate neighborhood in M and $\sigma : U \longrightarrow FM$ the section of FM given by

$$\sigma(x) = \{(\partial/\partial x^1)_x, \ldots, (\partial/\partial x^m)_x\}.$$

We set

$$\sigma^\star \omega^i_j = \Gamma^i_{jk} dx^j,$$

where the m^3 functions Γ^i_{jk} are called the **Christoffel components** of Γ.

Let $(U, x^i), (\bar{U}, \bar{x}^i)$ be two coordinate neighborhoods with $U \cap \bar{U} \neq \emptyset$. If Γ^k_{jk}, $\bar{\Gamma}^i_{jk}$ are the corresponding Christoffel components, a direct computation from (4.49) shows that

$$\bar{\Gamma}^\alpha_{\beta\gamma} = \Gamma^i_{jk} \frac{\partial x^j}{\partial \bar{x}^\beta} \frac{\partial x^k}{\partial \bar{x}^\gamma} \frac{\partial \bar{x}^\alpha}{\partial x^i} + \frac{\partial^2 x^i}{\partial \bar{x}^\beta \partial \bar{x}^\gamma} \frac{\partial \bar{x}^\alpha}{\partial x^i}. \qquad (4.50)$$

The components Γ^i_{jk} permit us to reconstruct the connection Γ. In fact, with respect to the induced coordinate system (x^i, X^i_j) on FM we have

$$\omega^i_j = Y^i_k(dX^k_j + \Gamma^k_{\alpha\beta} X^\beta_j dx^\alpha), \tag{4.51}$$

where (Y^i_j) is the inverse matrix of (X^i_j). Hence, if for each coordinate system (U, x^i) on M, we have a set of functions Γ^i_{jk} satisfying (4.50), then we can define ω^i_j by (4.51) and ω by $\omega = \omega^i_j E^j_i$.

Now, let Γ be a connection in FM. Since TM is a vector bundle associated with FM, then Γ induces a connection $\tilde{\Gamma}$ in TM. It is a straigthforward computation to prove that $\tilde{\Gamma}$ is in fact a linear connection on M. Conversely, let $\tilde{\Gamma}$ be a linear connection on M. Then, proceeding as above, we can construct a connection in FM from the Christoffel components Γ^i_{jk}. Thus, a **linear connection on M is a connection in FM.**

4.12 Exercises

4.12.1 Prove that the vertical lift X^v to TM of a vector field X on M is homogeneous of degree -1.
4.12.2 Prove that for any semispray ξ on TM and any vector field X on M $[X^c, \xi]$ is a vertical vector field and $[X^v, \xi]$ projects onto X.
4.12.3 Prove that if $\tilde{X}$ is a vector field on TM and ξ a semispray such that $[\tilde{X}, \xi] = 0$ and $J\tilde{X} = X^v$, then $\tilde{X}$ is the complete lift of X, i.e., $\tilde{X} = X^c$.
4.12.4 Let ξ be a semispray on TM. Prove that

$$[i_\xi, d_J] = i_\xi d_J + d_J i_\xi = L_C - i_{[\xi, J]}$$

4.12.5 Let $\Gamma = -L_\xi J$ be the connection in TM defined by an arbitrary semispray ξ on TM. We denote by t, $\mathbf{H}$ and T the weak torsion, tension and strong torsion of Γ, respectively. Prove that: (i) t vanishes; (ii) $\mathbf{H} = -(1/2)[\xi^\star, J]$; (iii) $T = (1/2)[\xi^\star, J]$, where $\xi^\star$ is the deviation of ξ.
4.12.6 Let Γ be a connection in a principal bundle $P(M, G)$.

(i) Prove that the horizontal lift X^H to P of a vector field X on M is invariant by G, i.e., $(TR_a)X^H = X^H$ for all $a \in G$.

(ii) Prove that the horizontal component of $[X^H, Y^H]$ is $[X, Y]^H$.

(iii) Let A be an element of the Lie algebra of G and $\tilde{X}$ a horizontal vector field on P. Prove that $[\tilde{X}, \lambda A]$ is horizontal. If, in particular, $\tilde{X}$ is the horizontal lift X^H of a vector field X on M, prove that $[X^H, \lambda A] = 0$.

4.12.7 Let $B_G(M)$ be a G–structure on M. Consider a connection Γ in $B_G(M)$. Then we can define a connection $\tilde{\Gamma}$ in FM (i.e., a linear connection on M) as follows. For each $p \in FM$, let be $q \in B_G(M)$ and $a \in G$ such that $p = qa$. Thus we define the horizontal subspace $\tilde{H}_p$ by $\tilde{H}_p = TRa(H_q)$, where H_q is the horizontal subspace at q defined by Γ. Prove that $\tilde{\Gamma}$ is in fact a connection in FM (Γ is called a G–**connection**).

4.12.8 (i) Let $P(M,G)$ be a principal bundle. Prove that there always exists a connection in P (*Hint*: use partitions of unity).

(ii) Prove that, if $B_G(M)$ is a G–structure on M, then there always exists a G–connection on M.

(iii) Let J be an almost complex structure on a 2n–dimensional manifold M. Prove that a linear connection Γ is a $Gl(n,C)$–connection if and only if $\nabla J = 0$, where ∇ is the covariant derivative associated with Γ.

(iv) Extend (iii) for other polynomial structures (almost tangent structures, almost product structures, etc...).æ

Chapter 5

Symplectic manifolds and cotangent bundles

5.1 Symplectic vector spaces

Let V be a real vector space of dimension m and ω a skew–symmetric bilinear form (2–form) on V, i.e., $\omega \in \wedge^2 V$. Consider the linear mapping $S_\omega : V \longrightarrow V^\star$, where $V^\star$ is the dual space of V, defined by

$$u \longrightarrow S_\omega(u) = i_u \omega. \tag{5.1}$$

Here $i_u\omega$ is the interior product of the vector u by the form ω, i.e.,

$$(i_u\omega)(v) = \omega(u, v), \; v \in V.$$

Let $Im S_\omega$ be the image of S_ω and $ker\, S_\omega = \{u \in V / i_u\omega = 0\}$ the kernel of S_ω. The **rank** of ω, denoted by rank ω, is the dimension of $Im\, S_\omega$. The dimension of $ker\, S_\omega$ is called **corank** of ω, denoted by corank ω. If in particular corank $\omega = 0$ then $dim\, V = rank\, \omega$. In such a case we say that ω is **non–degenerate**, of **maximal rank** or even **regular**. It is possible to show, without major difficulties the following assertion: "A necessary and sufficient condition for $\omega \in \wedge^2 V$ be non–degenerate is that $S_\omega : V \longrightarrow V^\star$ defined by (5.1) be an isomorphism".

Proposition 5.1.1 *Let $\omega \in \wedge^2 V$. Then there is a basis $\{u_i \mid 1 \leq i \leq m = dim\, V\}$ for V and an integer $2s \leq m$ such that*
(1) $\omega(u_i, u_{s+i}) = -\omega(u_{s+i}, u_i) = 1$, for all $i \leq s$,
(2) for all other values, $\omega(u_i, u_j) = 0$.

Proof: The proof is obtained by induction on dimension of V. The result is evident if $dim V = 2$. Let $u_1 \in V$ be an arbitrary vector. Then there is a vector u_2 such that $\omega(u_1, u_2) \neq 0$, since ω is non zero. We may choose u_2 in such a way that $\omega(u_1, u_2) = -\omega(u_2, u_1) = 1$ (because, if necessary, we may change u_2 by a numerical factor). Suppose that $m \geq 2$. We may choose two vectors u_1 and u_{s+1} such that $\omega(u_1, u_{s+1}) = -\omega(u_{s+1}, u_1) = 1$. Let F be the space spanned by these vectors. Let W defined by

$$W = \{v \in V / \omega(v, u_1) = \omega(v, u_{s+1}) = 0\}.$$

Then $V = F \oplus W$. Hence, by induction, there is an integer $2s \leq m$ and a basis $\{u_i, u_{s+i}, u_j / 2 \leq i \leq s,\ 2s+1 \leq j \leq m\}$ for W such that (1) and (2) hold. Then taking $\{u_i, u_{s+i}, u_j / 1 \leq i \leq s,\ 2s+1 \leq j \leq m\}$ as a basis for V one has the desired result. □

Let $\{u_i^\star / 1 \leq i \leq m\}$ be the corresponding dual basis for $V^\star$ of $\{u_i / 1 \leq i \leq m\}$. Then a direct computation obtained from the action of the forms $u_i^\star \wedge u_{s+i}^\star, 1 \leq i \leq s$, on the pair of vectors (u_j, u_l), shows that

$$\omega = u_1^\star \wedge u_{s+1}^\star + \ldots + u_s^\star \wedge u_{2s}^\star. \tag{5.2}$$

Thus, rank $\omega = 2s$. Also, if we take the s–exterior product of ω

$$\omega^s = \omega \wedge \ldots \wedge \omega = -(s)!\, u_1^\star \wedge u_2^\star \wedge \ldots \wedge u_{2s}^\star.$$

we see that $\omega^s \neq 0$ and $\omega^{s+1} = 0$. In fact, if these two last properties hold for $\omega \in \wedge^2 V$ then ω has rank $2s$. On the other hand, we have

$$S_\omega(u_i) = u_{s+i}^\star,\ S_\omega(u_{s+i}) = -u_i^\star,\ S_\omega(u_j) = 0,$$

$$1 \leq i \leq s,\ 2s+1 \leq j \leq m.$$

From these comments we have the following.

Theorem 5.1.2 *The rank of every 2–form ω on V is an even number. The following assertions are equivalent:*
(1) rank $\omega = 2s \leq m = dim\, V$,
(2) the power $\omega^s \neq 0$ and $\omega^{s+1} = 0$,
(3) there are $2s$ linear independent forms (1–forms) on V, denoted by $u_1^\star, \ldots, u_{2s}^\star$, such that

$$\omega = u_1^\star \wedge u_{s+1}^\star + \ldots + u_s^\star \wedge u_{2s}^\star.$$

Also, under these conditions, $\{u_1^\star, \ldots, u_{2s}^\star\}$ is a basis for $Im\, S_\omega$.

Definition 5.1.3 *A* **symplectic structure** *on a finite (real) vector space V is given by a 2–form ω on V such that ω is non–degenerate. The form ω is called* **symplectic** *and the pair (V, ω) a* **symplectic vector space**.

From the preceding considerations, we easily see that if (V, ω) is a symplectic vector space then dim V is an even number, say $2n$. The following proposition is a consequence of the results given above:

Proposition 5.1.4 *Let V be a vector space of even dimension $2n$, and ω a 2–form on V. Then the following assertions are equivalent:*
(1) (V, ω) is a symplectic vector space,
(2) $S_\omega : V \longrightarrow V^*$ is an isomorphism,
(3) ω^n is a volume form on V.

From Theorem 5.1.2 and equality (2) of Proposition 5.1.4 we see that if ω is a symplectic form on V there is a basis $\{u_i \mid 1 \leq i \leq 2n\}$ for V such that

$$\omega = \sum_{i=1}^{n} u_i^* \wedge u_{n+i}^*.$$

We call such basis **symplectic**. The matrix of ω with respect to a symplectic basis is

$$\mathcal{M} = \begin{bmatrix} 0 & I_n \\ -I_n & 0 \end{bmatrix}.$$

Definition 5.1.5 *Let U and V be vector spaces and $h : U \longrightarrow V$ a linear mapping. Suppose that α and ω are symplectic forms on U and V, respectively. We say that h is* **symplectic** *if the linear adjoint mapping $h^* : V^* \longrightarrow U^*$ is such that $h^*\omega = \alpha$, i.e.,*

$$\omega(hu, hv) = \alpha(u, v), \text{ for all } u, v \in U.$$

A symplectic linear mapping is injective. If dim U = dim V then h is called **symplectic isomorphism**. In particular, if $U = V$ then h is said **symplectic automorphism**, and we say that h **preserves** the symplectic form: $h^*\omega = \omega$.

Definition 5.1.6 *The group of the symplectic automorphisms of a symplectic vector space (V, ω), endowed with the rule of composition of maps, is called* **symplectic group** *and it is denoted by $Sp(V)$.*

Proposition 5.1.7 *If (V, ω) is a symplectic vector space and $h \in Sp(V)$ then det $h = 1$.*

Proof: We remark that if $h \in Sp(V)$ then h preserves the volume form ω^n, i.e., $h^*(\omega^n) = \omega^n$. Since $\omega^n = h^*\omega^n = (\det h)\omega^n$, then det $h = 1$ (with respect to the volume form ω^n). □

The reader is invited to show the following:

Proposition 5.1.8 *If h is an automorphism of V and H is its matrix with respect to a symplectic basis then h is a symplectic automorphism if and only if*

$$H^t \mathcal{M} H = \mathcal{M}.$$

From Proposition 5.1.7 one has that $Sp(V)$ is isomorphic to the group of matrices $Sp(n) = \{H \in Gl(2n, R) \mid H^t \mathcal{M} H = \mathcal{M}\}$ which is a closed subgroup of $Gl(2n, R)$. Thus $Sp(V)$ admits a Lie group structure.

Remark 5.1.9 Let $V = R^{2n}$ and

$$\omega_0 = \sum_{i=1}^{n} dx^i \wedge dx^{n+i},$$

where $(x^1, \ldots, x^{2n})$ are the canonical coordinates on R^{2n}. Then (R^{2n}, ω_0) is a symplectic vector space. We can easily check that

$$Sp(n) = \{H \in Gl(2n, R)/\omega_0(Hx, Hy) = \omega_0(x, y),\ x, y \in R^{2n}\}$$

where $H : R^{2n} \longrightarrow R^{2n}$. Then the Lie algebra of $Sp(n)$ is

$$sp(n) = \{H \in gl(2n, R)/\omega_0(Hx, y) + \omega_0(x, Hy) = 0,\ x, y \in R^{2n}\}.$$

Definition 5.1.10 *Let V be a vector space of finite dimension, $\omega \in \wedge^2 V$ and K a subspace of V. Then the subspace*

$$K^\perp = \{u \in V/\omega(u, v) = 0,\ for\ all\ v \in K\}$$

is called the **orthocomplement** *of K in V with respect to ω. (In particular, if $v \in V$, then we may define $v^\perp = \{u \in V/\omega(u, v) = 0\}$).*

One has:

$$\text{Ker } S_\omega = V^\perp \text{ implies corank } \omega = \dim V^\perp;$$
$$\dim V + \dim (V^\perp \cap K) = \dim K + \dim K^\perp;$$
$$\text{in particular, if } \omega \text{ is symplectic then } \dim V = \dim K + \dim K^\perp.$$

Definition 5.1.11 *Let (V, ω) be a symplectic vector space and K a subspace of V. We say that*
(1) K is **isotropic** *if $K \subset K^\perp$;*
(2) K is **coisotropic** *if $K^\perp \subset K$;*
(3) K is **Lagrangian** *if K is a maximal isotropic subspace of (V, ω);*
(4) K is **symplectic** *if $K \cap K^\perp = 0$.*

Proposition 5.1.12 *Let (V, ω) be a symplectic vector space. A necessary and sufficient condition for a vector subspace $K \subset V$ to be Lagrangian is $K = K^\perp$.*

Proof: If K is a proper subspace of $K^\perp$, $K \neq K^\perp$, then K is not maximal. In fact, we may choose $u \in K^\perp$ such that $u \notin K$ and so

$$\omega(v_1 + a_1 u, v_2 + a_2 u) = 0$$

for all $v_1, v_2 \in K$ and $a_1, a_2 \in R$. Then $K + < u >$ is also isotropic.□

It is clear that if K is Lagrangian in (V, ω) then its complement K' in V is also Lagrangian. Also from the proof of Proposition 5.1.12 one has that every finite dimensional symplectic vector space has a Lagrangian subspace.

Proposition 5.1.13 *Suppose that K is an isotropic subspace of a vector symplectic space (V, ω). K is a Lagrangian subspace if and only if dim $K =$ $(1/2)$ dim V.*

Proof: If V is symplectic then $\dim V = 2n$. If K is Lagrangian then $K = K^\perp$. Thus $\dim V = 2 \dim K$. Conversely, suppose $\dim K = (1/2) \dim V$, that is, $\dim K = n$. Then $n = \dim K^\perp$ (since $\dim V = \dim K + \dim K^\perp$), and so $K = K^\perp$ because K is isotropic.□

Now, let K be a vector space of dimension n. Consider $V = K \oplus K^*$, where K^* is the dual space of K. We may define a symplectic form on V as follows:

$$\omega(u + \alpha, v + \beta) = \alpha(v) - \beta(u), \; u, v \in K, \; \alpha, \beta \in K^*.$$

One easily proves that K is a Lagrangian subspace of (V, ω) and its Lagrangian complement is, precisely, K^*. Conversely, let (V, ω) be a symplectic vector space and K a Lagrangian subspace in (V,ω). By K' we represent the complement of K and consider the isomorphism from K' to $K^\star$ given by $\bar{S}_\omega(v) = i_v\omega$, i.e., $\bar{S}_\omega$ is the restriction of S_ω composed with the canonical projection of $V^\star$ onto $K^\star$. Then, for all $u, v \in K$, $\bar{u}, \bar{v} \in K'$

$$\omega(u+\bar{u}, v+\bar{v}) = \omega(u,\bar{v}) + \omega(\bar{u}, v) = \bar{S}_\omega(\bar{u})v - \bar{S}_\omega(\bar{v})u.$$

Set $\bar{S}_\omega(\bar{u}) = \alpha$, $\bar{S}_\omega(\bar{v}) = \beta$. Then a 2–form ω_K on $K \oplus K^\star$ is defined by

$$\omega_K(u+\alpha, v+\beta) = \alpha(v) - \beta(u).$$

Moreover, as the mapping $1 \oplus \overline{S_\omega}$ is an isomorphism from $K \oplus K'$ onto $K \oplus K^\star$, ω_K is a symplectic form on $K \oplus K^\star$ such that

$$(1 \oplus \bar{S}_\omega)^\star \omega_K = \omega, \text{ i.e.}$$

the following diagram

$$\begin{array}{ccc} (K \oplus K') \times (K \oplus K') & \xrightarrow{\ \omega\ } & \\ \big\downarrow (1 \oplus \bar{S}_\omega) \times (1 \oplus \bar{S}_\omega) & & R \\ (K \oplus K^\star) \times (K \oplus K^\star) & \xrightarrow{\ \omega_K\ } & \end{array}$$

is commutative. Hence $1 \oplus \overline{S_\omega} : V = K \oplus K' \longrightarrow K \oplus K^\star$ is a symplectic isomorphism and V may be identified to $K \oplus K^\star$.

Next, we extend these definition to vector bundles.

Definition 5.1.14 *Let (E,p,M) be a vector bundle. Suppose that, for each $x \in M$, there exists a symplectic form $\omega(x)$ on the vector space $E_x = p^{-1}(x)$ such that the assignement $x \longrightarrow \omega(x)$ is C^∞ (i.e., if s_1 and s_2 are two C^∞ sections of E, then $\omega(s_1,s_2)$ defined by $\omega(s_1,s_2)(x) = \omega(x)(s_1(x), s_2(x))$ is a C^∞ function on M). Then E is said to be* **symplectic**. *Now, let (E,p,M) be a symplectic vector bundle and K a subbundle of E. We define a new subbundle $K^\perp$ of E by*

$$(K^\perp)_x = \{e \in E_x / \omega(x)(e,e') = 0, \ for\ all\ e' \in E_x\},$$

$x \in M$. Then K is said to be **isotropic**, *resp.* **coisotropic**, *resp.* **Lagrangian**, *resp.* **symplectic** *if K_x is isotropic, resp. coisotropic, resp. Lagrangian, resp. symplectic, for every $x \in M$.*

Suppose now that V is a vector space and ω a 2–form on V of rank $2s$ but not necessarily of maximal rank. Let dim $V = 2s + r$, $r \geq 0$. Then (V, ω) is said to be a **presymplectic vector space** (if $r = 0$, then (V, ω) is symplectic). The form ω is said to be **presymplectic**. If we consider the linear map $S_\omega : V \longrightarrow V^\star$ then S_ω is not necessarily an isomorphism.

Definition 5.1.15 *A linear mapping $P : V \longrightarrow V$ is a* **projector** *on V if $P^2 = P$. If P is a projector then $V = Im\, P \oplus ker\, P$.*

If P is a projector then we may define its complement $Q = Id - P$. Then $Q^2 = Q$ and $PQ = QP = 0$. Also, $Im\, P = ker\, Q$, $ker\, P = Im\, Q$. Let us set

$$V_P = Im\, P,\ V_Q = Im\, Q.$$

Then $V = V_P \oplus V_Q$. If $P^\star$ and $Q^\star$ are the adjoint operators on $V^\star$ we also have $V^\star = V_{P^\star} \oplus V_{Q^\star}$, where $P^\star\alpha = \alpha \circ P$, $Q^\star\alpha = \alpha \circ Q$, $\alpha \in V^\star$.

Definition 5.1.16 *We say that the projector P is* **adapted** *to the presymplectic form ω on V if*

$$Ker\ P = V_Q = ker\ S_\omega.$$

Remark 5.1.17 If V is a vector space with an inner product $<,>$ and ω is a presymplectic form on V then we may take V_P as being the orthonormal complement of $ker\ S_\omega$ with respect to $<,>$. This gives an adapted projector on V.

Proposition 5.1.18 *Let (V, ω) be a presymplectic vector space with an adapted projector $P : V \longrightarrow V$. If $\alpha \in V^\star$, then there exists a unique vector $v \in V_P$ such that*

$$i_v\omega = P^\star\alpha.$$

In particular, if $\alpha \in V_{P^\star}$, then $i_v\omega = \alpha$.

Proof: As $ker\ S_\omega = V_Q$ one has that the restriction of S_ω to the subspace V_P is injective, and so

$$(S\omega/_{V_P})(V_P) = V_{P^\star}.$$

as $P^*S_\omega(v) = S_\omega(v)$ for all $v = P(v) \in V_P$. Let us show this assertion. If $u \in V$, since ker $S_\omega = V_Q$, one has

$$(S_\omega(v))u = \omega(v,u) = \omega(v, Pu + Qu) = \omega(v, Pu) + \omega(v, Qu)$$

$$= \omega(v, Pu) = (S_\omega(v))(Pu) = P^*(S_\omega(v))(u)$$

and so $S_\omega(v) = P^*(S_\omega(v))$ (recall that $Qu \in Ker\ S\omega$). Therefore S_ω induces an isomorphism from V_P to V_{P^*}.□

5.2 Symplectic manifolds

Let S be a C^∞ manifold of dimension m, TS (resp. T^*S) its tangent (resp. cotangent) bundle with canonical projections $\tau_S : TS \longrightarrow S$ (resp. $\pi_S : T^*S \longrightarrow S$). Let ω be a 2–form on S. The **rank** (resp. **corank**) **of ω at a point x of S** is the rank (resp. corank) of the form $\omega(x) \in \wedge^2(T_xS)$. We say that ω is **non–degenerate** or of **maximal rank** if for every point $x \in S$, $\omega(x)$ is non–degenerate.

Definition 5.2.1 *An* **almost symplectic form** *(or* **almost symplectic structure***) on a manifold S is a non–degenerate 2–form ω on S. The pair (S,ω) is called an* **almost symplectic manifold**. *Then S has even dimension, say $2n$.*

Let (S,ω) be an almost symplectic manifold of dimension $2n$. Then, for each $x \in S$, $(T_xS, \omega(x))$ is a symplectic vector space. Thus there exists a symplectic basis $\{e_1, \ldots, e_{2n}\}$ for T_xS, which is called a **symplectic frame** at x. Let B be the set of all symplectic frames at all the points of S. If $\{e_i\}, \{e_i'\}$ are two symplectic frames at x, then they are related by a matrix $A \in Sp(n)$. Further, by using an argument as in section 1, we can find a local section of FM over a neighborhood of each point of M which takes values in B. Hence B is a $Sp(n)$–structure on S. Conversely, let $B_{S_{p(n)}}$ be a $Sp(n)$–structure on S. Then we can define a 2–form ω on S as follows:

$$\omega(x)(X,Y) = \omega_0(z^{-1}X, z^{-1}Y),\ X, Y \in T_xM,$$

where $z \in B_{Sp(n)}$ is a linear frame at x. (Obviously, $\omega(x)$ is independent on the choice of the linear frame $z \in B_{Sp(n)}$ at x). Since ω_0 is non–degenerate, then ω is an almost symplectic form on S.

Summing up, we have proved the following.

Proposition 5.2.2 *Giving a symplectic structure is the same as giving a $Sp(n)$-structure.*

Let (S, ω) be an almost symplectic manifold of dimension $2n$. Then

$$\omega^n = \omega \wedge \ldots \wedge \omega \ (n \text{ times})$$

is a volume form on S. Thus we have

Proposition 5.2.3 *Every almost symplectic manifold is orientable.*

Next we define a vector bundle homomorphism

$$S_\omega : TS \longrightarrow T^*S$$

by

$$S_\omega(X) = i_X(\omega(x)), \ X \in T_xS, \ x \in M.$$

Proposition 5.2.4 *S_ω is a vector bundle isomorphism.*

Proof: Let (u, x^i) be a coordinate neighborhood of S. Then we have induced local coordinates (x^i, v^i), (x^i, p_i) on TU, T^*U, respectively. Suppose that

$$\omega = \sum_{1 \leq i,j \leq 2n} \omega_{ij} dx^i \wedge dx^j,$$

where $\omega_{ij} = -\omega_{ji}$. Hence we have

$$S_\omega(\partial/\partial x^i) = \omega_{ij} dx^j.$$

Thus the map S_ω is locally given by

$$S_\omega(x^i, v^i) = (x^i, v^i \omega_{ij}).$$

Then S_ω is C^∞ and rank S_ω = rank $\omega = 2n$. Therefore we have the required result. □

Furthermore, ω defines a linear mapping (also denoted by S_ω)

$$S_\omega : \chi(S) \longrightarrow \wedge^1 S$$

given by

$$S_\omega(X) = i_X \omega.$$

An easy computation shows that S_ω is, in fact, an isomorphism of $C^\infty(S)$-modules.

Definition 5.2.5 *An almost symplectic form (or structure) ω on a manifold S is said to be* **symplectic** *if it is closed, i.e., $d\omega = 0$. Then the pair (S, ω) is called a* **symplectic manifold**.

Remark 5.2.6 If (S, ω) is an almost symplectic manifold, then TS is a symplectic vector bundle. If, in addition, ω is closed, then TS is a symplectic vector bundle such that $d\omega = 0$.

The reader may take notice of the study developed in the preceding section for vector spaces to reobtain some results in terms of the vector bundle structure of the tangent bundle of a given C^∞ finite dimensional manifold. For example, a submanifold K of a symplectic manifold (S, ω) is called **isotropic**, resp. **coisotropic**, resp. **Lagrangian**, resp. **symplectic** in (S, ω) if $T_xK \subset (T_xK)^\perp$, resp. $(T_xK)^\perp \subset T_xK$, resp. if it is a maximal isotropic submanifold of S, resp. if $(T_xK) \cap (T_xK)^\perp = 0$ for each $x \in K$. We have $dim\ K \leq n$, resp. $dim\ K \geq n$, resp. $dim\ K = n$, if $dim\ S = 2n$. We will return to Lagrangian submanifolds in the next chapter.

Remark 5.2.7 Obviously, K is an isotropic, resp. coisotropic, resp. Lagrangian, resp. symplectic submanifold of (S, ω) if and only if the tangent bundle TK is an isotropic, resp. coisotropic, resp. Lagrangian, resp. symplectic, vector subbundle of TS.

Definition 5.2.8 *Let (S, ω) and (W, α) be symplectic manifolds of same dimension, say $2n$. A differentiable mapping $h : S \longrightarrow W$ is called* **symplectic transformation** *if $h^\star \alpha = \omega$, i.e.,*

$$\alpha(dh(x)X_1, dh(x)X_2) = \omega(X_1, X_2),$$

for all $x \in S$ and $X_1, X_2 \in T_xS$.

For a symplectic mapping $h : S \longrightarrow W$ one has that $dh(x) : T_xS \longrightarrow T_{h(x)}W$ is a symplectic isomorphism. Thus h is a local diffeomorphism. If h is a global diffeomorphism then h is said to be C^∞**symplectic diffeomorphism** (or **symplectomorphim**). In particular, when $S = W$ then a symplectic map $h : S \longrightarrow W$ preserves the symplectic form ω on S, i.e., $h^\star \omega = \omega$. In such a case h is said to be a **canonical transformation**. This definition is more general than the ones adopted in Classical Mechanics (which states that a transformation is canonical if it preserves the Hamilton equations, see Arnold [4]).

Definition 5.2.9 *Let (S,ω) be a symplectic manifold. A vector field X on S is called a* **symplectic vector field** *(or an* **infinitesimal symplectic transformation***) if its flow consists of symplectic transformations.*

Proposition 5.2.10 *The following assertions are equivalent:*
(1) X is a symplectic vector field;
(2) the Lie derivative $L_X\omega = 0$;
(3) $i_X\omega = df$ (locally) for some function f, i.e., $d(i_X\omega) = 0$.

Proof: The equivalence of (1) and (2) follows from the definition of Lie derivative and from the fact that φ_t, the flow of X, is symplectic:

$$L_X\omega = \frac{d}{dt}(\varphi_t^\star\omega)/_{t=0} = \lim_{t\longrightarrow 0}\left(\frac{\varphi_t^\star\omega - \omega}{t}\right) = 0.$$

The equivalence of (2) and (3) follows from the H. Cartan formula

$$L_X\omega = (i_X d + d i_X)\omega = d i_X\omega$$

and the Poincaré lemma. ⌐

5.3 The canonical symplectic structure on the cotangent bundle

In this section we shall prove that the cotangent bundle of a manifold carries a natural symplectic structure.

Let M be an n–dimensional manifold, $T^\star M$ its cotangent bundle and $\pi_M : T^\star M \longrightarrow M$ the canonical projection. We define a canonical 1–form λ_M on $T^\star M$ as follows:

$$\lambda_M(p)(X) = p(x)(d\pi_M(p)X),\ X \in T_p(T^\star M),\ p \in T_x^\star M.$$

If (q^i) are coordinates in M and (q^i, p_i) are the induced coordinates in $T^\star M$, we obtain

$$\lambda_M(q^j, p_j)(\partial/\partial q^i) = \left(\sum_j p_j dq^j\right)(\partial/\partial q^i) = p_i,$$

$$\lambda_M(q^j, p_j)(\partial/\partial p_i) = 0.$$

Then λ_M is locally expressed by

$$\lambda_M = \sum_i p_i dq^i$$

Definition 5.3.1 *λ_M is called the* **Liouville form** *on T^*M.*

The following proposition gives an important property of the Liouville form.

Proposition 5.3.2 *The Liouville form λ_M on T^*M is the unique 1–form on T^*M such that*

$$\beta^* \lambda_M = \beta, \tag{5.3}$$

for any 1–form β on M.

Proof: Suppose that β is locally given by

$$\beta = \beta_i dq^i, \ i.e.,$$

$$\beta : M \longrightarrow T^*M, \ \beta(q^i) = (q^i, p_i).$$

Then we have

$$\beta^* (\lambda_M) = \beta^* (p_i dq^i) = (p_i \circ \beta) d(q^i \circ \beta) = \beta_i dq^i = \beta.$$

Furthermore, let λ be a 1–form on T^*M such that (5.3) holds for λ. If λ is locally given by

$$\lambda = \sum_i a_i dq^i + \sum_i b_i dp_i,$$

and $\beta = \sum_i \beta_i dq^i$ is an arbitrary 1–form on M. We obtain

$$\beta^* \lambda = \sum_i a_i (q^j \circ \beta_j) dq^i + \sum_i b_i (q^j \circ \beta_j) d\beta_i$$

$$= \sum_i \left(a_i + b_j \frac{\partial \beta_j}{\partial q^i} \right) dq^i$$

$$= \sum_i \beta_i dq^i$$

Hence

$$a_i + b_j \frac{\partial \beta_j}{\partial q^i} = \beta_i,$$

which implies

$$a_i = \beta_i, \ b_i = 0.$$

Thus $\lambda = \lambda_M$.□

If we now set

$$\omega_M = -d\lambda_M,$$

we locally have

$$\omega_M = \sum_i dq^i \wedge dp_i \tag{5.4}$$

From (5.4) one easily deduces that ω_M is a symplectic form on T^*M, which is called the **canonical symplectic form on** T^*M.

Remark 5.3.3 Since T^*M carries a symplectic structure ω_M we deduce that it is orientable.

Now, let $F : M \longrightarrow M$ be a diffeomorphism. We may define a diffeomorphism

$$T^*F : T^*M \longrightarrow T^*M$$

as follows:

$$(T^*F)(\alpha)(X) = \alpha(dF(x)X), \ \alpha \in T^*_xM, \ X \in T_{F^{-1}(x)}M.$$

Since F is a diffeomorphism, we may choose coordinates in M such that F is locally given by the identity map, i.e.,

$$F : (q^i) \longrightarrow (q^i)$$

Then T^*F is also given by the identity map:

$$T^*F : (q^i, p_i) \longrightarrow (q^i, p_i).$$

Thus, we have

$$(T^*F)^* \lambda_M = \lambda_M \text{ and then } (T^*F)^* \omega_M = \omega_M.$$

Hence we have,

Proposition 5.3.4 *T^*F is a symplectomorphism.*

We will return to symplectic manifolds in Section 5.6.

5.4 Lifts of tensor fields to the cotangent bundle

Let M be an m–dimensional manifold, T^*M its cotangent bundle and $\pi_M : T^*M \longrightarrow M$ the canonical projection.

Vertical lifts

If f is a function on M, then the **vertical lift** of f to T^*M is the function f^v on T^*M defined by

$$f^v = f \circ \pi_M.$$

In local coordinates (q^i, p_i) we have

$$f^v(q^i, p_i) = f(q^i) \tag{5.5}$$

As in the case of the tangent bundle we can consider the **vertical bundle** $V(T^*M)$ defined by

$$V(T^*M) = Ker\{T\pi_M : TT^*M \longrightarrow TM\},$$

i.e.,

$$V(T^*M) = \bigcup_{z \in T^*M} V_z(T^*M),$$

where $V_z(T^*M) = Ker\{d\pi_M(z) : T_z(T^*M) \longrightarrow T_{\pi_M(z)}M\}$, for all $z \in T^*M$. A tangent vector v to T^*M at z such that $v \in V_z(T^*M)$ is called **vertical**. A

vertical vector field X is a vector field on T^*M such that $X(z) \in V_z(T^*M)$ for all $z \in T^*M$.

Now, let α be a 1–form on M. The **vertical lift** of α to T^*M is the vertical vector field α^v on T^*M defined by

$$\alpha^v = -(S_{\omega_M})^{-1}(\pi_M^*\alpha).$$

If α is locally given by $\alpha = \alpha_i(q)dq^i$, then we have

$$\alpha^v = \alpha_i(q)\partial/\partial p_i \tag{5.6}$$

From (5.5) and (5.6) we obtain

$$\alpha^v f^v = 0, \ (\alpha+\beta)^v = \alpha^v + \beta^v,$$

$$(f\alpha)^v = f^v\alpha^v, \ [\alpha^v, \beta^v] = 0,$$

for all function f and all 1–forms α, β on M.

The operator $\imath$

If X is a vector field on M, we define a function $\imath X$ on T^*M by

$$(\imath X)(\alpha) = \alpha(X(x)),$$

for all $\alpha \in T_x^*M$.

If $X = X^i\, \partial/\partial q^i$, then we have

$$(\imath X)(q^i, p_i) = p_i X^i \tag{5.7}$$

Now, let F be a tensor field of type (1,1) on M. We define a 1–form $\imath F$ on T^*M by

$$(\imath F)_\alpha(\tilde{X}) = \alpha_{\pi_M(\alpha)}(F_{\pi_M(\alpha)}(T\pi_M(\tilde{X}))),$$

where $\alpha \in T^*M$, $\tilde{X} \in T_\alpha(T^*M)$.

If $F = F_i^j\, \partial/\partial q^j \otimes dq^i$, then we have

$$\imath F = p_i F_j^i dq^j \tag{5.8}$$

If S is a tensor field of type $(1, s)$ on M, we define a tensor field $\imath S$ of type $(0, s)$ on T^*M by

$$(iS)_\alpha(\tilde{X}_1,\ldots,\tilde{X}_s) = \alpha_{\pi_M(\alpha)}(S_{\pi_M(\alpha)}(T\pi_M(\tilde{X}_1),\ldots,T\pi_M(\tilde{X}_s))),$$

where $\alpha \in T^*M$, $\tilde{X}_1,\ldots,\tilde{X}_x \in T_\alpha(T^*M)$.

If

$$S = S^i_{j_1\ldots j_s}\partial/\partial q^i \otimes dq^{j_1} \otimes \ldots \otimes dq^{j_s}, \tag{5.9}$$

then we have

$$iS = p_i S^i_{j_1\ldots j_s} dp^{j_1} \otimes \ldots \otimes dp^{j_s}. \tag{5.10}$$

Thus we obtain an operator

$$i : \mathcal{T}^1_s(M) \longrightarrow \mathcal{T}^0_s(T^*M)$$

The operator γ

As we have seen, the canonical symplectic structure ω_M on T^*M induces an isomorphism

$$S_{\omega_M} : \chi(T^*M) \longrightarrow \wedge^1(T^*M).$$

This isomorphism may be extended to an isomorphism (also denoted by S_{ω_M})

$$S_{\omega_M} : \mathcal{T}^1_{s-1}(T^*M) \longrightarrow \mathcal{T}^0_s(T^*M)$$

as follows:

$$(S_{\omega_M}K)(\tilde{X}_1,\ldots,\tilde{X}_s) =< K(\tilde{X}_1,\ldots,\tilde{X}_{s-1}),\ S_{\omega_M}(\tilde{X}_s) >,$$

for all $\tilde{X}_1,\ldots,\tilde{X}_s \in \chi(T^*M)$.

Hence, if S is a tensor field of type $(1,s)$ on M, we define a tensor field γS of type $(1,s-1)$ on T^*M given by

$$\gamma S = -(S_{\omega_M})^{-1}(iS).$$

If S is locally given by (5.9), then we obtain

$$\gamma S = p_i S^i_{j_1\ldots j_k}\partial/\partial p_{j_k} \otimes dq^{j_1} \otimes \ldots \otimes dq^{j_s-1} \tag{5.11}$$

Hence we have an operator

$$\gamma : \mathcal{T}_s^1(M) \longrightarrow \mathcal{T}_{s-1}^1(T^*M)$$

A direct computation from (5.11) shows that

$$\gamma(S+T) = \gamma S + \gamma T.$$

If F (resp. S) is a tensor field of type (1,1) (resp. (1,2)) then (5.11) becomes

$$\gamma F = p_i F_j^i \partial/\partial p_j \tag{5.12}$$

(resp.

$$\gamma S = p_i S_{jk}^i \partial/\partial p_j \otimes dq^k) \tag{5.13}$$

since (5.8) and (5.10).

From (5.6), (5.12) and (5.13) we easily deduce the following

Proposition 5.4.1 *Let be $\alpha \in \wedge^1 M$, $F, G \in \mathcal{T}_1^1(M)$ and $S \in \mathcal{T}_2^1(M)$. Then we have*

$$(\gamma S)(\alpha^v) = (\gamma S)(\gamma F) = 0,$$

$$[\alpha^v, \gamma F] = (\alpha \circ F)^v,$$

$$[\gamma F, \gamma G] = \gamma[F, G],$$

where $\alpha \circ F$ is a 1–form on M defined by $(\alpha \circ F)(X) = \alpha(F(X))$.

Complete lifts of vector fields

Let X be a vector field on M. Then the **complete lift** of X to T^*M is the vector field X^c on T^*M defined by

$$X^c = (S_{\omega_M})^{-1}(d(iX))$$

If $X = X^i \partial/\partial q^i$, then we have

$$X^c = X^i \partial/\partial q^i - p_j (\partial X^j/\partial q^i) \partial/\partial p_i \tag{5.14}$$

From (5.5), (5.6), (5.12), (5.13) and (5.14), we obtain the following

Proposition 5.4.2 *Let be $f \in C^\infty(M)$, $X, Y \in \chi(M)$, $\alpha \in \wedge^1 M$, $F \in \mathcal{T}_1^1(M)$ and $S \in \mathcal{T}_2^1(M)$. Then we have*

$$(X+Y)^c = X^c + Y^c, (fX)^c = f^v X^c - (iX)(df)^v,$$

$$[X^c, \alpha^v] = (L_X \alpha)^v,\ [X^c, \gamma F] = \gamma(L_X F),$$

$$[X^c, Y^c] = [X, Y^c],\ (\gamma S)X^c = \gamma(S_X),$$

where S_X is the tensor field of type (1,1) on M defined by $S_X(Z) = S(X, Z)$ for any $Z \in \chi(M)$.

Complete lifts of tensor fields of type (1,1)

Now, let F be a tensor field of type (1,1) on M. Then the **complete lift** of F to T^*M is the tensor field F^c of type (1,1) on T^*M given by

$$F^c = (S_{\omega_M})^{-1}(d(iF)).$$

If F is locally given by $F = F^i_j\, \partial/\partial q^i \otimes dq^j$, then we have

$$\begin{aligned} F^c &= F^i_j\, \partial/\partial q^i \otimes dq^j \\ &+ p_k(\partial F^k_j/\partial q^i - \partial F^k_i/\partial q^j)\, \partial/\partial q^i \otimes dp_j \\ &+ F^i_j\, \partial/\partial p_i \otimes dp_j, \end{aligned} \tag{5.15}$$

since (5.8).

From (5.6), (5.12) and (5.15) we obtain the following

Proposition 5.4.3 *Let be $\alpha \in \wedge^1 M$, $X \in \chi(M)$ and $F, G \in \mathcal{T}_1^1(M)$. Then we have*

$$F^c \alpha^v = (\alpha \circ F)^v,\ F^c(\gamma G) = \gamma(GF),$$

$$F^c X^c = (FX)^c + \gamma(L_X F).$$

Complete lifts of tensor fields of type (1,2)

Suppose now that S is a skew–symmetric tensor field of type $(1,2)$ on M. Then it is not hard to prove that $\imath S$ is a 2–form on T^*M. Then we define the **complete lift** of S to T^*M by

$$S^c = (S_{\omega_M})^{-1}(d(\imath S)).$$

Thus S^c is a tensor field of type $(1,2)$ on T^*M. By a straightforward computation from (5.10) we obtain the following

Proposition 5.4.4 *Let be* $X, Y \in \chi(M)$, $\alpha, \beta \in \wedge^1 M$, $F, G \in \mathcal{T}_1^1(M)$ *and* $S \in \mathcal{T}_2^1(M)$. *Then we have*

$$S^c(\alpha^v, \beta^v) = 0, \; S^c(\alpha^v, \gamma G) = 0,$$

$$S^c(\alpha^v, Y^c) = -(\alpha \circ S_Y)^v, \; S^c(\gamma F, \gamma G) = 0,$$

$$S^c(\gamma F, Y^c) = -\gamma(F S_Y),$$

$$S^c(X^c, Y^c) = (S(X,Y))^c - \gamma((L_X S)_Y - (L_Y S)_X + S_{[X,Y]})$$

5.5 Almost product and almost complex structures on the cotangent bundle

In this section we apply the constructions of lifts of tensor fields to obtain some interesting structures on T^*M.

Before proceeding further we prove the following lemma.

Lemma 5.5.1 *Let* $\tilde{S}_1$ *and* $\tilde{S}_2$ *be tensor fields of type* $(0,s)$ *(or (1,s)) on* T^*M *such that*

$$\tilde{S}_1(S_1^c, \ldots, X_s^c) = \tilde{S}_2(X_1^c, \ldots, X_s^c)$$

for any $X_1, \ldots, X_s \in \chi(M)$. *Then* $\tilde{S}_1 = \tilde{S}_2$.

Proof: It is sufficient to show that if $\tilde{S}(X_1^c, \ldots, X_s^c) = 0$ for any $X_1, \ldots, X_s \in \chi(M)$, then $\tilde{S} = 0$.

We only prove the case of tensor fields of type (1,1). The general case may be proved in a similar way.

Let $\tilde{F}$ be a tensor field of type (1,1) on T^*M such that $\tilde{F}X^c = 0$ for any vector field X on M. Then

$$\tilde{F}(\partial/\partial q^i)^c = \tilde{F}(\partial/\partial q^i) = \tilde{F}_i^j \partial/\partial q^j + \tilde{F}_i^{\bar{j}} \partial/\partial p_j = 0$$

implies

$$\tilde{F}_i^j = \tilde{F}_i^{\bar{j}} = 0.$$

Now suppose that X is a vector field on M locally given by $X = X^i \partial/\partial q^i$. From (5.14) we have

$$\tilde{F}(X^c) = -p_j(\partial X^j/\partial q^i)\tilde{F}(\partial/\partial p_i)$$

$$= -p_j(\partial X^j/\partial q^i)\tilde{F}_{\bar{i}}^j \partial/\partial q^j$$

$$-p_j(\partial X^j/\partial q^i)\tilde{F}_{\bar{i}}^{\bar{j}} \partial/\partial p_j = 0$$

which implies

$$p_j(\partial X^j/\partial q^i)\tilde{F}_{\bar{i}}^j = p_j(\partial X^j/\partial q^i)\tilde{F}_{\bar{i}}^{\bar{j}} = 0.$$

Hence $\tilde{F}_{\bar{i}}^j = \tilde{F}_{\bar{i}}^{\bar{j}} = 0$ except on the zero–section. Since $\tilde{F}_{\bar{i}}^j$ and $\tilde{F}_{\bar{i}}^{\bar{j}}$ are continous it follows that $\tilde{F}_{\bar{i}}^j = \tilde{F}_{\bar{i}}^{\bar{j}} = 0$ at all T^*M.□

Proposition 5.5.2 *Let F be a tensor field of type (1,1) on M. Then we have*

$$(F^c)^2 = (F^2)^c + \gamma N_F,$$

where N_F is the Nijenhuis tensor of F.

Proof: In fact, we have

$$(F^c)^2 X^c = F^c((FX)^c + \gamma(L_X F))$$

$$= (F^2 X)^c + \gamma L_{FX} F + \gamma((L_X F)F)$$

$$= (F^2 X)^c + \gamma\{L_{FX} F + (L_X F)F\},$$

since Proposition 5.4.3.

On the other hand we obtain

$$((F^2)^c + \gamma N_F)X^c = (F^2 X)^c + \gamma(L_X F^2) + \gamma(N_F)_X$$

$$= (F^2 X)^c + \gamma\{L_X F^2 + (N_F)_X\}$$

since Propositions 5.4.2 and 5.4.3.

Now we have

$$(L_{FX} F + (L_X F)F)(Y) = [FX, FY] - F[FX, Y]$$

$$+[X, F^2 Y] - F[X, FY]$$

$$= [X, F^2 Y] - F^2[X, Y] + [FX, FY] - F[FX, Y] - F[X, FY] + F^2[X, Y]$$

$$= (L_X F^2)Y + N_F(X, Y)$$

$$= (L_X F^2 + (N_F)_X)Y,$$

for any $Y \in \chi(M)$. Hence

$$(F^c)^2 X^c = ((F^2)^c + \gamma N_F)X^c,$$

for any $X \in \chi(M)$, from which we have the required result since Lemma 5.5.1. □

By a straightforward computation from Propositions 5.4.1, 5.4.2, 5.4.3, 5.4.4 and 5.5.2, we easily obtain the following.

Proposition 5.5.3 *Let F be a tensor field of type (1,1) on M. Then we have*

$$N_{F^c} = (N_F)^c.$$

Complete lifts of almost product structures

Let F be an almost product structure on M. Since $(F^c)^2 = (F^2)^c + \gamma N_F = Id + \gamma N_F$, we have

Proposition 5.5.4 *F^c is an almost product structure on T^*M if and only if F is integrable.*

From Proposition 5.5.3, we obtain

Proposition 5.5.5 *If F is an integrable almost product structure on M then F^c is an integrable almost product structure on T^*M.*

Now let F be an integrable almost product structure on M with projection operators P and Q, i.e.,

$$P = \frac{1}{2}(I + F),\ Q = \frac{1}{2}(I - F).$$

Since $P^2 = P$ and $Q^2 = Q$, we have

$$(P^c)^2 = (P^2)^c + \gamma N_P = (P^2)^c = P^c,$$

$$(Q^c)^2 = (Q^2)^c + \gamma N_Q = (Q^2)^c = Q^c,$$

since Proposition 3.1.4. Hence P^c and Q^c are the projection operators corresponding to F^c, i.e.,

$$P^c = \frac{1}{2}(I + F^c),\ Q^c = \frac{1}{2}(I - F^c).$$

Suppose that rank $P = r$, rank $Q = s$, $r + s = m$. Since F is integrable, then exists for each point of M a coordinate neighborhood U with local coordinates (q^i) such that

$$Im\, P =< \partial/\partial q^1, \ldots, \partial/\partial q^r >,$$

$$Im\, Q =< \partial/\partial q^{r+1}, \ldots, \partial/\partial q^m > .$$

An easy computation shows that

$$Im\, P^c =< (\partial/\partial q^1)^c, \ldots, (\partial/\partial q^r)^c, (dq^1)^v, \ldots, (dq^r)^v >$$

$$=< \partial/\partial q^1, \ldots, \partial/\partial q^r, \partial/\partial p_1, \ldots, \partial/\partial p_r >,$$

$$Im\, Q^c =< (\partial/\partial q^{r+1})^c, \ldots, (\partial/\partial q^m)^c,\ (dq^{r+1})^v, \ldots, (dq^m)^v >$$

$$=< \partial/\partial q^{r+1}, \ldots, \partial/\partial q^m,\ \partial/\partial p_{r+1}, \ldots, \partial/\partial p_m > .$$

Complete lifts of almost complex structures

Let F be an almost complex structure on M. From Propositions 5.5.2 and 5.5.3 we have

Proposition 5.5.6 *(1) F^c is an almost complex structure on T^*M if and only if F is integrable, i.e., F is a complex structure on M. (2) If F is a complex structure on M, then F^c is a complex structure on T^*M.*

5.6 Darboux Theorem

Regarding (5.4) we see that ω_M has a local expression similar to the expression for symplectic forms on vector spaces. We show in this section a fundamental result due to Darboux which stablish the non–linear analogue of symplectic vector spaces.

In order to prove the Darboux Theorem, we first introduce the notion of time–dependent vector field.

Definition 5.6.1 *A* **time–dependent vector field** *on a manifold M is a C^∞ map $X : R \times M \longrightarrow TM$ such that*

$$X(t, x) \in T_xM.$$

We remark that all the results obtained in Chapter 1 still hold for time–dependent vector fields. Thus, we define $\Phi_{t,s}(x)$ to be the integral curve of X_t through time $t = s$, i.e.,

$$\frac{d}{dt}(\phi_{t,s}(x)) = X_t(\phi_{t,s}(x))$$

and

$$\phi_{t,s}(x) = x, \quad t = s,$$

where X_t is the vector field on M given by $X_t(x) = X(t, x)$. In fact, $\phi_{t,s}$ is the **(time–dependent) local 1–parameter group generated by** X_t. We have

$$\phi_{t,s} \circ \phi_{s,r} = \phi_{t,r}, \; \phi_{t,t} = Id.$$

The proof of the following result is left to the reader as an exercise.

Proposition 5.6.2 *Let X be a time–dependent vector field on M. Then we have*

$$(d/dt)(\phi^*_{t,s}\alpha) = \phi^*_{t,s}(L_{X_t}\alpha),$$

for every p–form α on M.

Theorem 5.6.3 (Darboux Theorem). *Let ω be an almost symplectic form on a $2n$–dimensional manifold S. Then $d\omega = 0$ if and only if for each $x \in S$ there exists a coordinate neighborhood U with local coordinates $(x^1, \ldots, x^{2n})$ such that*

$$\omega = \sum_{i=1}^{n} dx^i \wedge dx^{n+i}$$

on U.

Proof: We will use an idea of Moser [98] adapted by Weinstein known as the "path method". (An alternative proof using the notion of class of a form may be founded in Godbillon [63]; we also remit to Arnold [4] for a beautiful and geometrical proof).

First, we note that if $\omega = \sum_i dx^i \wedge dx^{n+i}$, then ω is closed. In order to prove the converse, we can suppose that $S = R^{2n}$ and $x = 0 \in R^{2n}$. Let ω_1 be the constant 2–form on R^{2n} defined by

$$\omega_1(y) = \omega_0 = \sum_i dx^i \wedge dx^{n+i}, \; y \in R^{2n},$$

where $(x^1, \ldots, x^{2n})$ are the canonical coordinates in R^{2n}. We put

$$\omega_t = \omega + t(\omega_1 - \omega), \; t \in [0,1].$$

For each $t \in [0,1]$, $\omega_t(0) = \omega_0$. Thus $\omega_t(0)$ is non–degenerate for each t. Hence there exists a neighborhood U of 0 such that ω_t is non–degenerate for all $t \in [0,1]$, since $Gl(2n, R)$ is an open set of $gl(2n, R)$. We can suppose that U is an open ball centered at $0 \in R^{2n}$. Then, from the Poincaré lemma, $\omega_1 - \omega = d\alpha$, for some 1–form α on U, with $\alpha(0) = 0$ (since $d(\omega_1 - \omega) = 0$). Now, let X_t be a vector field defined by

$$i_{X_t}\omega_t = -\alpha.$$

So, X_t is a time–dependent vector field such that $X_t(0) = 0$, since

$$i_{X_t(0)}\omega_0 = -\alpha(0) = 0.$$

Then there exists an open ball $V \subset U$ centered at $0 \in R^{2n}$ such that the (time–dependent) 1–parameter group $\phi_{t,0}$ is defined for all $t \in [-1, 1]$. By Proposition 5.6.2, we have

$$d/dt(\phi^{\star}_{t,0}\omega_t) = \phi^{\star}_{t,0}(L_{X_t}\omega_t) + \phi^{\star}_{t,0}(d/dt(\omega_t))$$

$$= \phi^{\star}_{t,0}(di_{X_t}\omega_t) + \phi^{\star}_{t,0}(\omega_1 - \omega)$$

$$= -\phi^{\star}_{t,0}(d\alpha - \omega_1 + \omega) = 0.$$

Thus

$$\phi^{\star}_{1,0}\,\omega_1 = \phi^{\star}_{0,0}\,\omega = \omega_0 = \omega.$$

Therefore $\phi_{1,0}$ gives the required change of coordinates which transforms ω to ω_1.□

The coordinate neighborhoods given by Darboux Theorem are called **symplectic** and its coordinate functions $(x^1, \ldots, x^{2n})$ are called **symplectic** (or **canonical**) **coordinates**. (If we set $x^i = q^i$, $x^{n+i} = p_i$, $1 \leq i \leq n$, then ω is locally given by (5.4)).

In terms of the theory of G–structures, the Darboux theorem can be rewritten as follows.

Corollary 5.6.4 *Let $B_{Sp(n)}$ be a $Sp(n)$– structure on S with almost symplectic form ω. Then $B_{Sp(n)}$ is integrable if and only if ω is symplectic.*

Definition 5.6.5 *Let S be a manifold of dimension $2n + r$ and ω a closed 2–form on S of constant rank $2n$. Then ω is said to be a* **presymplectic form** *(or* **structure***) on S and the pair (S, ω) is called a* **presymplectic manifold**.

Let (S, ω) be a presymplectic manifold. Suppose that dim $S = 2n+r$ and rank $\omega = 2n$. Let x be a point of S and (U, φ) a coordinate neighborhood at x such that $\varphi(x) = 0 \in R^{2n+r}$. If we shrink U, if necessary, we may suppose that there are neighborhoods $V \subset R^{2n}$, $W \subset R^r$ at the origin

$0 \in R^{2n}$, $0 \in R^r$, respectively, such that $\varphi(U) = V \times W$. Consider the pull-back $(\varphi^{-1})^*\omega$ on $\varphi(U)$ and denote by ω_V, resp. ω_W, the 2-forms on V, resp. W, defined by

$$\omega_V = ((\varphi^{-1})^*\omega)/_V, \ \omega_W = ((\varphi^{-1})^*\omega)/_W.$$

Since $d\omega_V = 0$ and ω_V is of maximal rank $2n$, from Darboux theorem, there is a coordinate system $(x^1, \ldots x^{2n})$ such that

$$\omega_V = \sum_{i=1}^{n} dx^i \wedge dx^{n+i}.$$

Therefore we have the following.

Theorem 5.6.6 (**Generalized Darboux Theorem**).- *Suppose that S is a $(2n+r)$-dimensional manifold and ω a 2-form on S of constant rank $2n$. Then ω is closed (i.e., ω is a presymplectic form on S) if and only if for each point $x \in S$ there exists a coordinate neighborhood U with local coordinates $(x^1, \ldots, x^{2n}, y^1, \ldots, y^r)$ such that*

$$\omega = \sum_{i=1}^{n} dx^i \wedge dx^{n+i}$$

on U.

Remark 5.6.7 We may prove directly the above generalized theorem using the path method and then obtain Darboux Theorem as a Corollary. Pulling back the form ω to $V \times W$ and applying Frobenius theorem, we may prove that the form on $V \times W$ has a local expression of type

$$\sum_{i=1}^{n} a_{i,n+i}(y^1, \ldots, y^{2n+r}) dy^i \wedge dy^{n+i}$$

Now, along V, defined by the local equations $y^{2n+1} = \ldots = y^{2n+r} = 0$, we may also suppose that at the origin the matrix $(a_{i,n+i})$ is of type

$$\begin{pmatrix} 0 & I_n \\ -I_n & 0 \end{pmatrix}.$$

As R^{2n} is a symplectic vector space, we consider a symplectic form on V with canonical expression and then we may apply the same procedure as in Darboux Theorem.

Remark 5.6.8 Another point of view for proving the generalized theorem consists in adopting induction on r. If $r = 0$, the result is just Darboux Theorem. Suppose that the assertion is true for $r - 1$. Then there is a $(2n + r - 1)$–dimensional subspace of T_xS, $x \in S$, on which $\omega(x)$ has rank $2n$. If we apply Frobenius theorem for vector fields one obtain a submanifold N of S of dimension $2n + r - 1$. By induction we have a coordinate system on N such that ω has the expression $\sum_{j=1}^{n} dy^j \wedge dy^{n+j}$ on this coordinate system. We finish the proof by considering the distribution $Ker\ S$ and choosing a vector field $Y \in Ker\ S_\omega$ such that $Y(x) \in T_xN$, $x \in N$. Using the flow of such Y one obtains a coordinate system $(x^1, \ldots, x^{2n}, z^1, \ldots, z^r)$ such that ω assumes the above expression (for further details see Robinson [106], for example).

5.7 Almost cotangent structures

The concept of an almost cotangent structure was introduced by Bruckheimer [10] and interpreted by Clark and Goel [18] as a certain type of G–structure.

Let T^*M be the cotangent bundle of an m–dimensional manifold M and $\pi_M : T^*M \longrightarrow M$ the canonical projection. Let (U, q^i), $(\bar{U}, \bar{q}^i)$ be coordinate neigborhoods with $U \cap \bar{U} \neq \emptyset$ and (T^*U, q^i, p_i), $(T^*\bar{U}, \bar{q}^i, \bar{p}_i)$ the induced coordinate neighborhoods on T^*M. Then (q^i, p_i), $(\bar{q}^i, \bar{p}_i)$ are related by a change of coordinates whose Jacobian matrix has the form

$$\begin{bmatrix} A & 0 \\ B & (A^{-1})^t \end{bmatrix}, \tag{5.16}$$

where $A = (\partial q^j/\partial \bar{q}^i)$, $B = [(\partial^2 \bar{q}^k/\partial q^j \partial q^l)(\partial q^l/\partial \bar{q}^i)\bar{p}_k]$.

This suggests the following definition.

Definition 5.7.1 *Let N be a $2m$–dimensional manifold carrying a G–structure whose group G consists of all $2m \times 2m$ matrices of the form (5.16), where $A \in Gl(m, R)$ and $A^tB = B^tA$. Such a structure is called an* **almost cotangent structure**, *and such a manifold N is called an* **almost cotangent manifold**.

Now, let B be an almost cotangent structure on a $2m$–dimensional manifold N. We define a 2–form ω on N by specifying its components to be

$$\omega_0 = \begin{bmatrix} 0 & -I_m \\ I_m & 0 \end{bmatrix}$$

relative to any adapted frame $\{X_1,\ldots,X_m,X_{m+1},\ldots,X_{2m}\}$ at x. Then ω is well–defined and determines an almost symplectic structure on N. In fact, if $\{\theta^1,\ldots,\theta^m,\theta^{m+1},\ldots,\theta^{2m}\}$ is the dual coframe, then we have

$$\omega = \sum_{i=1}^{m} \theta^i \wedge \theta^{m+i}.$$

Furthermore let V be an m–dimensional distribution on N defined by

$$V_x = < X_{m+1},\ldots,X_{2m} > .$$

Thus V is a Lagrangian distribution with respect to ω, i.e.,

$$\omega(X,Y) = 0 \text{ for all } X,Y \in V.$$

Conversely, suppose that N is a $2m$– dimensional manifold endowed with an almost symplectic form ω, together with a Lagrangian distribution V. Let x be a point of N and $\{Y_1,\ldots,Y_m,Y_{m+1},\ldots,Y_{2m}\}$ a frame at x which is adapted for V, i.e.,

$$V_x = < Y_{m+1},\ldots,Y_{2m} > .$$

Then the matrix of ω relative to $\{Y_a,\ 1 \leq a \leq 2m\}$ is

$$\begin{bmatrix} P & -Q^t \\ Q & 0 \end{bmatrix},$$

where $P^t = -P$ and $\det Q \neq 0$. We now construct a new frame $\{X_a\}$ at x as follows:

$$\{X_a\} = \{Y_a\} \begin{bmatrix} A & 0 \\ B & C \end{bmatrix},$$

where $A = Q^{-1}$, $C = I_m$ and $B = \frac{1}{2}(Q^{-1})^t P Q^{-1}$. Then

$$V_x = < X_{m+1},\ldots,X_{2m} >$$

and the matrix of ω relative to $\{X_a\}$ is ω_0. Now, let $\{X_a\}$, $\{\bar{X}_a\}$ be two frames at x as above. Since $V_x = < X_{m+1},\ldots,X_{2m} > = < \bar{X}_{m+1},\ldots,\bar{X}_{2m} >$, we deduce that they are related by a matrix of the form

$$\begin{bmatrix} A & 0 \\ B & C \end{bmatrix}. \tag{5.17}$$

Moreover, since they are adapted to ω, we deduce that

$$C = (A^{-1})^t \text{ and } A^t B = B^t A.$$

Hence the matrix (5.17) belongs to G. So the set of such a frames $\{X_a\}$ at all points of N defines a G–structure on M.

Summing up, we have

Proposition 5.7.2 *Giving an almost cotangent structure on N is the same as giving an almost symplectic form ω, together with a Lagrangian distribution V.*

Example

We have seen that the cotangent bundle T^*M of an m–dimensional manifold M carries a canonical almost cotangent structure. Actually, we easily see that the corresponding almost symplectic form is precisely ω_M and the corresponding Lagrangian distribution is $V(T^*M)$.

Next, we stablish the integrability conditions of an almost cotangent structure.

Theorem 5.7.3 *An almost cotangent structure (ω, V) on a 2m–dimensional manifold N is integrable if and only if ω is symplectic and V is involutive.*

Proof: Obviously, if (ω, V) is integrable then V is involutive and $d\omega = 0$. We prove that this is also sufficient. In fact, let x be a point of N. Since V is involutive, then, from the Frobenius theorem, local coordinates (q^i, y_i) may be introduced such that $\{\partial/\partial y_i\}$ span V. Choose a coframe field $\{\psi^1, \ldots, \psi^{2m}\}$ at x adapted to the almost cotangent structure (ω, V). Then we have

$$\psi^i = A^i_j dq^j, \ \det A \neq 0.$$

We now define a new coframe field $\{\theta^1, \ldots, \theta^{2m}\}$ at x by

$$\theta^i = dq^i, \ \theta^{m+i} = A^i_j \psi^{m+j}, \quad 1 \leq i, j \leq m.$$

Then $\{\theta^1, \ldots, \theta^{2m}\}$ is adapted to (ω, V). Suppose that

$$\theta^{m+i} = P^i_j dq^j + Q^i_j dy_j.$$

Since $d\omega = 0$, then we have

$$d\omega = d(\theta^i \wedge \theta^{m+i}) = 0.$$

Thus,

$$dq^i \wedge \{[(\partial P^i_j/\partial q^k)dq^k + (\partial P^i_j/\partial y_k)dy^k] \wedge dq^j$$

$$+[(\partial Q^i_j/\partial q^k)dq^k + (\partial Q^i_j/\partial y_k)dy_k] \wedge dy_j\} = 0.$$

and therefore we obtain

$$\partial Q^i_j/\partial y_k = \partial Q^i_k/\partial y_j$$

It follows that the equations

$$\partial F^i/\partial y_j = Q^i_j$$

admit differentiable solutions $F^i = F^i(q, y)$ on a neighborhood of x. Then we may construct a new local coordinates (q^i, z_i) at x by setting

$$z_i = F^i(q, y).$$

Therefore we have

$$\theta^i = dq^i, \; \theta^{m+i} = (P^i_j - \partial F^i/\partial q^j)dq^j + dz_j$$

$$= \bar{P}^i_j dq^j + dz_j,$$

where $\bar{P}^i_j = P^i_j - \partial F^i/\partial q^j$.

The condition $d\omega = 0$ implies that

$$(\partial \bar{P}^i_j/\partial q^k) + (\partial \bar{P}^k_i/\partial q^j) + (\partial \bar{P}^j_k/\partial q^i) = 0, \tag{5.18}$$

$$\partial(\bar{P}^i_j - \bar{P}^j_i)/\partial z_k = 0. \tag{5.19}$$

Now, consider the equations

$$\partial H^i/\partial q^j - \partial H^j/\partial q^i = \bar{P}^i_j - \bar{P}^j_i \tag{5.20}$$

From (5.19) we deduce that the right–side of (5.20) depends only on (q^i), and from (5.18) we deduce that there exist differentiable solutions $H^i(q)$ of (5.20) on a neighborhood of x. Next, we define functions p_i by

$$p_i = z_i + H^i(q),\ 1 \le i \le m.$$

Therefore we have

$$\theta^i = dq^i,$$

$$dp_i = \theta^{m+i} + (\partial H^i/\partial q^j - \bar{P}^i_j)\theta^j.$$

Then (q^i, p_i) is a local coordinate system at x. Furthermore, a simple computation shows that $\{dq^i, dp_i\}$ is a coframe field adapted to (ω, V) since

$$\partial H^i/\partial q^j - \bar{P}^i_j$$

is symmetric in i, j. Hence

$$V = \langle \partial/\partial p_1, \ldots, \partial/\partial p_m \rangle$$

and

$$\omega = \sum_{i=1}^{m} dq^i \wedge dp_i.$$

This ends the proof. □

Remark 5.7.4 Notice that the example $(\omega_M, V(T^*M))$ is an integrable almost cotangent structure.

Let (ω, V) be an integrable almost cotangent structure on a $2m$–dimensional manifold N. Then ω is a symplectic form and V an involutive distribution. Since Exercises 3.8.2 and 5.9.1, it is easy to prove the following.

Proposition 5.7.5 *If (ω, V) is an integrable almost cotangent structure on M, then there exists a symmetric linear connection ∇ on M such that $\nabla\omega = 0$ and $\nabla_X Y \in V$ for all $Y \in V$, i.e., ∇ is an almost cotangent connection.*

5.8 Integrable almost cotangent structures which define fibrations

As we have seen in Section 5.7, the integrability of an almost cotangent structure implies that it is locally equivalent to the cotangent bundle T^*M of a differentiable manifold M. In this section we shall prove that an integrable almost cotangent structure verifying some global hypotheses is (globally) diffeomorphic to some cotangent bundle. This result is due to Thompson [115].

Definition 5.8.1 *Let (ω, V) be an integrable almost cotangent structure on a 2m–dimensional manifold N. Since V is involutive then V determines a foliation (also denoted by V). Let M be the space of leaves and $\pi : N \longrightarrow M$ the canonical projection. We say that (ω, V)* **defines a fibration** *if M has a differentiable structure of manifold such that π is a surjective submersion (then M is a quotient manifold of N).*

Suppose that (ω, V) is a integrable almost cotangent structure on N which defines a fibration $\pi : N \longrightarrow M$. We now show that there is a construction which generalizes the vertical lift construction on a cotangent bundle.

Let α be a 1–form on M. We define the **vertical lift** α^v of α to N by

$$\alpha^v = -(S_\omega)^{-1}(\pi^*\alpha).$$

In adapted coordinates (q^i, p_i), we have

$$\alpha^v = \alpha_i(q)\partial/\partial p_i, \tag{5.21}$$

where $\alpha = \alpha_i(q)dq^i$.

Proposition 5.8.2 *For all 1–forms α, β on M, we have*

$$[\alpha^v, \beta^v] = 0.$$

Proof: Directly from (5.21).□

Since (ω, V) is integrable, there exists a symmetric almost cotangent connection ∇. We have the following.

Proposition 5.8.3 *∇ induces, by restriction, a flat connection on each leaf of V.*

We omit the proof which is similar to these of Proposition 2.10.3.

Theorem 5.8.4 *Suppose that the flat connection induced from ∇ on each leaf of V is geodesically complete and that each leaf is connected and simply connected. Then N is diffeomorphic to T^*M. Moreover the diffeomorphism, F say, can be chosen such that $\omega = F^*(\omega_M + \pi_M^*\phi)$, where ϕ is a closed 2–form on M.*

Proof: By similar arguments to those used in the proof of Theorem 2.10.4 we prove that $\pi : N \longrightarrow M$ is an affine bundle modeled on T^*M. Now we choose a global section s of N over M. Then N may be identified with T^*M, with s playing the role of zero section. We call the resulting diffeomorphism $F : N \longrightarrow T^*M$ and consider the 2–form

$$\Omega = \omega - F^*\omega_M$$

Then Ω is closed and verify $i_X\Omega = 0$ for any vertical vector field $X \in V$. Then there exists a closed 2–form ϕ on N such that

$$\Omega = \pi^*\phi.$$

Since $\pi_M \circ F = \pi$, we have

$$\omega = F^*(\omega_M + \pi_M^*\phi). \square$$

Corollary 5.8.5 *Suppose that (ω, V) verifies all the hypotheses of the Theorem 5.8.4 except that the leaves of V are simply connected. Then, if the leaves of V are assumed to be mutually diffeomorphic, T^*M is a covering space of N and the leaves of V are of the form $T^k \times R^{m-k}$, $0 \leq k \leq m$. Moreover, if the leaves of V are compact then T^*M is a covering space of N and this leaves are diffeomorphic to T^m.*

The following definition was also introduced by Thompson [115].

Definition 5.8.6 *We say that an almost cotangent structure (N, ω, V) is* **regular** *if it verifies all the hypotheses of the Theorem 5.8.4. In such a case, (N, π, M, ω, V) is called a* **regular almost cotangent structure**. *If (N, π, M, ω, V), $(\bar{N}, \bar{\pi}, M, \bar{\omega}, \bar{V})$ are two regular almost cotangent structures, they will be said to be* **equivalent** *if there exists a bundle morphism $F : N \longrightarrow \bar{N}$ over the identity of M such that*

$$F^*\bar{\omega} - \omega = \pi^*(d\alpha),$$

for some 1–form α on M, i.e., $F^*\bar{\omega} - \omega$ is cohomologous to zero.

Proposition 5.8.7 *There is a one–to–one correspondence between the set of equivalence classes of regular almost cotangent structures and elements of* $H^2(M, R)$.

Proof: Let (N, π, M, ω, V) be a regular almost cotangent structure. From Theorem 5.8.4, there exists a diffeomorphism $F_s : N \longrightarrow T^*M$ such that

$$\omega = F_s^* \omega_M - \pi^* \phi_s,$$

where ϕ_s is a closed 2–form on M and s is the section of N which is used to define F_s. Then $F_s \circ s = s_0$ (zero–section of T^*M) and we have

$$s^* \omega = \phi_s.$$

Let $\bar{s}$ be another section of N and $F_{\bar{s}}$ the corresponding diffeomorphism such that

$$\omega = F_{\bar{s}}^* \omega_M - \pi^* \phi_{\bar{s}}.$$

Then there exists a section σ of T^*M (i.e., a 1–form on M) such that $\bar{s} = s + \sigma$, i.e., $\bar{s}(x) = s(x) + \sigma(x)$, for each point $x \in M$. A simple computation shows that

$$F_s \circ \bar{s} = \sigma.$$

Then

$$\phi_{\bar{s}} = \bar{s}^* \omega = \bar{s}^* (F_s^* \omega_M + \pi^* \phi_s)$$

$$= (F_s \circ \bar{s})^* \omega_M + (\pi \circ \bar{s})^* \phi_s = \sigma^* \omega_M + \phi_s$$

$$= \phi_s + d\sigma.$$

Thus $\phi_{\bar{s}}$ and ϕ_s define the same cohomology class of $H^2(M, R)$.

Now, we prove that the mapping defined above is surjective. Let $[\phi] \in H^2(M, R)$. Then the corresponding regular almost cotangent structure is $(T^*M,\ \pi_M,\ M,\ \omega_M + \pi_M^* \phi, V(T^*M))$.

Finally, we prove that the mapping is injective. Suppose that (N, π, M, ω, V), $(\bar{N}, \bar{\pi}, M, \bar{\omega}, \bar{V})$ are regular almost cotangent structures and that

F and $\bar{F}$ are the respective diffeomorphism with T^*M corresponding to the sections s and $\bar{s}$. Then we have

$$\bar{s}^*\omega - s^*\omega = \phi_{\bar{s}} - \phi_s = d\alpha,$$

for some closed 1–form α on M. Now, a direct computation shows that

$$((\bar{F})^{-1} \circ F)^*\bar{\omega} = \omega - \pi^* d\alpha$$

Hence $(\bar{F})^{-1} \circ F$ is an equivalence of regular almost cotangent structures. This ends the proof. □

The last result of this section shows that the vanishing of the element of $H^2(M, R)$ characterizes T^*M as a regular almost cotangent structure up to equivalence.

Proposition 5.8.8 *Suppose that (N, π, M, ω, V) is a regular almost cotangent structure. Then (N, π, M, ω, V) is equivalent to $(T^*M, \pi_M, M, \omega_M, V(T^*M))$ if and only if the element of $H^2(M, R)$ it determines is zero. In such a case ω is exact, say $\omega = -d\lambda$, and the equivalence F verifies $F^*\lambda_M = \lambda$.*

The proof is left to the reader as an exercise.

Remark 5.8.9 In de León et al. [35], [36] we introduce the concept of p–almost cotangent structures which generalizes almost cotangent structures and prove some results similar to those proved in this section.

5.9 Exercises

5.9.1 (i) Let (S, ω) be an almost symplectic manifold. Define a linear connection ∇ on S by

$$2\omega(\nabla_X Y, Z) = X\omega(Y, Z) + Y\omega(X, Z) - Z\omega(X, Y)$$

$$+\omega([X, Y], Z) + \omega([Z, X], Y) + \omega(X, [Z, Y]),$$

for all $X, Y, Z \in \chi(S)$. Prove that ∇ is an **almost symplectic connection**, i.e., $\nabla\omega = 0$.

(ii) Prove that, if ω is symplectic, then ∇ is a locally Euclidean connection, i.e., $T = 0$ and $R = 0$, where T and R are the torsion and curvature tensors of ∇, respectively (*Hint*: compute the Christoffel components of ∇ in canonical coordinates).

5.9.2 Show that if (V, ω) is a symplectic vector space and K is a subspace of V then $dim\, V = dim\, K + dim\, K^{\perp}$.

5.9.3 Show that $(K^{\perp})^{\perp} = K$. If K_1 and K_2 are subspaces of V then $K_1^{\perp} \cap K_2^{\perp} = (K_1 + K_2)^{\perp}$ and $(K_1 \cap K_2)^{\perp} = K_1^{\perp} + K_2^{\perp}$.

5.9.4 Let K be a coisotropic subspace of V. Then the symplectic form ω induces a symplectic form on $K/K^{\perp}$.

5.9.5 Is it true that every symplectic vector space admits a Lagrangian subspace?

5.9.6 Let (V_1, ω_1) and (V_2, ω_2) be symplectic vector spaces and $\pi_1 : V_1 \times V_2 \longrightarrow V_1$, $\pi_2 : V_1 \times V_2 \longrightarrow V_2$ the canonical projections. Show that $\pi_1^* \omega_1 - \pi_2^* \omega_2$ is a symplectic form on $V_1 \times V_2$.

5.9.7 Show that $h \in Sp(V)$ if and only if the graph of h

$$\{(x, h(x))/x \in V\}$$

is a Lagrangian subspace of $V \times V$.

5.9.8 Show that every closed 2–form ω on a manifold S of dimension $2s$ is symplectic if and only if ω^s is a volume form.æ

Chapter 6

Hamiltonian systems

6.1 Hamiltonian vector fields

Suppose that T^*M is the cotangent bundle of a manifold M and $H : T^*M \longrightarrow R$ a function on T^*M. If ω_M is the canonical symplectic structure on T^*M, then there exists a unique vector field X_H on T^*M such that $i_{X_H}\omega_M = dH$; X_H is called in the literature Hamiltonian vector field of energy H. In this section we extend this definition to arbitrary symplectic manifolds.

Definition 6.1.1 *Let (S,ω) be a symplectic manifold and $H : S \longrightarrow R$ a function on S. Since the map $S_\omega : \chi(S) \longrightarrow \wedge^1 S$ is an isomorphism, there exists a unique vector field X_H on S such that*

$$i_{X_H}\omega = dH.$$

We call X_H a **Hamiltonian vector field** *with* **energy** *(or* **Hamiltonian energy***) H. The triple (S,ω,X_H) (or (S,ω,H)) is called a* **Hamiltonian system**.

From Proposition 5.2.10, we deduce that a Hamiltonian vector field on (S,ω) is symplectic. Conversely, a vector field X on (S,ω) is said to be **locally Hamiltonian** if for every point $x \in S$ there is a neighborhood U of x and a function H on U such that $X = X_H$ on U. From Proposition 5.2.10, we easily deduce that a vector field X on S is locally Hamiltonian if and only if X is a symplectic vector field.

Example

A classical example of a local Hamiltonian vector field which cannot be Hamiltonian is the following. Consider the 2–torus T^2 with local coordinates (x, y). Then

$$\omega = dx \wedge dy$$

is a well–defined symplectic form on T^2. Let

$$X = \partial/\partial x + \partial/\partial y$$

Then

$$i_X\omega = dy - dx,$$

which is closed in T^2 (but not exact!). Then X is locally Hamiltonian. But X cannot be Hamiltonian. In fact, if $X = X_H$ for some function H defined on T^2, then, since T^2 is compact, H has a critical point and at this point dH vanishes. Hence X would correspondingly have a zero.

Let (S, ω) be a symplectic manifold of dimension $2n$ and X_H a Hamiltonian vector field with energy H. Let (q^i, p_i) be canonical coordinates in S. Suppose that $\sigma : I = (-\epsilon, \epsilon) \longrightarrow S$ is an integral curve of X_H, i.e.,

$$X_H(\sigma(t)) = \dot{\sigma}(t),\ t \in I. \tag{6.1}$$

In local coordinates we have

$$\sigma(t) = (q^i(t),\ p_i(t)),$$

$$\dot{\sigma}(t) = \frac{dq^i}{dt}\frac{\partial}{\partial q^i} + \frac{dp_i}{dt}\frac{\partial}{\partial p_i}.$$

Consider the isomorphism

$$S_\omega : X \in \chi(S) \longrightarrow S_\omega(X) = i_X\omega \in \wedge^1 S.$$

A simple computation shows that

$$S_\omega(\partial/\partial q^i) = dp_i,\ s_\omega(\partial/\partial p_i) = -dq^i. \tag{6.2}$$

Hence we obtain

$$S_\omega^{-1}(dq^i) = -\partial/\partial p_i,\ S_\omega^{-1}(dp_i) = \partial/\partial q^i. \tag{6.3}$$

From (6.2) and (6.3), we deduce that if X is a vector field on S with local expression

$$X = X^i\partial/\partial q^i + \bar{X}^i\partial/\partial p_i,$$

then

$$S_\omega(x) = -\bar{X}^i dq^i + X^i dp_i.$$

Also, if α is a 1–form on S locally given by

$$\alpha = \alpha_i dq^i + \bar{\alpha}_i dp_i,$$

then

$$S_\omega^{-1}(\alpha) = \bar{\alpha}_i\partial/\partial q^i - \alpha_i\partial/\partial p_i.$$

Then, since

$$dH = \frac{\partial H}{\partial q^i}dq^i + \frac{\partial H}{\partial p_i}dp_i,$$

we obtain

$$X_H = S_\omega^{-1}(dH) = (\partial H/\partial p_i)\,\partial/\partial q^i - (\partial H/\partial q^i)\,\partial/\partial p_i. \tag{6.4}$$

From (6.1) and (6.4), we have

$$\frac{dq^i}{dt} = \frac{\partial H}{\partial p_i},\ \frac{dp_i}{dt} = -\frac{\partial H}{\partial q^i},\ 1 \leq i \leq n, \tag{6.5}$$

which are called the **Hamilton** (or **canonical**) **equations**.

The equation

$$i_{X_H}\omega = dH$$

is called the **symplectic** or **intrinsical form** of Hamilton equations.

Example

In Classical Mechanics the **phase space of momenta** is the cotangent bundle of the **configuration manifold** M. Consider a function H on T^*M given by

$$H(\alpha_x) = \frac{1}{2} g_x(\alpha_x, \alpha_x) - U \circ \pi_M, \text{ for all } \alpha_x \in T_x^* M,$$

where g is a metric in T^*M and $U : M \longrightarrow R$ is a C^∞ function on M. Then in local coordinates (q^i, p_i) we have

$$H(q^i, p_i) = \frac{1}{2} g^{ij} p_i p_j + U(q^i),$$

where $g^{ij} = g(dp_i, dp_j)$.

Now, suppose that $M = R^3$ and consider the metric g on $T^* R^3 \simeq R^3 \times R^3$ defined by

$$g((q^i, p_i), (q^i, \bar{p}_i)) = \frac{1}{m} \sum_{i=1}^{3} p_i \bar{p}_i;$$

where $m > 0$. Thus the Hamiltonian H is given by

$$H(q^1, q^2, q^3, p_1, p_2, p_3) = \frac{1}{2m} \sum_{i=1}^{3} (p_i)^2 + U(q^1, q^2, q^3).$$

and the Hamilton equations (6.5) are

$$dq^i/dt = p_i/m, \; dp_i/dt = -\partial U/\partial q^i, \; 1 \leq i \leq 3,$$

Hence we have

$$m\ddot{q}^i = m\frac{d^2 q^i}{dt^2} = -\partial U/\partial q^i, \; 1 \leq i \leq 3,$$

which is Newton's second law for a particle of mass m moving in a potential $U(q^1, q^2, q^3)$ in R^3.

Proposition 6.1.2 *Let (S, ω) and (W, α) be symplectic manifolds of same dimension and let h be a symplectic transformation from (S, ω) to (W, α). Then*

$$(Th) X_{F \circ h} = X_F,$$

for any function F on W.

Proof: Recall that for all k–form θ on W we have

$$h^\star(d\theta) = d(h^\star\theta)$$

and

$$h^\star(i_Z\theta) = i_{Th^{-1}(Z)}(h^\star\theta).$$

Hence, if $\theta = \omega$ and $Z = X_F$, where $F : W \longrightarrow R$ is a function on W, we have

$$i_{(Th^{-1})X_F}\omega = h^\star(i_{X_F}\omega) = h^\star(dF) = d(h^\star F) = d(F \circ h)$$

since $h^\star\alpha = \omega$. But, as ω is non- degenerate, the map $Z \longrightarrow i_Z\omega$ is an isomorphism. Since $i_{X_{F\circ h}}\omega = d(F \circ h)$ we deduce that $(Th^{-1})X_F = X_{F\circ h}$.□

6.2 Poisson brackets

Poisson brackets is the most important operation given by the symplectic structure.

Definition 6.2.1 *Let (S, ω) be a symplectic manifold of dimension $2n$. Let F and G be C^∞ functions on S. Then the* **Poisson bracket** *of F and G is defined by*

$$\{F, G\} = \omega(X_F, X_G) \tag{6.6}$$

From (6.6) we deduce

$$\{F, G\} = \omega(X_F, X_G) = (i_{X_F}\omega)(X_G) = i_{X_G}i_{X_F}\omega.$$

Now, we observe that $(i_{X_F}\omega)(Y) = (dF)Y$ implies $\omega(X_F, Y) = YF$. Thus, if we suppose $Y = X_G$ for some function G on S the following equality holds:

$$L_{X_G}F = X_G(F) = \omega(X_F, X_G) = \{F, G\}.$$

The reader is invited to prove the following:

Proposition 6.2.2 *For all C^∞ functions F, G and H on a symplectic manifold (S,ω) one has:*
(1) $\{F,G\} = -\{G,F\}$;
(2) $\{F,GH\} = \{F,G\}H + G\{F,H\}$;
(3) $\{F,\{G,H\}\} + \{G\{H,F\}\} + \{H,\{F,G\}\} = 0$ *(Jacobi identity)*
(4) $\{aF,G\} = a\{F,G\}$, *for all* $a \in R$;
(5) $\{F+G,H\} = \{F,H\} + \{G,H\}$.

We recall that a real vector space endowed with an internal operation g such that $g(a,b) = -g(b,a)$ and $g(a,g(b,c)) + g(b,g(c,a)) + g(c,g(a,b)) = 0$ is called Lie algebra. Therefore, if we consider the real vector space $C^\infty(S)$ of all C^∞ functions on the symplectic manifold (S,ω) then $C^\infty(S)$ is a Lie algebra with the Poisson bracket being the product (see Proposition 6.2.2).

Next, we define Poisson brackets for 1–forms.

Definition 6.2.3 *Let (S,ω) be a symplectic manifold and $\alpha,\beta \in \wedge^1 S$. The* **Poisson bracket** *$\{\alpha,\beta\}$ is the 1–form given by*

$$\{\alpha,\beta\} = -S_\omega([X_\alpha, X_\beta]),$$

where $i_{X_\alpha}\omega = \alpha$ and $i_{X_\beta}\omega = \beta$.

Hence the following diagram

$$\begin{array}{ccc} \chi(S)\times\chi(S) & \xrightarrow{-[\,,\,]} & \chi(S) \\ \downarrow{\scriptstyle S_\omega\times S_\omega} & & \downarrow{\scriptstyle S_\omega} \\ \wedge^1 S\times\wedge^1 S & \xrightarrow{\{\,,\,\}} & \wedge^1 S \end{array}$$

is commutative. Therefore $\wedge^1 S$ endowed with the Poisson bracket $\{,\}$ of 1–forms is a Lie algebra.

Proposition 6.2.4 *We have*

$$\{\alpha,\beta\} = -L_{X_\alpha}\beta + L_{X_\beta}\alpha + d(i_{X_\alpha} i_{X_\beta}\omega).$$

Proof: In fact, we have

$$0 = (d\omega)(X_\alpha, X_\beta, Z)$$

$$= X_\alpha(\omega(X_\beta, Z)) + X_\beta(\omega(Z, X_\alpha)) + Z(\omega(X_\alpha, X_\beta))$$

$$-\omega([X_\alpha, X_\beta], Z) - \omega([X_\beta, Z]), X_\alpha) - \omega([Z, X_\alpha], X_\beta).$$

Now, since

$$\omega(X_\alpha, Z) = (i_{X_\alpha}\omega)(Z) = \alpha(Z),$$

$$\omega(X_\beta, Z) = (i_{X_\beta}\omega)(Z) = \beta(Z),$$

we obtain

$$0 = X_\alpha(\beta(Z)) - X_\beta(\alpha(Z)) + Z(\omega(X_\alpha, X_\beta))$$

$$+\{\alpha, \beta\}(Z) + \alpha([X_\beta, Z]) + \beta([Z, X_\alpha])$$

$$= X_\alpha(\beta(Z)) - \beta([X_\alpha, Z]) - X_\beta(\alpha(Z))$$

$$+\alpha([X_\beta, Z]) + \{\alpha, \beta\}(Z) + Z(\omega(X_\alpha, X_\beta))$$

$$= (L_{X_\alpha}(Z) - (L_{X_\beta}\alpha)(Z) + \{\alpha, \beta\}(Z)$$

$$+d(\omega(X_\alpha, X_\beta))(Z).$$

Therefore we deduce

$$\{\alpha, \beta\} = -L_{X_\alpha}\beta + L_{X_\beta}\alpha + d(i_{X_\alpha}i_{X_\beta}\omega). \square$$

Corollary 6.2.5 *If α and β are closed then $\{\alpha, \beta\}$ is exact.*

Proof: If α and β are closed we have

$$L_{X_\alpha}\beta = i_{X_\alpha}d\beta + di_{X_\alpha}\beta = di_{X_\alpha}\beta,$$

$$L_{X_\beta}\alpha = i_{X_\beta}d\alpha + di_{X_\beta}\alpha = di_{X_\beta}\alpha.$$

Then, from Proposition 6.2.4, we obtain

$$\{\alpha, \beta\} = -di_{X_\alpha}\beta + di_{X_\beta}\alpha + d(i_{X_\alpha}i_{X_\beta}\omega)$$

$$= d(-i_{X_\alpha}\beta + i_{X_\beta}\alpha + i_{X_\alpha}i_{X_\beta}\omega).\square$$

The following result relates the Poisson brackets of functions and 1–forms.

Corollary 6.2.6 *We have*

$$d\{F, G\} = \{dF, dG\}$$

Proof: From Proposition 6.2.4 we obtain

$$\{dF, dG\} = -L_{X_F}(dG) + L_{X_G}(dF) + d(i_{X_F}i_{X_G}\omega)$$

$$= -d(X_F G) + d(X_G F) - d\{F, G\}$$

$$= -d(L_{X_F}G) + d(L_{X_G}G) - d\{F, G\}$$

$$= -d\{G, F\} + d\{F, G\} - d\{F, G\}$$

$$= d\{F, G\}.\square$$

Corollary 6.2.7 *We have*

$$X_{\{F,G\}} = -[X_F, X_G].$$

Proof: In fact, we have

$$i_{X_{\{F,G\}}}\omega = d\{F,G\} = \{dF, dG\}$$

$$= -S_\omega([X_F, X_G]) = -i_{[X_F,X_G]}\omega. \square$$

Let us now recall that if (S,ω) is a symplectic manifold of dimension $2n$ then ω^n is a volume form on S.

Theorem 6.2.8 (**Liouville Theorem**). *Let (S,ω) be a symplectic manifold of dimension $2n$ and φ_t the flow of a Hamiltonian vector field. Then $\varphi_t^\star$ preserves the volume form ω^n for all t, i.e., $\varphi_t^\star\,\omega^n = \omega^n$, for all t.*

Proof: Since $\varphi_t^\star\omega = \omega$ it follows that

$$\varphi_t^\star\omega^n = \varphi_t^\star(\omega \wedge \ldots \wedge \omega) = (\varphi_t^\star\omega) \wedge \ldots \wedge (\varphi_t^\star\omega)$$

$$= (\varphi_t^\star\omega)^n = \omega^n. \square$$

To end this section we develop some computations in local coordinates. Let (S,ω) be a symplectic manifold and $(q^1,\ldots,q^n,p_1,\ldots,p_n)$ canonical coordinates in S. From (6.4), we deduce that the Poisson bracket of two functions F and G is given by

$$\{F,G\} = L_{X_G}F = (\partial F/\partial q^i)(\partial G/\partial p_i) - (\partial F/\partial p_i)(\partial G/\partial q^i) \tag{6.7}$$

From (6.7) we obtain the Poisson brackets of the canonical coordinates:

$$\{q^i,q^j\} = \{p_i,p_j\} = 0,\ \{q^i,p_j\} = \delta^i_j.$$

Furthermore, if F is a function on S, we have

$$\{F,q^i\} = -\{q^i,F\} = -\partial F/\partial p_i,$$

$$\{F,p_i\} = -\{p_i,F\} = \partial F/\partial q^i. \tag{6.8}$$

Using (6.8), Hamilton equations may be written as

$$\frac{dq^i}{dt} = \{q^i,H\},\ \frac{dp_i}{dt} = -\{p_i,H\}.$$

Let us now show that a canonical transformation $h : (S, \omega) \longrightarrow (S, \omega)$ preserves Hamilton equations. To see this it is sufficient to show that the Poisson brackets are invariant under the action of h. Invariant here means

$$h^*\{F, G\} = \{h^*F, h^*G\}, \text{ i.e.,}$$

$$\{F, G\} \circ h = \{F \circ h, G \circ h\}.$$

In fact, we have

$$h^*\{F, G\} = \{F, G\} \circ h = (X_G F) \circ h$$

$$= (((Th)X_{G\circ h})F) \circ h \ \text{ (by Proposition 6.1.2)}$$

$$= (X_{G\circ h})(F \circ h)) = \{F \circ h, G \circ h\} = \{h^*F, h^*G\}.$$

In particular, if $h : (q, p) \longrightarrow (\bar{q}.\bar{p})$, where (q^i, p_i) and $(\bar{q}^i, \bar{p}_i)$ are canonical coordinates, we have

$$h^*\{q^i, H\} = \{q^i \circ h, H \circ h\} = \{\bar{q}^i, K\} = \frac{d\bar{q}^i}{dt},$$

$$h^*\{p_i, H\} = \{p_i \circ h, H \circ h\} = \{\bar{p}_i, K\} = -\frac{d\bar{p}_i}{dt},$$

where $K = h^*H = H \circ h$. This shows that canonical transformations preserve the form of Hamilton equations.

6.3 First integrals

We say that a 1–form α on a manifold M is a **first integral** of a vector field X if $i_X\alpha = \alpha(X) = 0$. Also, a function F on M such that $XF = 0$ is called a **first integral** of X. Obviously, if F is a first integral of X then $dF(X) = XF = 0$ and thus dF is a first integral of X. Now, let (S, ω) be a symplectic manifold. Then for every 1–form α on S there exists a unique vector field X_α such that $\alpha = i_{X_\alpha}\omega$. In such as case, as

$$i_{X_\alpha}\alpha = i_{X_\alpha} i_{X_\alpha}\omega = 0,$$

α is a first integral of X_α. In particular, if $\alpha = dH$ then H is a first integral of $X_H = X_\alpha$. If follows that the Hamiltonian function H is constant along the integral curves of X_H; in fact, if $\sigma(t)$ is an integral curve of X_H, we have

$$X_H(H)(\sigma(t)) = \frac{d}{dt}H(\sigma(t)) = dH(\sigma(t))(X_H(\sigma(t)) = 0.$$

This gives the well–known principle of "**energy conservation law**". A first integral F of X_H is usually called **a constant of motion**. Then H is a constant of motion. More generally, let F and G be C^∞ functions on (S, ω). Then there is a vector field X_F, resp. X_G such that

$$i_{X_F}\omega = dF, \text{ resp. } i_{X_G}\omega = dG.$$

Hence

$$L_{X_F}G = i_{X_F}dG + di_{X_F}G = i_{X_F}dG$$

$$= i_{X_F}i_{X_G}\omega = -i_{X_G}i_{X_F}\omega = -i_{X_G}dF$$

$$= -L_{X_G}F,$$

i.e.,

$$X_F(G) = -X_G(F),$$

Thus we have

Proposition 6.3.1 *If F and G are functions on a symplectic manifold (S, ω) such that $X_F(G) = 0$ then $X_G(F) = 0$.*

Definition 6.3.2 *If α and β are 1–forms on a symplectic manifold (S, ω) such that $\omega(X_\alpha, X_\beta) = 0$ then they are said* **in involution**. *Two functions F and G on S are* **in involution** *if dF and dG are in involution.*

Let us recall that if α and β are closed then

$$\{\alpha, \beta\} = d(-i_{X_\alpha}\beta + i_{X_\beta}\alpha + i_{X_\beta}\omega)$$

(see Corollary 6.2.5). Hence, if α and β are in involution we have $\{\alpha, \beta\} = 0$. Also we have:

Proposition 6.3.3 *Two functions F and G on S are in involution if and only if* $\{F,G\}=0$.

Proof: Since $X_{dF}=S_F$, $X_{dG}=X_G$, we obtain

$$\{F,G\}=\omega(X_F,X_G)=\omega(X_{dF},X_{dG}).$$

Hence $\{F,G\}=0$ if and only if $\omega(X_{dF},X_{dG})=0$, or, equivalently, F and G are in involution. □

We have seen before that

$$X_G(F)=\omega(X_F,X_G)=\{F,G\}=-X_F(G).$$

So, if F and G are in involution, then $X_G(F)=X_F(G)=0$, that is, F (resp. G) is a first integral of X_G (resp. X_F) and conversely. As $i_{X_\alpha}\omega=\alpha$ and $i_{X_\beta}\omega=\beta$ we have

$$i_{X_\beta}i_{X_\alpha}\omega=i_{X_\beta}\alpha=-i_{X_\alpha}\beta.$$

Thus a similar result holds for 1–forms.

Proposition 6.3.4 *Let X_F be a Hamiltonian vector field on a symplectic manifold (S,ω). Then $i_{X_F}\omega$ is a first integral of the Hamiltonian vector field X_H if and only if* $[X_F,X_H]=0$.

Proof: As we know

$$(i_{X_F}\omega)(X_H)=\omega(X_F,X_H)=\{F,H\},$$

$$d\{F,H\}=\{dF,dH\}=-i_{[X_F,X_H]}\omega.$$

Therefore $(i_{X_F}\omega)(X_H)=0$ implies $\{F,H\}=0$ and then $[X_F,X_H]=0$. Conversely $[X_F,X_H]=0$ implies $\{dF,dH\}=0$ and so F and H are in involution (see Proposition 6.3.3). Hence $(i_{X_F}\omega)(X_H)=0$. □

Suppose that β is a closed 1–form on a symplectic manifold (S,ω) such that for all closed 1–form α on S, α and β are in involution. Then $\beta=0$. In particular, if $\beta=dF$ such that for all function G, F and G are in involution then F is constant ($dF=0$) (in fact, F is locally constant if S is not supposed a connected manifold). This is called a **regular condition** for the Poisson brackets.

6.4 Lagrangian submanifolds

A set of C^∞ functions $f_1, \ldots, f_k$ on a symplectic manifold (S, ω) is said to be **independent** if the corresponding Hamiltonian vector fields $X_{f_1}, \ldots, X_{f_k}$ are linearly independent (this is equivalent to say that the 1–forms $df_1, \ldots, df_k$ are linearly independent).

Let K be a submanifold of codimension k of (S, ω) locally defined by the independent functions $f_1 = \ldots = f_k = 0$, $k \leq n$.

Lemma 6.4.1 *K is coisotropic if and only if $\{f_i, f_j\} = 0$ on $f_1 = \ldots = f_k = 0$, $1 \leq i, j \leq k$.*

Proof: If $x \in K$ then the tangent space $T_x K$ is the orthonormal complement of the vector space spanned by $X_{f_1}(x), \ldots, X_{f_k}(x)$ with respect to $\omega(x)$. In fact, if $Y \in T_x S$, then

$$\omega(X_{f_i}(x), Y) = (i_{X_{f_1(x)}}\omega)(Y) = df_i(x)(Y) = Y(f_i), \quad 1 \leq i \leq k.$$

So $Y \in T_x K$ if and only if $df_i(x)(Y) = 0$, for all $1 \leq i \leq k$. Therefore

$$< X_{f_1}(x), \ldots, X_{f_k}(x) > \subset (T_x K)^\perp. \tag{6.9}$$

Now, suppose that K is coisotropic. Then $(T_x K)^\perp \subset T_x K$ for all $x \in K$. So X_{f_i} is tangent to K and thus $X_{f_i}(f_j) = \{f_j, f_i\} = 0$ for any $1 \leq i, j \leq k$. Conversely, if $\{f_i, f_j\} = 0$ for any $1 \leq i, j \leq k$, then X_{f_i} is tangent to K for all i and thus (6.9) holds. Hence we deduce that $(T_x K)^\perp \subset T_x K$, for all $x \in K$.□

Coisotropic submanifolds are also known as **first class constrained manifolds** in Dirac terminology (see Chapter 8).

Let us recall that a submanifold N of (S, ω) is called symplectic if $(T_x N) \cap (T_x N)^\perp = 0$ for all $x \in N$.

Lemma 6.4.2 *Let K_0 be a submanifold of codimension k in $K \subset (S, \omega)$. Then K_0 is symplectic if and only if*

$$(T_x K_0) \cap (T_x K)^\perp = 0, \; x \in K_0 \tag{6.10}$$

Proof: If K_0 is symplectic then

$$(T_x K_0) \cap (T_x K_0)^\perp = 0.$$

As $T_xK_0 \subset T_xK$, we have $(T_xK)^\perp \subset (T_xK_0)^\perp$. Then (6.10) holds. Conversely, if (6.10) holds then taking into account exercise (2) of the present chapter, we have

$$((T_xK_0) \cap (T_xK)^\perp)^\perp = O^\perp = T_xS \Longrightarrow (T_xK_0)^\perp + T_xK = T_xS$$

$$\Longrightarrow dim\,[(T_xK_0)^\perp \cap (T_xK)] = k \Longrightarrow$$

$$(T_xK_0)^\perp \cap (T_xK) = (T_xK)^\perp, \text{ since dim } (T_xK)^\perp = k,$$

and so

$$(T_xK_0) \cap (T_xK_0)^\perp = 0. \square$$

Theorem 6.4.3 (**Jacobi's Theorem**). *Suppose that $\mathcal{C} = \{f_1, \ldots, f_k\}$ are C^∞ functions on a neighborhood of a point $x \in S$, where (S, ω) is a 2n–dimensional symplectic manifold. If they are in involution, then $k \leq n$ and there exists a neighborhood of x on which there is defined a set of C^∞ functions $f_{k+1}, \ldots, f_n$ such that $\bar{\mathcal{C}} = \{f_1, \ldots, f_n\}$ is in involution.*

Proof: (See Duistermaat [50], p. 100). □
Now, we show an important result (we follow Duistermaat [50]).

Theorem 6.4.4 *Let K be a submanifold of codimension k in a 2n–dimensional symplectic manifold (S, ω). Through each point $x \in K$ there passes a Lagrangian submanifold $L \subset K$ if and only if K is coisotropic.*

Proof: If through each $x \in K$ passes a Lagrangian submanifold $L \subset K$ then

$$T_xK \supset T_xL = (T_xL)^\perp \supset (T_xK)^\perp. \tag{6.11}$$

Thus K is coisotropic. Let us see the converse. As K is coisotropic, from Lemma 6.4.1 $\{f_i, f_j\} = 0$. Thus (see Corollary 6.2.7) $[X_{f_i}, X_{f_j}] = 0$, $1 \leq i, j \leq k$. Then $D : x \longrightarrow D(x) = (T_xK)^\perp$ gives an integrable distribution according to Frobenius Theorem.

Let K_0 be a submanifold of codimension k in K transversal to the integral manifolds of D, i.e.,

$$T_xK = T_xK_0 \oplus T_xN,$$

where N is an integral manifold of D through x. Then $(T_xK_0) \cap (T_xK)^\perp = 0$ and so from Lemma 6.4.2 K_0 is a symplectic manifold. Let L_0 be a Lagrangian submanifold of K_0. Let $x_0 \in L_0$ and U a sufficiently small neighborhood of x_0 on which K is defined by the above independent functions f^i. Let $(\varphi_i)_t$ be the flow of the vector field X_{f_i}. Define

$$L_1 = \{(\varphi_1)_{t_1}(x)/x \in U \cap L_0, \text{ for all } t_1 \in I_1\},$$

where I_1 is a small neighborhood of $0 \in R$. Then we have

$$T_xL_1 = T_xL_0 + < X_{f_1}(x) > \quad ((\varphi_1)_0(x) = x)$$

and so

$$(T_xL_1)^\perp = (T_xL_0)^\perp \cap < X_{f_1}(x) >^\perp .$$

But $T_xL_0 \subset T_xK$ implies $(T_xK)^\perp \subset (T_xL_0)^\perp = T_xL_0$. Thus

$$(T_xL_1)^\perp \supset (T_xL_0) + < X_{f_1}(x) > = T_xL_1,$$

i.e., T_xL_1 is isotropic if $x \in U \cap L_0$. As the map

$$d(\varphi_1)_t(x) : (T_xS, \omega(x)) \longrightarrow (T_{(\varphi_1)_t(x)}S, \; ((\varphi_1)_t)^*\omega(x))$$

is symplectic then $d(\varphi_1)_t(x)(T_xL_1)$ is also isotropic showing that L_1 is isotropic with dimension $n - k + 1$. If we repeat the argument for

$$L_i = \{((\varphi_1)_{t_i} \circ \ldots \circ (\varphi_1)_{t_1})(x)/x \in U \cap L_0, \; (t_1, \ldots, t_i) \in I_1 \times \ldots \times I_i\}$$

we find that L_i is an isotropic submanifold of dimension $n - k + i$. Thus, taking $i = k$, one obtains a Lagrangian submanifold $L = L_k$ through $x_0 \in L_0$. If for such x_0 there is (locally) only one Lagrangian submanifold then the assertion of the theorem is proved. So let us now show the (local) unicity of L.

If $\tilde{L}$ is any Lagrangian submanifold of K then from (6.11) the integral manifolds of D are tangent to $\tilde{L}$ and, if in addition $L_0 \subset \tilde{L}$ then $\tilde{L}$ contains (locally) the integral manifolds passing through points of L_0. Now, L was defined by

$$L = \{((\varphi_k)_{t_k} \circ \ldots \circ (\varphi_1)_{t_1})(x)/x \in U \cap L_0, \; (t_1, \ldots, t_k) \in I_1 \times \ldots \times I_k\}$$

and as the differential of the mapping

$$(t_1, \ldots, t_k, x) \longrightarrow ((\varphi_k)_{t_k} \circ \ldots \circ (\varphi_1)_{t_1})(x)$$

has rank m ($T_x L_0$ and $(T_x K)^\perp$ are transversal) one deduces that for sufficiently small neighborhoods U and $I_1 \times \ldots \times I_k$, the mapping is an embedding and L has dimension n, $U \cap L_0 \subset L \subset \tilde{L}$ (locally). Thus $\tilde{L} = L$ since $\tilde{L}$ is Lagrangian and $dim \tilde{L} = n$.□

Proposition 6.4.5 *If (S, ω) is a symplectic manifold then a submanifold K of S is Lagrangian if and only if there is a fiber bundle E such that $TK \oplus E = TS/_K$ with $T_x K$ and E_x being isotropic subspaces of $T_x S$.*

Proof: One direction is obvious. The other direction is obtained by using the fact that on every symplectic manifold there is an almost complex structure J. Thus, for each $x \in S$, $J_x(T_x K) = E_x$ is a Lagrangian complement of the Lagrangian subspace $T_x K$.□

Proposition 6.4.6 *Let α be a 1–form on M and $L \subset T^* M$ its graph. Then L is Lagrangian if and only if α is closed.*

Proof: Let $\alpha : M \longrightarrow T^* M$ be a 1–form on M locally given by $\alpha(q^i) = \alpha_i dq^i$. Let (q^i, p_i) be the induced coordinates in $T^* M$. We recall that the canonical symplectic structure on $T^* M$ is given by $\omega_M = -d\lambda_M$, where λ_M is the Lioville form. Then we have

$$\omega_M = dq^i \wedge dp_i.$$

Hence

$$\alpha^*(\omega_M) = \alpha^*(dq^i \wedge dp_i) = d(q^i \circ \alpha) \wedge d(p_i \circ \alpha)$$

$$= dq^i \wedge d\alpha_i = -d\alpha$$

Since the graph L of α is given by

$$L = \{(x, \alpha(x))/x \in M\}$$

we have dim $L = \frac{1}{2}$ dim $(T^* M)$. Furthermore, since α is closed if and only if $\alpha^* \omega_M = 0$, we deduce that L is Lagrangian if and only if α is closed.□

The following result is given as an exercise.

Proposition 6.4.7 *Let $f : S \longrightarrow W$ be a symplectomorphism from a symplectic manifold (S, ω) to the symplectic manifold (W, Ω). Then $S \times W$ is a symplectic manifold with symplectic form*

$$\beta = \pi_1^\star \omega - \pi_2^\star \Omega,$$

where the π's are the obvious canonical projections. The graph of f is a Lagrangian submanifold of $S \times W$.

Definition 6.4.8 *Let (S, ω) be a symplectic manifold and L a Lagrangian submanifold. If L is the graph of a closed 1-form α then locally there exists a function F such that $\alpha = dF$. We call F the* **generating function** *of L.*

Let (q, p) be symplectic coordinates in S (we are omitting the index i for q, p for simplicity), $f : S \longrightarrow S$ a symplectomorphism and set

$$f(q,p) = (\bar{q}(q,p), \bar{p}(q,p)).$$

Then $(\bar{q}, \bar{p})$ are symplectic coordinates and $\omega = d\bar{q} \wedge d\bar{p}$. Let us now suppose that $(q, \bar{q})$ are independent coordinates, that is, the matrix $(\partial(q, \bar{q})/\partial(q, p))$ is non–singular. Then the graph of f is given by

$$(q, p, \bar{q}(q,p), \bar{p}(q,p))$$

and may be expressed as the image of a closed 1–form $\alpha = dF$ where $(q, \bar{q})$ has the form

$$(q,\ (\partial F/\partial q)(q, \bar{q}), q,\ (\partial F/\partial \bar{q})(q, \bar{q}))$$

If $H : T^\star M \longrightarrow R$ is a Hamiltonian function then

$$H(q,p) = H(q, \frac{\partial F}{\partial q}) = H(\bar{q},\ (\partial F/\partial \bar{q})).$$

The **Hamilton - Jacobi method** for solving the Hamiltonian equations consists in showing that H is independent of p by the use of F. In fact, the generating function F satisfies the equation

$$H(\bar{q},\ (\partial F/\partial \bar{q})(q, \bar{q})) = G.$$

The Hamiltonian equations are

$$\frac{dq}{dt} = 0, \quad \frac{d\bar{p}}{dt} = \frac{\partial G}{\partial \bar{q}}$$

and a solution is given by $q(t) = q(0)$, $\bar{p}(t) = p(0) + t(\partial G/\partial \bar{q})$.

Proposition 6.4.9 *A neccessary and sufficient condition for the (autonomous) change of coordinates* $(q^i, p_i) \longrightarrow (\bar{q}^i, \bar{p}_i)$ *be canonical is that*

$$\frac{\partial q}{\partial \bar{q}} = \frac{\partial \bar{p}}{dp}, \; \frac{\partial q}{\partial \bar{p}} = -\frac{\partial \bar{q}}{\partial p},$$

$$\frac{\partial p}{\partial \bar{q}} = -\frac{\partial \bar{p}}{\partial q}, \; \frac{\partial p}{\partial \bar{p}} = \frac{\partial \bar{q}}{\partial q}.$$

Proof: We want a transformation from the set of variables (q, p) to another set $(\bar{q}, \bar{p})$ such that

$$H(q, p) = G(\bar{q}(q, p), \bar{p}(q, p))$$

$$\frac{d\bar{q}}{dt} = \frac{\partial G}{\partial \bar{p}}, \; \frac{d\bar{p}}{dt} = -\frac{\partial G}{\partial \bar{q}} \quad \text{(Hamilton equations)}$$

The symplectic form of the Hamilton equations are (for each Hamiltonian):

$$i_{X_H}\omega = dH, \; i_{X_G}\bar{\omega} = dG,$$

where $\bar{\omega}$ is ω expressed in the new coordinates.

Let us suppose that the change of coordinates is canonical. Then $X_H = X_G$ and

$$X_H = (\partial H/\partial p)\, \partial/\partial q - (\partial H/\partial q)\, \partial/\partial p = \dot{q}\, \partial/\partial q + \dot{p}\, \partial/\partial p$$

(the dot means derivative with respect to t). But

$$X_H(q(t)) = \dot{q}(t) = \dot{\bar{q}}(t)\, \frac{\partial q}{\partial \bar{q}} + \dot{\bar{p}}(t)\, \frac{\partial q}{\partial \bar{p}} = X_G(q(t)) \tag{6.12}$$

$$X_H(p(t)) = \dot{p}(t) = \dot{\bar{q}}(t)\, \frac{\partial p}{\partial \bar{q}} + \dot{\bar{p}}(t)\, \frac{\partial p}{\partial \bar{p}} = X_G(p(t)) \tag{6.13}$$

On the other hand

$$X_H = \left(\frac{\partial G}{\partial \bar{q}} \frac{\partial \bar{q}}{\partial p} + \frac{\partial G}{\partial \bar{p}} \frac{\partial \bar{p}}{\partial p} \right) \frac{\partial}{\partial q} - \left(\frac{\partial G}{\partial \bar{q}} \frac{\partial \bar{q}}{\partial q} + \frac{\partial G}{\partial \bar{p}} \frac{\partial \bar{p}}{\partial q} \right) \frac{\partial}{\partial p}$$

$$= \left(-\frac{\partial \bar{q}}{\partial p}\, \dot{\bar{p}} + \frac{\partial \bar{p}}{\partial p}\, \dot{\bar{q}} \right) \frac{\partial}{\partial q} - \left(-\frac{\partial \bar{q}}{\partial q}\, \dot{\bar{p}} + \frac{\partial \bar{p}}{\partial q}\, \dot{\bar{q}} \right) \frac{\partial}{\partial p}$$

The substitution of (6.12) and (6.13) into X_H and the comparison with the above expression gives the desired result. We leave to the reader the proof of the other direction. □

Corollary 6.4.10 *The above change of coordinates is canonical if and only if*

$$\{F,G\}_{q,p} = \{F,G\}_{\bar{q},\bar{p}}$$

Proof: We have

$$\{F,G\}_{\bar{q},\bar{p}} = \frac{\partial F}{\partial \bar{q}}\frac{\partial G}{\partial \bar{p}} - \frac{\partial F}{\partial \bar{p}}\frac{\partial G}{\partial \bar{q}}$$

$$= \left(\frac{\partial F}{\partial q}\frac{\partial q}{\partial \bar{q}} + \frac{\partial F}{\partial p}\frac{\partial p}{\partial \bar{q}}\right)\left(\frac{\partial G}{\partial q}\frac{\partial q}{\partial \bar{p}} + \frac{\partial G}{\partial p}\frac{\partial p}{\partial \bar{p}}\right)$$

$$- \left(\frac{\partial F}{\partial q}\frac{\partial q}{\partial \bar{p}} + \frac{\partial F}{\partial p}\frac{\partial p}{\partial \bar{p}}\right)\left(\frac{\partial G}{\partial q}\frac{\partial q}{\partial \bar{q}} + \frac{\partial G}{\partial p}\frac{\partial p}{\partial \bar{q}}\right)$$

If the change of coordinates is canonical, from Proposition 6.4.9, we deduce

$$\{F,G\}_{\bar{q},\bar{p}} = \{F,G\}_{q,p}.$$

The converse is proved by a similar procedure. □

Now, let us consider the generating function $F = F(q,\bar{q})$. One has

$$p = \frac{\partial F}{\partial q}, \ \bar{p} = -\frac{\partial F}{\partial \bar{q}}, \ G = H + \frac{\partial F}{\partial t}$$

and F is the generating function of a canonical transformation.

The use of Proposition 6.1.2 and Jacobi's Theorem gives the following result (see also Weber [124]).

Proposition 6.4.11 *Let (S,ω) be a symplectic manifold of dimension $2n$ and $C = \{f_1, \ldots, f_n\}$ a set of C^∞ independent functions in involution on a neighborhood U of a point $x \in X$. Suppose that for every $1 \leq i,j \leq n$ we have rank $(\partial f_i/\partial p_j) = n$, where (q^i, p_i) are canonical coordinates for S. Then there is a local canonical transformation $g : U \longrightarrow g(U) \subset S$ such that $(Tg)X_{f_i} = X_{p_i}$.*

6.5 Poisson manifolds

Let (S,ω) be a symplectic manifold. Then to (S,ω) corresponds an operator $\{\ ,\ \}$ on the algebra of C^∞ functions on S, such that $\{\ ,\ \}$ is a skew-symmetric bilinear mapping defined by $\{F,G\} = \omega(X_f, X_G)$, where X_F and X_G are vector fields on S defined by $i_{X_F}\omega = dF$, $i_{X_G}\omega = dG$. This result suggested Lichnerowicz [90] to study manifolds on which is defined an operator $\{\ ,\ \}$ giving a structure of Lie algebra on the space of C^∞ functions on such manifolds.

Definition 6.5.1 *A* **Poisson structure** *on a manifold P is defined by a bilinear map*

$$C^\infty(P) \times C^\infty(P) \longrightarrow C^\infty(P),$$

where $C^\infty(P)$ is the space of C^∞ functions on P, noted by $(F,G) \longrightarrow \{F,G\}$ such that the following properties are verified:
(1) $\{F,G\} = -\{G,F\}$ (skew-symmetry)
(2) $\{F,\{G,H\}\} + \{G,\{H,F\}\} + \{H,\{F,G\}\} = 0$ (Jacobi identity)
(3) $\{F,GH\} = G\{F,H\} + H\{F,G\}$, $\{FG,H\} = \{F,H\}G + F\{G,H\}$.
We call $\{\ ,\ \}$ **Poisson bracket** *and the pair $(P,\{\ ,\ \})$* **Poisson manifold**.

So, on every symplectic manifold (S,ω) there is defined a Poisson structure, canonically associated to the symplectic structure defined by ω.

Let $\{\ ,\ \}$ be a Poisson structure on a manifold P. From (3) of Definition 6.5.1 we see that the map

$$\{F,\ \} : C^\infty(P) \longrightarrow C^\infty(P)$$

$$G \longrightarrow \{F,G\}$$

is a derivation. Therefore there is a unique vector field X_F on P such that $X_F(G) = \{F,G\}$; X_F is called the **Hamiltonian vector field of F**. One can easily check that the Hamiltonian vector field $X_{\{F,G\}}$ of the function $\{F,G\}$ is $[X_F, X_G]$.

Suppose that $F, G \in C^\infty(P)$ and x is a point of P. Then we have

$$\{F,G\}(x) = (X_F(G))(x) = dG(x)(X_F(x)).$$

Therefore, $\{F,G\}(x)$ is a function depending on $dG(x)$, for F fixed. The same reasoning gives that, for each G fixed, $\{F,G\}(x)$ is a function depending on $dF(x)$ (one has $\{G, F\}(x) = dF(x)(X_G(x)))$. So, for each $x \in P$, there is a bilinear map

$$\Omega(x) : T_x^\star P \times T_x^\star P \longrightarrow R$$

such that

$$\Omega(x)(dF(x),\ dG(x)) = \{F, G\}(x)$$

with $\Omega(x)$ being skew–symmetric. Then $x \longrightarrow \Omega(x)$ defines a skew–symmetric tensor field of type $(2,0)$ on P. Therefore one has (Libermann and Marle [88])

Theorem 6.5.2 *Let $(P, \{\ ,\ \})$ be a Poisson manifold. Then there is a unique 2–form Ω on P such for all $F, G \in C^\infty(P)$ and $x \in P$,*

$$\{F, G\}(x) = \Omega(x)(dF(x),\ dG(x)).$$

We call Ω the "**Poisson tensor field**".

Remark 6.5.3 We may try to show that if it is given a skew–symmetric tensor field Ω of type *(2,0)* on *P*, then is defined on P a Poisson structure. Lichnerowicz showed that this is only possible if Ω verifies the identity

$$[\Omega, \Omega] = 0,$$

where $[\ ,\ ]$ is the **Schouten bracket**. These bracket are characterized by the following properties. Suppose that A (resp. B) is a skew–symmetric tensor field of type *(a,0)* (resp. *(b,0))*. The skew–symmetric tensor field $[\Omega, \Omega]$ of type $(a + b - 1, 0)$ is defined by

$$i_{[A,B]}\beta = (-1)^{ab+b} i_A d i_B \beta + (-1)^a i_B d i_A \beta,$$

where β is a closed $(a + b - 1)$–form. If $a = 1$, then $[A, B] = L_A B$. One has

$$[A, B] = (-1)^{ab}[B, A].$$

If C is a skew–symmetric tensor field of type $(c,0)$ then

$$(-1)^{ab} \Big[[B, C], A] + (-1)^{bc}[[C, A], B] + (-1)^{ca}[[A, B], C\Big] = 0$$

(Jacobi identity).

Let us consider the particular case where $a, b = 2$. Take

$$A = \Sigma\, X_i \wedge Y_i, \quad B = \Sigma\, Z_j \wedge W_j.$$

Then

$$[A, B] = \sum_{i,j} ([X_i, Z_j] \wedge Y_i \wedge W_j$$

$$+[Y_i, W_j] \wedge X_i \wedge Z_j - [X_i, W_j] \wedge Y_i \wedge Z_j$$

$$-[Y_i, Z_j] \wedge X_i \wedge W_j + (div\, X_i) Y_i \wedge Z_j \wedge W_j$$

$$-(div\, Y_i) X_i \wedge Z_j \wedge W_j + (div\, Z_j) X_i \wedge Y_i \wedge W_j$$

$$-(div\, W_j) X_i \wedge Y_i \wedge Z_j).$$

Thus, $[\Omega, \Omega] = 0$ if and only if Poisson brackets satisfy Jacobi identity.

From these comments we may say now that a "**Poisson manifold**" is a pair (P, Ω) where P is a manifold of dimension m and Ω is a skew–symmetric tensor field of type (2,0) on P of rank $2n \leq m$ verifying $[\Omega, \Omega] = 0$" (If $2n = m$, then P is a symplectic manifold).

From the above Theorem we see that if (P, Ω) is a Poisson manifold, then for all $x \in P$ and all 1–form α on P, there is a map

$$\bar{\rho}_{\Omega(x)} : T_x^* P \longrightarrow T_x P$$

such that

$$\beta(\bar{\rho}_{\Omega(x)}(\alpha)) = \Omega(x)(\alpha, \beta), \text{ for all} \beta \in T_x^* P.$$

Therefore, for every point $x \in P$ one obtains a mapping $\bar{\Omega} = \rho_\Omega : T^* P \longrightarrow TP$ given by

$$\beta(\bar{\Omega}(\alpha)) = \Omega(\alpha, \beta),$$

for all $\alpha, \beta \in \wedge^1 P$. It is clear now that the Hamilton vector field X_F is given by $X_F = \bar{\Omega}(dF)$ and

$$\{F,G\} = \bar{\Omega}(dF)G = -\bar{\Omega}(dG)F = \Omega(dF,dG). \tag{6.14}$$

Therefore, we have an analogous construction to the symplectic case. We observe that if (S,ω) is a symplectic manifold one obtains trivially a Poisson structure Ω on S, defining $\bar{\Omega} = (S_\omega)^{-1}$, where $S_\omega : TS \longrightarrow T^*S$. In such a case $\bar{\Omega}$ is an isomorphism.

Let x be an arbitrary point of P. Then $Im\,\bar{\Omega}(x)$ (resp. ker $\bar{\Omega}(x)$) is a vector subspace of T_xP and its dimension is called the **rank** (resp **corank**) of $\Omega(x)$. The rank (corank) depends on x. As Ω is of type $(2,0)$, the rank of $\Omega(x)$ is an even number. If it is constant and equal to the dimension of P, then Ω is said to be **non–degenerate**.

Proposition 6.5.4 *Let (P,Ω) be a Poisson manifold such that the induced map $\bar{\Omega} : T^*P \longrightarrow TP$ is an isomorphism. For all $x \in P$ and $X,Y \in T_xP$, we put*

$$\omega(x)(X,Y) = \Omega(x)((\bar{\Omega})^{-1}(X),\ (\bar{\Omega})^{-1}(Y)).$$

Then ω is a closed 2–form.

Proof: Let us first recall the following facts: Suppose that X_{F_1}, X_{F_2} are Hamiltonian vector fields, that is,

$$\bar{\Omega}(dF_1(x)) = X_{F_1}(x),\ \bar{\Omega}(dF_2(x)) = X_{F_2}(x).$$

Then

$$[X_{F_1}, X_{F_2}] = X_{\{F_1,F_2\}} = \bar{\Omega}(d\{F_1,F_2\}).$$

So, if F_3 is a third function such that $\bar{\Omega}(dF_3(x)) = X_{F_3}(x)$, one has, from the definition of ω,

$$\omega(x)([X_{F_1}, X_{F_2}], X_{F_3}) = \omega(x)(\bar{\Omega}(x)d\{F_1,F_2\}, \bar{\Omega}(dF_3))$$

$$= \Omega(x)(d\{F_1,F_2\}, dF_3)$$

$$= \{\{F_1,F_2\},F_3\}(x),$$

where in the last equality we have used again (6.14). Also, we have

$$X_{F_1}(\omega(X_{F_2}, X_{F_3})) = \bar{\Omega}(dF_1)\omega(\bar{\Omega}(dF_2), \bar{\Omega}(dF_3))$$

$$= \bar{\Omega}(dF_1)\Omega(dF_2, dF_3)$$

$$= \{F_1, \{F_2, F_3\}\},$$

where in the last equality we have again used (6.14). From these equalities we may prove that

$$d\omega(X_{F_1}, X_{F_2}, X_{F_3}) = X_{F_1}\omega(X_{F_2}, X_{F_3})$$

$$+X_{F_2}\omega(X_{F_3}, X_{F_1}) + X_{F_3}\omega(X_{F_1}, X_{F_2})$$

$$-\omega([X_{F_1}, X_{F_2}], X_{F_3}) - \omega([X_{F_3}, X_{F_1}], X_{F_2})$$

$$-\omega([X_{F_2}, X_{F_3}], X_{F_1}).$$

Therefore, we have

$$3d\omega(X_{F_1}, X_{F_2}, X_{F_3}) = 2[\{F_1, \{F_2, F_3\}\}$$

$$+\{F_2, \{F_3, F_1\}\} + \{F_3, \{F_1, F_2\}\}] = 0 \text{ (from Jacobi identity).}$$

Then ω is closed.□

Theorem 6.5.5 *Suppose that (P, Ω) is a Poisson manifold of even dimension $2p$. If Ω is non–degenerate then there is defined a symplectic structure on P.*

Proof: In fact, if Ω is non–degenerate, then ω is a non–degenerate closed 2–form on P.□

From this theorem we see that the non–degenerate Poisson structure on a manifold of even dimension is equivalent to the symplectic structure. In the general situation, a Poisson structure defines a **morphism** from T^*P to TP and, in the symplectic case, an **isomorphism** from TS to T^*S. We remark that we may also consider situations where rank of $\bar{\Omega}$ is constant equal to $2n < dim\, P$, which generates a study similar to presymplectic manifolds. We suggest the paper of S. Benenti [5] in this direction.

6.6 Generalized Liouville dynamics and Poisson brackets

Marmo et al. [94] proposed an extension of volume forms preserving vector fields to arbitrary manifolds having, as a particular case, some results obtained from the symplectic formalism. This generalization offers a possibility of re–obtaining Poisson brackets in a different way as presented up to now.

Definition 6.6.1 *Suppose that N is a manifold of dimension n (even or odd) and θ a volume form on N. We say that a vector field X on N has a* **Liouville property** *with respect to θ if the Lie derivative $L_X\theta$ vanishes, i.e., if θ is invariant under X.*

The motivation of such definition is clear: suppose that (S, ω), is a symplectic manifold of dimension $2n$. Then ω^n is a volume form and if X_F is a Hamiltonian vector field with energy F, then we have seen that $L_{X_F}\omega^n = 0$.

If we develop $L_X\theta = 0$, then one has $di_X\theta = 0$. So, we shall say that X is **locally Liouville**, if for each $x \in N$, there is a neighborhood U of x and a $(n-2)$–form λ on U such that

$$i_X\theta/_U = d\lambda.$$

We shall say that X is **Liouville** if λ is globally defined.

For a local Liouville vector field X one has that $d\lambda$ is invariant under X. Furthermore, if the $(n-3)$–form $i_X\lambda$ is closed then λ is also invariant since

$$L_X\lambda = i_X d\lambda + di_X\lambda = i_X i_X\theta = 0.$$

Therefore we may say that λ plays the role that a function H plays in the Hamiltonian formalism. We will see now some extensions previously presented for the present generalization. For example, suppose that

$$i_X\theta = \alpha.$$

Then a necessary and sufficient condition to X be a locally Liouville vector field is α to be closed:

$$L_X\theta = 0 \Longleftrightarrow di_X\theta = 0.$$

From this, one obtains the following simple result. Suppose that $F_1, \ldots,$ F_{n-1} are C^∞ functions on N such that

$$\lambda = dF_1 \wedge \ldots \wedge dF_{n-1} \neq 0.$$

Then λ is closed and X is locally Liouville. Therefore each F_i, $1 \leq i \leq n-1$, is a constant of motion, since $i_X\lambda = 0$ if and only if $X(F_i) = 0$, $1 \leq i \leq n-1$. The dynamics here is therefore characterized by a set of Hamiltonian constants of motion. This kind of generalization goes to a type of Mechanics called Nambu Mechanics (Marmo et al. [94]).

We show now the relation of Liouville vector fields and Poisson brackets has an equivalent form (see Flanders [55], p. 180):

$$\{F, G\}\, \omega^n = n\,(dF \wedge dG) \wedge \omega^{n-1},$$

where ω is a symplectic form and ω^n the corresponding volume form. In such a case, if X_H is a Hamiltonian vector field, $i_{X_H}\omega = dH$, then

$$(L_{X_H}F)\,\omega^n = (dF) \wedge (i_{X_H}\omega^n)$$

$$= n(dF) \wedge (i_{X_H}\omega) \wedge \omega^{n-1}$$

$$= n\, dF \wedge dH \wedge \omega^{n-1}$$

$$= \{F, H\}\, \omega^n$$

(and so $L_{X_H}F = \{F, H\}$). We observe that the first equality in the above expression is obtained from

$$0 = i_X(dF \wedge \omega^n) = (i_X dF)\,\omega^n - dF \wedge i_X\omega^n$$

$$= (L_X F)\omega^n - dF \wedge i_X\omega^n.$$

Suppose now that X is Liouville on a manifold N. Then X is defined (at least locally) by a $(n-2)$–form λ through the equation $L_X\theta = 0$, i.e., $i_X\theta = d\lambda$. Hence

$$0 = i_X(dF \wedge \theta) = (L_X F)\theta - dF \wedge d\lambda,$$

that is,

$$(L_X F)\theta = dF \wedge d\lambda,$$

and if we define the Poisson bracket of a C^∞ function F on N with respect to λ by

$$\{F, \lambda\}\theta = dF \wedge d\lambda,$$

then obviously we have an analogous result for the Lie derivative:

$$L_X F = \{F, \lambda\}.$$

We may extend the definition of Poisson brackets for $(n-2)$–forms as we did for 1–forms (see section 6.2). For this, suppose that X and Y are Liouville vector fields (at least locally). Then there are $(n-1)$–forms λ_X, λ_Y such that

$$i_X\theta = \lambda_X, \; i_Y\theta = \lambda_Y,$$

with $d\lambda_X = d\lambda_Y = 0$. Put

$$\{\lambda_X, \lambda_Y\} = i_{[X,Y]}\theta.$$

Then it is not difficult to see that this brackets verify the same properties of Poisson brackets. Moreover, we have

$$L_X\lambda_Y = L_X i_Y\theta = i_{[X,Y]}\theta = \{\lambda_X, \lambda_Y\}.$$

6.7 Contact manifolds and non–autonomous Hamiltonian systems

Let M be a $(2n+1)$–dimensional manifold and ω a closed 2–form on M of rank $2n$. From the Generalized Darboux Theorem (with $r = 1$) there is a coordinate system (x^i, y^i, u), $1 \leq i \leq n$, on each point of M such that

$$\omega = \sum_{i=1}^{n} dx^i \wedge dy^i.$$

In particular, let α be a contact form on M, i.e., α is a 1–form on M such that $\alpha \wedge (d\alpha)^n \neq 0$. We set

$$\omega = -d\alpha.$$

Since ω has rank $2n$, there exist on each point of M a coordinate system (x^i, y^i, u) such that

$$\omega = \sum_{i=1}^{n} dx^i \wedge dy^i.$$

Then

$$0 = d\alpha + \sum_{i=1}^{n} dx^i \wedge dy^i = d(\alpha - \sum_{i=1}^{m} y^i dx^i).$$

Therefore, there exists a locally defined function z such that

$$\alpha - \sum_{i=1}^{n} y^i dx^i = dz,$$

and, so,

$$\alpha = dz + \sum_{i=1}^{n} y^i dx^i.$$

If we put $Y = \partial/\partial z$, then we have

$$i_Y \alpha = \alpha(Y) = 1,$$

$$i_Y d\alpha = -i_Y (\sum_{i=1}^{n} dx^i \wedge dy^i) = 0.$$

This shows the existence and unicity of such vector field Y which is called the **Reeb vector field** (see Godbillon [63], Marle [93]).

Let us return to the general situation, i.e., M is a $(2n+1)$–dimensional manifold and ω is a closed 2–form of rank $2n$ on M. We denote by Δ_ω the 1–dimensional distribution on M defined by

$$\Delta_\omega = \{X / i_X \omega = 0\}$$

We notice that Δ_ω is involutive. In fact, for two vector fields $X, Y \in \Delta_\omega$, we have

$$i_{[X,Y]}\omega = L_X i_Y \omega - i_Y L_X \omega$$

$$= -i_Y L_X \omega \quad (\text{since } i_Y \omega = 0)$$

$$= -i_Y((i_X d + d i_X)\omega)$$

$$= -i_Y d i_X \omega \quad (\text{since } \omega \text{ is closed})$$

$$= 0 \quad (\text{since } i_X \omega = 0).$$

Alternatively, Δ_ω may be viewed as a vector bundle over M of rank 1; in fact, Δ_ω is a vector subbundle of TM. A vector field X such that $X \in \Delta_\omega$ (i.e., $i_X \omega = 0$) is called a **characteristic vector field.**

Let now (S, ω) be a symplectic manifold of dimension $2n$ and consider the product manifold $R \times S$. Let $p_S : R \times S \longrightarrow S$ be the canonical projection on the second factor, i.e.,

$$p_S(t, x) = x.$$

Set $\omega' = (p_S)^\star \omega$. Then ω' is a closed 2–form on $R \times S$ of rank $2n$. Consider the distribution $\Delta_{\omega'}$ and define a vector field X on $R \times S$ by

$$X(t, x) = \partial/\partial t \in T_{(t,x)}(R \times S) \simeq T_t R \oplus T_x S.$$

We have

$$\omega'(X(t, x), Z) = \omega'(\partial/\partial t, Z)$$

$$= (p_S^\star \omega)(\partial/\partial t, Z)$$

$$= \omega(0, Z)$$

$$= 0,$$

for all $Z \in T_{(t,x)}(R \times S)$. This shows that $X \in \Delta_{\omega'}$. Furthermore, if $Y \in \Delta_{\omega'}$, i.e., $i_Y\omega' = 0$, we have

$$0 = \omega'(Y, Z) = \omega'(a\,\partial/\partial t + u,\ b\,\partial/\partial t + u')$$

$$= (p_S^*\omega)(a\,\partial/\partial t + u,\ b\,\partial/\partial t + u')$$

$$= \omega(u, u'),$$

where $Y = a\,\partial/\partial t + u$, $Z = b\,\partial/\partial t + v$, $a, b \in C^\infty(R \times S)$, $u, v \in \chi(S)$. Hence, $u = 0$ and so $Y = a\,\partial/\partial t$. Therefore, $\Delta_{\omega'}$ is globally spanned by $\partial/\partial t$.

Now, let $H : R \times S \longrightarrow R$ be a function on $R \times S$. For each $t \in R$, we define $H_t : S \longrightarrow R$ by

$$H_t(x) = H(t, x).$$

We consider the Hamiltonian vector field X_{H_t} on S with energy H_t, i.e.,

$$i_{X_{H_t}}\omega = dH_t.$$

For simplicity, we set $X_t = X_{H_t}$. Define a mapping $X : R \times S \longrightarrow TS$ by

$$X(t, x) = X_t(x) \in T_xS,\ t \in R,\ x \in S.$$

Then there is a vector field X_H on $R \times S$ given by

$$X_H(t, x) = \partial/\partial t + X(t, x), \text{ i.e.,}$$

$$X_H(t, x) = \partial/\partial t + X_t(x).$$

If (q^i, p_i) are canonical coordinates in S, i.e., $\omega = \sum_{i=1}^n dq^i \wedge dp_i$, then ω' is locally given by the same expression:

$$\omega' = \sum_{i=1}^n dq^i \wedge dp_i.$$

Let $\sigma : I = (-\epsilon, \epsilon) \longrightarrow R \times S$ be an integral curve of X_H, with $\epsilon > 0$. Then we have

$$\sigma(t) = (a(t),\ q^i(t),\ p_i(t)).$$

Since σ is an integral curve of X_H,

$$\dot{\sigma}(t) = X_H(\sigma(t))$$

Thus $dq/dt = 1$, i.e., $a(t) = t$. Therefore

$$\sigma(t) = (t,\ q^i(t), p_i(t)).$$

So we obtain

$$\dot{\sigma}(t) = X_H(t, q^i(t), p_i(t)) = \partial/\partial t + X_t(q^i(t), p_i(t)).$$

Since $i_{X_t}\omega = dH_t$, we deduce

$$X_t = \frac{\partial H_t}{\partial p_i}\frac{\partial}{\partial q^i} - \frac{\partial H_t}{\partial q^i}\frac{\partial}{\partial p_i}.$$

$$= \frac{\partial H}{\partial p_i}\frac{\partial}{\partial q^i} - \frac{\partial H}{\partial q^i}\frac{\partial}{\partial p_i}.$$

Therefore, σ is an integral curve of X_H if and only if

$$\frac{dq^i}{dt} = \frac{\partial H}{\partial p_i},\ \frac{dp_i}{dt} = -\frac{\partial H}{\partial q^i},\ 1 \leq i \leq m$$

which are the **Hamilton equations** for a non–autonomous Hamiltonian H.

Proposition 6.7.1 *We have*

$$L_{X_H} H = \frac{\partial H}{\partial t}.$$

Proof: Since $X_H(t,x) = \partial/\partial t + X_t(x)$, we have

$$(L_{X_H}H)(t,x) = X_H(t,x)H = \frac{\partial H}{\partial t} + X_t(x)H_t$$

But $X_t(x)H_t = (L_{X_t}H_t)(x) = 0$. Then

$$L_{X_H} H = \frac{\partial H}{\partial t}.\square$$

Now, let us define a 2–form ω_H on $R \times S$ by

$$\omega_H = \omega' + dH \wedge dt.$$

Proposition 6.7.2 *(1) ω_H is a closed 2-form of rank 2n; (2) X_H is the unique vector field on $R \times S$ such that*

$$i_{X_H}\omega_H = 0, \; i_{X_H}dt = 1. \tag{6.15}$$

Proof: (1) Obviously, ω_H is closed. Furthermore, $dt \wedge (\omega_H)^n = dt \wedge (\omega')^n \neq 0$, since ω' is of rank $2n$. Then ω_H has rank $2n$.
(2) We have

$$dt(X_H(t,x)) = dt(\partial/\partial t) + X_t(x)) = dt(\partial/\partial t) = 1.$$

Moreover, let Y be a tangent vector at (t,x). We obtain

$$(i_{X_H}\omega')(t,x)(Y) = \omega'(X_H(t,x),Y) = \omega(dp_S(t,x)X_H(t,x),\; dp_S(t,x)(Y))$$

$$= \omega(X_t(x),\; dp_S(t,x)(Y)) = (i_{X_t}\omega)(x)(dp_S(t,x)(Y))$$

$$= dH_t(x)(dp_S(t,x)(Y)),$$

$$i_{X_H}dH = L_{X_H}H = \frac{\partial H}{\partial t},$$

$$i_{X_H}dt = 1.$$

Then we have

$$(i_{X_H}\omega_H)(t,x)(Y) = (i_{X_H}\omega')(t,x)(Y) + (i_{X_H}dH)(t,x)dt(Y)$$

$$-(i_{X_H}dt)(t,x)dH(t,x)(Y)$$

$$= dH_t(x)(dp_S(t,x)(Y)) + \frac{\partial H}{\partial t}(t,x)dt(Y)$$

$$-dH(t,x)(Y) = 0,$$

since

$$dH_t(x)(dp_S(t,x)(Y)) + \frac{\partial H}{\partial t}(t,x)dt(Y) = dH(t,x)Y.$$

Finally, since Δ_{ω_h} is of dimension 1, then X_H is unique. □

Now, suppose that (X,ω) is a symplectic manifold of dimension $2n$ such that $\omega = -d\theta$ (this is the case when S is the cotangent bundle of a manifold). We set

$$\theta' = (p_S)^*\theta,\ \theta_H = \theta' + Hdt.$$

Proposition 6.7.3 *If H is a non-vanishing function on $R \times S$, then θ_H is a contact form.*

Proof: In fact, we have

$$\theta_H(X_H) = \theta'(X_H) + Hdt(X_H)$$

$$= (p_S)^*\theta(X_H) + H,$$

that is,

$$\theta_H(X_H)(t,x) = \theta(x)(X_t(x)) + H(t,x).$$

Then θ_H does not vanish on Δ_{ω_H}. Therefore, one easilly deduces that $\theta_H \wedge (d\theta_H)^n \neq 0$. □

6.8 Hamiltonian systems with constraints

In this section we examine some topics in the geometric theory of constraints, inspired in Weber [124]. This study will be continued in Section 7.13.

Let (S,ω) be a symplectic manifold of dimension $2n$.

Definition 6.8.1 *A non-zero 1-form α on S is called a* **constraint** *on S. A set $C = \{\alpha_1, \ldots, \alpha_r\}$ of r linearly independent 1-forms on S is called a* **system of constraints** *on S. We say that a curve σ in S* **satisfies the constraints** *if*

$$\alpha_a(\dot{\sigma}(t)) = 0,\ 1 \leq a \leq r.$$

Now we consider a Hamiltonian system (S, ω, H) together with a system C of constraints on S. We call (S, ω, H, C) a **Hamiltonian system with constraints.**

It is clear that a curve σ satisfying the Hamilton equations for H will not satisfy the constraints in general. For a curve σ to satisfy also the constraints it is necessary that some additional forces (called **canonical constraint forces**) act on the system besides the "force" dH. Then the equation of motion becomes

$$i_X \omega = dH + \alpha, \tag{6.16}$$

where α is a 1–form on S which is the resultant of the canonical constraint forces. We must require that if $\alpha_a(Y) = 0$, for all $a, 1 \leq a \leq k$, then $\alpha(Y) = 0$.

But this condition holds if and only if

$$\alpha = \wedge^a \alpha_a$$

Hence the equation (6.16) becomes

$$i_X \omega = dH + \wedge^a \alpha_a, \; \alpha_a(X) = 0. \tag{6.17}$$

(Here the $\wedge^a$ are Lagrange multipliers).

If we denote by X_a the vector fields on S given by

$$i_{X_a} \omega = \alpha_a, \; 1 \leq a \leq r$$

and by X_H the Hamiltonian vector field with energy H, i.e.,

$$i_{X_H} \omega = dH,$$

then we easily deduce from (6.17) that

$$X = X_H + \wedge^a X_a.$$

Now, suppose that α_a is locally given by

$$\alpha_a = (A_a)_i dq^i + (B_a)_i dp_i,$$

where (q^i, p_i) are canonical coordinates for (S, ω).

It is easy to prove that a curve $\sigma(t) = (q^i(t), p_i(t))$ in S is an integral curve of X if and only if it satisfy

$$\left.\begin{array}{rcl} dq^i/dt &=& \partial H/\partial p_i + \wedge^a(B_a)_i, \\ dp_i/dt &=& -\partial H/\partial q^i - \wedge^a(A_a)_i, \\ (A_a)_i(dq^i/dt) + (B_a)_i(dp_i/dt) &=& 0, \end{array}\right\} \quad (6.18)$$

$1 \leq i \leq n$, $1 \leq a \leq r$.

Next, we shall explain the geometrical meaning of constraints.

Let C be a system of constraints on S. Then we define a distribution D on S as follows. For each $x \in S$, $D(x)$ is given by

$$D(x) = \{X \in T_xS/\alpha_a(X) = 0, \text{ for all } a,\ 1 \leq a \leq r\}.$$

Thus D is a $(2n - r)$–dimensional distribution on S.

Definition 6.8.2 *A system of constraints C is called* **holonomic** *if D is integrable; otherwise we call C* **anholonomic**.

Hence C is holonomic if and only if the ideal $\mathcal{Y}$ of $\wedge S$ generated by C is a differential ideal, i.e., $d\mathcal{Y} \subset \mathcal{Y}$. Obviously (6.18) holds for holonomic as well as anholonomic constraints. For a system of holonomic constraints the motion lies on a specific leaf of the foliation defined by D.

Remark 6.8.3 Note that the energy H for a solution $\sigma(t)$ of (6.18) is conserved. In fact, from (6.17) we have

$$0 = (i_X\omega)(X) = dH(X) = X(H).$$

6.9 Exercises

6.9.1 Show that a submanifold W of codimension r of a coisotropic manifold K of a symplectic manifold (S, ω) is Lagrangian if and only if it is transverse to the integral manifolds of the distribution $x \longrightarrow (T_xK)^{\perp}$.

6.9.2 Show that a transformation $f : (S, \omega) \longrightarrow (S, \omega)$ is canonical if the 1–form $\alpha = pdq - \bar{p}d\bar{q}$ is exact, where

$$f(q,p) = (\bar{q}(q,p), \bar{p}(q,p)).$$

Show that the converse is also true locally.

6.9.3 Let (S, ω) be a symplectic manifold with local coordinates $(x, \bar{y})$, $F : S \longrightarrow R$ a C^∞ map with Hessian matrix regular. Set

$$x = x, \quad \bar{x} = (\partial F / \partial \bar{y}).$$

$$y = \frac{\partial F}{\partial x}, \quad \bar{y} = \bar{y}.$$

Then the map $\varphi : S \longrightarrow S$ defined by $(x, y) \longrightarrow (\bar{x}, \bar{y})$ is a canonical transformation. The map F is called "**generating function of a canonical transformation**". (*Hint*: compute $d(\bar{y}\bar{x} - F)$).

6.9.4 Let $\{P_1, \Omega_1\}$ and $\{P_2, \Omega_2\}$ be Poisson manifolds and $h : P_1 \longrightarrow P_2$ a C^∞ map. Let $\{\ ,\ \}_1$, $\{\ ,\ \}_2$ be the corresponding Poisson brackets. We say that h is a **Poisson map** if

$$(*) \qquad \{f, g\}_2 \circ h = \{f \circ h, g \circ h\}_1$$

for all C^∞ functions f, g on P_2. Let $f : P_2 \longrightarrow R$ be a C^∞ function. We say that the vector fields $\bar{\Omega}_1(d(f \circ h))$ on P_1 and $\bar{\Omega}_2(df)$ on P_2 are **h–related** if

$$(**) \qquad T_x h(\bar{\Omega}_1(d(f \circ h)(x)) = \bar{\Omega}_2(df(h(x)))$$

Show that (*) holds if and only if (**) holds.

6.9.5 Let $\{P_1, \Omega_1\}$, $\{P_2, \Omega_2\}$ and $\{P_3, \Omega_3\}$ be Poisson manifolds and

$$h_{12} : P_1 \longrightarrow P_2, \quad h_{23} : P_2 \longrightarrow P_3$$

C^∞ mappings. Show that if h_{12} and h_{23} are Poisson maps then $h_{23} \circ h_{12}$ is also a Poisson map.

6.9.6 If h_{12} and $h_{23} \circ h_{12}$ are Poisson maps and h_{12} is surjective then h_{23} is a Poisson map?

6.9.7 Suppose that X is not a Liouville vector field with respect to a given volume form θ. Is it possible to find another volume form $\bar{\theta}$ such that X is Liouville with respect to $\bar{\theta}$? (Note that the set of n–forms on a n–dimensional manifold N is of dimension one and therefore there exists a function f on N such that $\bar{\theta} = f\theta$, $f \neq 0$).

6.9.8 Show that in local coordinates (x^A) the components X^A of a Liouville form are given by

$$X^A = (-1)^{A+B+1} \sum_B \frac{\partial \lambda_{BA}}{\partial x^B},$$

where λ is such that $i_X \theta = d\lambda$.

<u>6.9.9</u> Let M be a configuration manifold, ω_M the canonical symplectic form on T^*M and α a 2–form on M. Consider the form $\bar{\omega} = \omega + \pi_M^* \alpha$ and suppose that X and Y are Hamiltonian vector fields with respect to $\bar{\omega}$ and ω, respectively. Show that X, Y can be Liouville with respect to the same volume form.æ

Chapter 7

Lagrangian systems

7.1 Lagrangian systems and almost tangent geometry

In this section, we shall study a Lagrangian formalism for Classical Mechanics due to J. Klein [77], [78], [79].

Let M be an m–dimensional manifold and TM its tangent bundle with canonical projection $\tau_M : TM \longrightarrow M$. TM is called the **phase space of velocities** of the **configuration manifold** M. Let $L : TM \longrightarrow R$ be a differentiable function on TM called the **Lagrangian** function. We consider the following closed 2–form on TM:

$$\omega_L = -dd_J L. \tag{7.1}$$

The function

$$E_L = CL - L \ \ (\text{or } E_L = L_C L - L)$$

on TM is called the **energy function** associated to L.

Proposition 7.1.1 *We have*

$$i_J \omega_L = 0$$

Proof: In fact,

$$i_J \omega_L = -i_J dd_J L = i_J d_J dL \ \ (\text{since } dd_J = -d_J d)$$

$$= d_J i_J dL \text{ (since } i_J d_J = d_J i_J)$$

$$= d_J^2 L$$

$$= 0. \square$$

Consider the equation

$$i_X \omega_L = dE_L \tag{7.2}$$

Then we shall see that (7.2) under a certain condition on X is the intrinsical expression of the Euler–Lagrange equations of motion.

Let us first show the following.

Proposition 7.1.2 *The form ω_L is a symplectic form on TM if and only if, for any coordinate system (q^i, v^i), the Hessian matrix*

$$\left(\frac{\partial^2 L}{\partial v^i \partial v^j} \right)$$

is of maximal rank (i.e., an inversible matrix).

Proof: A straightforward computation shows that

$$\omega_L = \frac{\partial^2 L}{\partial q^j \partial v^i} dq^i \wedge dq^j + \frac{\partial^2 L}{\partial v^i \partial v^j} dq^i \wedge dv^j,$$

and so the m–times exterior product $\omega_L^m = \omega_L \wedge \ldots \wedge \omega_L$ gives

$$\omega_L^m = c \det \left(\frac{\partial^2 L}{\partial v^i \partial v^j} \right) dq^1 \wedge \ldots \wedge dq^m \wedge dv^1 \wedge \ldots \wedge dv^m,$$

where c is a non–zero constant function. So ω_L is symplectic if and only if ω_L^m is a volume form if and only if $\det \left(\frac{\partial^2 L}{\partial v^i \partial v^j} \right) \neq 0. \square$

If ω_L is symplectic we say that $L : TM \longrightarrow R$ is **regular** or **non–degenerate**. Otherwise L is said to be **singular, irregular** or **degenerate**.

Suppose that L is regular. Then the equation (7.2) admits a unique solution ξ, i.e., $i_\xi \omega_L = dE_L$, because ω_L is symplectic. Moreover, we have

Proposition 7.1.3 *The vector field ξ given by (7.2) is a semispray (or second–order differential equation).*

Proof: It is sufficient to show that $J\xi = C$. One has

$$\begin{aligned} i_C\omega_L &= -i_C dd_J L = i_C d_J dL \ (\text{since } dd_J = -d_J d) \\ &= i_J dL - d_J i_C dL \ (\text{since } i_C d_J + d_J i_C = i_J) \\ &= d_J L - d_J(CL) \\ &= -d_J(CL - L) \\ &= -d_J E_L. \end{aligned}$$

On the other hand,

$$\begin{aligned} i_J i_\xi \omega_L &= i_\xi i_J \omega_L - i_{J\xi}\omega_L \ (\text{since } i_\xi i_J - i_J i_\xi = i_{J\xi}) \\ &= -i_{J\xi}\omega_L, \end{aligned}$$

and

$$i_J dE = d_J E.$$

Therefore, one deduces

$$i_{J\xi}\omega_L = i_C\omega_L$$

and then

$$J\xi = C,$$

since ω_L is symplectic. □

Now, if ξ is the semispray given by (7.2), we locally have

$$\xi = v^i\partial/\partial q^i + \xi^i\partial/\partial v^i.$$

Therefore we obtain

$$i_\xi\omega_L = \left(\frac{\partial^2 L}{\partial q^i \partial v^j} v^i - \frac{\partial^2 L}{\partial q^j \partial v^i} v^j - \frac{\partial^2 L}{\partial v^i \partial v^j}\xi^j\right) dq^i$$

$$+\frac{\partial^2 L}{\partial v^i \partial v^j} v^j dv^i$$

On the other hand, we have

$$E_L = v^j \frac{\partial L}{\partial v^j} - L$$

and thus

$$dE_L = \left(\frac{\partial^2 L}{\partial q^i \partial v^j} v^j - \frac{\partial L}{\partial q^i} \right) dq^i + \frac{\partial^2 L}{\partial v^i \partial v^j} v^j dv^i.$$

Since $i_\xi \omega_L = dE_L$, we deduce

$$\frac{\partial^2 L}{\partial q^j \partial v^i} v^j + \frac{\partial^2 L}{\partial v^i \partial v^j} \xi^j - \frac{\partial L}{\partial q^i} = 0, \ 1 \leq i \leq m.$$

Let $\sigma : R \longrightarrow M$ be a path of ξ, i.e., $\dot{\sigma} : R \longrightarrow TM$ is an integral curve of ξ. If $\sigma(t) = (q^i(t))$, then $\dot{\sigma}(t) = (q^i(t), (dq^i/dt))$. Thus, σ is a path of ξ if and only if σ verifies

$$\frac{\partial^2}{\partial q^j \partial \dot{q}^i} \dot{q}^j + \frac{\partial^2 L}{\partial \dot{q}^i \partial \dot{q}^j} \ddot{q}^j - \frac{\partial L}{\partial q^i} = 0, \ 1 \leq i \leq m$$

(here the dots mean derivatives with respect to the time). That is,

$$\frac{d}{dt} \left(\frac{\partial L}{\partial \dot{q}^i} \right) - \frac{\partial L}{\partial q^i} = 0, \ 1 \leq i \leq m. \tag{7.3}$$

Hence, we have.

Theorem 7.1.4 *The paths of ξ are the solutions of the* **Euler–Lagrange equations** *(7.3).*

This suggests the following.

Definition 7.1.5 *The unique semispray ξ such that $i_\xi \omega_L = dE_L$, where $L : TM \longrightarrow R$ is a regular Lagrangian, is called* **Euler–Lagrange vector field** *for L and it is sometimes represented by ξ_L.*

We have $L_{\xi_L} E_L = \xi_L E_L = 0$ and E_L is constant along the integral curves of ξ_L (energy conservation law).

Example

Let $L : TR^2 \longrightarrow R$ be defined by

$$L(x, y, u, v) = (1/2)[(u^2 - v^2) - x^2 - y^2].$$

Then

$$d_J L = u\,dx - v\,dy$$

and

$$\omega_L = -dd_J L = dx \wedge du - dy \wedge dv.$$

The Hessian matrix of L is

$$\begin{bmatrix} 1 & 0 \\ 0 & -1 \end{bmatrix}$$

and hence, L is regular and ω_L is a symplectic form on TR^2.

A simple computation shows that the Euler–Lagrange vector field for L is

$$\xi_L = u\partial/\partial x + v\partial/\partial y - x\partial/\partial u + y\partial/\partial v$$

and so

$$J\xi_L = u\partial/\partial u + v\partial/\partial v = C.$$

The Euler–Lagrange equations for L are

$$\frac{d^2x}{dt^2} = x(t), \ \frac{d^2y}{dt^2} = y(t).$$

Definition 7.1.6 *A* **mechanical system** $\mathcal{M}$ *is a triple* (M, F, ρ), *where* M *is an* n*–dimensional manifold,* F *is a differentiable function on* TM *and* ρ *is a semibasic form on* TM, *called the* **force field**.

Proposition 7.1.7 *Let* $\mathcal{M} = (M, F, \rho)$ *be a mechanical system. Suppose that the 2–form* $\omega_F = -dd_J F$ *is symplectic. Then there is a unique semispray* ξ *satisfying the equation*

$$i_\xi \omega_F = dE_F + \rho, \tag{7.4}$$

where $E_F = CF - F$ *is the energy of* F. *The paths of* ξ *are the solutions of the system of equations*

$$\frac{d}{dt}\left(\frac{\partial F}{\partial \dot{q}^i}\right) - \frac{\partial F}{\partial q^i} = -\rho_i, \ 1 \le i \le m,$$

where $\rho = \rho_i dq^i$.

Proof: By a similar procedure used above we deduce that $J\xi = C$. Furthermore, from (7.4), one easily obtains

$$\frac{\partial^2 L}{\partial q^j \partial v^i} v^j + \frac{\partial^2 L}{\partial v^i \partial v^j} \xi^j - \frac{\partial L}{\partial q^i} = \rho_i, \ 1 \leq i \leq m. \tag{7.5}$$

Then if $\sigma(t) = (q^i(t))$ is a path of ρ (7.5) becomes

$$\frac{d}{dt}\left(\frac{\partial L}{\partial \dot{q}^i}\right) - \frac{\partial L}{\partial q^i} = \rho_i, \ 1 \leq i \leq m. \square \tag{7.6}$$

Definition 7.1.8 *A mechanical system* $\mathcal{M} = (M, F, \rho)$ *is said to be* **conservative** *if the force field* ρ *is a closed semibasic form.*

Let $\mathcal{M} = (M, F, \rho)$ be a conservative mechanical system. Then $L_\xi \omega_F = di_\xi \omega_F = d(dE_F + \rho) = 0$, since ρ is closed. Also we have (at least locally) $\rho = dV$ and thus $L_\xi(E_F + V) = 0$. So we deduce the energy conservation law.

Suppose that $\mathcal{M} = (M, F, \rho)$ is a conservative mechanical system such that there exists a differentiable function $U : M \longrightarrow R$ with $\rho = \tau_M^\star(dU) = d(U \circ \tau_M)$. Then $\mathcal{M}$ is said to be a **Lagrangian system**. If we set $L = F + U \circ \tau_M$ then is not hard to see that (7.4) and (7.6) become

$$i_\xi \omega_L = dE_L,$$

$$\frac{d}{dt}\left(\frac{\partial L}{\partial \dot{q}^i}\right) - \frac{\partial L}{\partial q^i} = 0.$$

So, conservative mechanical systems are Lagrangian systems. **Non-conservative mechanical systems** are mechanical systems for which the force field ρ **is not** a closed semibasic form.

7.2 Homogeneous Lagrangians

In this section, we develop the Klein formalism for homogeneous Lagrangians.

Definition 7.2.1 *A Lagrangian* $L : TM \longrightarrow R$ *is said to be* **homogeneous** *if the function* L *is homogeneous of degree 2, i.e.,* $CL = 2L$.

Consequently, if L is an homogeneous Lagrangian, we have

$$v^i \frac{\partial L}{\partial v^i} = 2L(q,v).$$

Thus, the energy function E_L associated to L coincides with L, i.e., $E_L = L$.

Proposition 7.2.2 *Let L be an homogeneous Lagrangian. Then: (1) ω_L is homogeneous of degree 1; and (2) the Euler–Lagrange vector field ξ_L is a spray.*

Proof: (1) In fact, if L is homogeneous then $\omega_L = -dd_J L$ verifies

$$L_C \omega_L = \omega_L,$$

since Proposition 4.2.9.
(2) We have

$$i_{[C,\xi_L]}\omega_L = L_C i_{\xi_L}\omega_L - i_{\xi_L} L_C \omega_L$$

$$= L_C dE_L - i_{\xi_L}\omega_L = L_C dL - dE_L$$

$$= d(L_C L) - dL = d(CL - L) = dE_L.$$

Since $i_{\xi_L}\omega_L = dE_L$, and ω_L is symplectic, it follows that $[C,\xi_L] = \xi_L$.□
For mechanical systems we have the following result.

Proposition 7.2.3 *Let $\mathcal{M} = (M,F,\rho)$ be a mechanical system such that F and ρ are homogeneous of degree k. Then the semispray ξ satisfying the equation*

$$i_\xi \omega_F = dE_F + \rho$$

is a spray.

Proof: We have

$$i_{[C,\xi]}\omega_F = L_C i_\xi \omega_F - i_\xi L_C \omega_F;$$

$$E_F = CF - F = kF - F = (k-1)F$$

$$L_C i_\xi \omega_F = L_C(dE_F + \rho) = d(L_C E_F) + L_C \rho$$

$$= k(k-1)dF + k\rho,$$

$$L_C \omega_F = (k-1)\omega_F$$

$$i_\xi L_C \omega_F = (k-1) i_\xi \omega_F = (k-1)dE_F + (k-1)\rho = (k-1)^2 dF + (k-1)\rho.$$

Hence

$$i_{[C,\xi]}\omega_F = k(k-1)dF + k\rho - (k-1)^2 dF - (k-1)\rho$$

$$= (k-1)dF + \rho = dE_F + \rho = i_\xi \omega_F.$$

Since ω_F is symplectic it follows that $[C, \xi] = \xi$.□

7.3 Connection and Lagrangian systems

Let ξ be an arbitrary semispray on TM. Then the endomorphism $\Gamma = -L_\xi J$ defined by

$$\Gamma(Y) = -[\xi, JY] + J[\xi, Y],$$

for all vector field Y on TM, is a connection in TM. Let

$$h = \frac{1}{2}(Id + L_\xi J), \; v = \frac{1}{2}(Id - L_\xi J)$$

be the horizontal and vertical projectors associated to Γ. Then we have

$$T(TM) = Im\, h \oplus Im\, v$$

(we remark that dim $Im\, h = \frac{1}{2}$dim$\,(TM) = m$, dim $M = m$). So, at every point $z \in TM$, the tangent vectors are decomposed in vertical and horizontal components belonging to the vertical and horizontal subspaces of the tangent space of TM at z. An interesting result about the above decomposition is given by the following theorem due to Crampin [21].

Theorem 7.3.1 *Let L be a regular Lagrangian on TM. Then there is a connection Γ_L in TM such that the vertical and horizontal subspaces of the tangent space of each point of TM, determined by the associated projectors of Γ are both Lagrangian subspaces for ω_L.*

We first prove the following lemma.

Lemma 7.3.2 *Let Ω be a 2-form and F a tensor field of type (1,1) on a manifold N. Then*

$$(X,Y) \longrightarrow (F\rfloor\Omega)(X,Y) = \Omega(FX,Y)$$

is a tensor field of type (0,2) on N. Moreover, if X is an arbitrary vector field on N, then

$$L_X(F\rfloor\Omega) = (L_X F)\rfloor\Omega + F\rfloor(L_X\Omega).$$

Proof: In fact,

$$(F\rfloor\Omega)(fX,gY) = (fg)(F\rfloor\Omega)(X,Y),$$

for any functions f and g on N. Hence $F\rfloor\Omega$ is a tensor field of type $(0,2)$ on N. Moreover, we have

$$(L_X(F\rfloor\Omega))(Y,Z) = X((F\rfloor\Omega)(Y,Z)) - (F\rfloor\Omega)([X,Y],Z)$$

$$-(F\rfloor\Omega)(Y,[X,Z])$$

$$= X(\Omega(FY,Z)) - \Omega(F[X,Y],Z) - \Omega(FY,[X,Z]),$$

$$((L_X F)\rfloor\Omega)(Y,Z) = \Omega((L_X F)Y,Z) = \Omega([X,FY],Z) - \Omega(F[X,Y],Z),$$

$$(F\rfloor(L_X\Omega))(Y,Z) = (L_X\Omega)(FY,Z)$$

$$= X(\Omega(FY,Z)) - \Omega([X,FY],Z) + \Omega(FY,[X,Z]).$$

This ends the proof. □

Proof of Theorem 7.3.1: Consider the symplectic 2–form $\omega_L = -dd_J L$ and the Euler–Lagrange vector field ξ_L for L. By Lemma 7.3.1, we have

$$L_{\xi_L}(J\rfloor\omega_L) = (L_{\xi_L}J)\rfloor\omega_L,$$

since $L_{\xi_L}\omega_L = 0$. We set $\Gamma_L = -L_{\xi_L}J$. Then

$$L_{\xi_L}(J\rfloor\omega_L) = -\Gamma_L\rfloor\omega_L.$$

Now, since $i_J\omega_L = 0$, we deduce that $J\rfloor\omega_L$ is symmetric, and, hence, $L_{\xi_L}(J\rfloor\omega_L)$ is also symmetric. Thus $\Gamma_L\rfloor\omega_L$ is symmetric, or equivalently, $i_{\Gamma_L}\omega_L = 0$. Then, for all vector fields X and Y on TM we obtain

$$\omega_L(\Gamma_L X, Y) + \omega_L(X, \Gamma_L Y) = 0.$$

So, if $h = \frac{1}{2}(Id + L_{\xi_L}J)$ and $v = \frac{1}{2}(Id - L_{\xi_L}J)$ are the horizontal and vertical projectors associated to Γ_L, we have

$$\omega_L(hX, Y) + \omega_L(X, hY) = \omega_L(X, Y),$$

$$\omega_L(vX, Y) + \omega_L(X, vY) = \omega_L(X, Y),$$

$$\omega_L(hX, Y) - \omega_L(X, vY) = 0.$$

Therefore

$$\omega_L(hX, hY) = \omega_L(vX, vY) = 0. \square$$

Next, we shall prove that there is an almost Hermitian structure on TM such that its Kähler form is precisely ω_L.

Proposition 7.3.3 *Let L be a regular Lagrangian on TM. Then*

$$\bar{g}(JX, JY) = -\omega_L(JX, Y)$$

defines a metric on the vertical bundle $V(TM)$.

Proof: (1) In fact $\bar{g}$ is well–defined. If Y and Y' are vector fields on TM such that $JY = JY'$, then $J(Y - Y') = 0$ implies that $Y - Y'$ is vertical. Thus $\omega_L(JX, Y) = \omega_L(JX, Y')$.
(2) $\bar{g}$ is symmetric. In fact,

$$\bar{g}(JY, JX) = -\omega_L(JY, X) = \omega_L(X, JY)$$

$$= -\omega_L(JX, Y) = \bar{g}(JX, JY).$$

(3) $\bar{g}$ is non–degenerate, since if $\bar{g}(JX, JY) = 0$ for any vector field JY on TM, it follows that $\omega_L(JX, Y) = 0$ for any vector field Y on TM. Since ω_L is non–degenerate, we deduce that $JX = 0$.□

In local coordinates (q^i, v^i) we have

$$\bar{g}(\partial/\partial v^i, \partial/\partial v^j) = -\omega_L(\partial/\partial v^i, \partial/\partial q^j)$$

$$= \frac{\partial^2 L}{\partial v^i \partial v^j}.$$

Hence, if Γ is an arbitrary connection in TM, we can prolongate $\bar{g}$ to a Riemannian metric g_Γ on TM. Consider the almost Hermitian structure (TM, F, g_Γ). We have

Proposition 7.3.4 *Let K_Γ be the Kähler form of (TM, F, g_Γ). Then*

$$K_\Gamma = i_v \omega_L.$$

Proof: In fact,

$$(i_v \omega_L)(X, Y) = \omega_L(vX, Y) + \omega_L(X, vY)$$

$$= \omega_L(vX, Y) - \omega_L(vY, X) = -g_\Gamma(vX, JY) + g_\Gamma(vY, JX)$$

$$= -g_\Gamma(X, JY) + g_\Gamma(JX, Y) = K_\Gamma(X, Y).□$$

Corollary 7.3.5 *Let $\Gamma_L = -L_{\xi_L} J$. Then the Kähler form K_{Γ_L} of (TM, F, g_{Γ_L}) coincides with ω_L.*

Proof: In fact,

$$i_v \omega_L = \omega_L.$$

by Theorem 7.3.1.□

Next we prove that there is a canonical connection in TM whose paths are precisely the solutions of the Euler–Lagrange equations.

We have seen in Chapter 4 that the connection $\Gamma_L = -L_{\xi_L} J$ verifies:

- its associated semispray is $\xi_L + \frac{1}{2}\xi_L^\star$,
- its tension is $\mathbf{H} = -\frac{1}{2}[\xi_L^\star, J]$,
- its weak torsion vanishes,
- its strong torsion is $T = \frac{1}{2}[\xi_L^\star, J]$.

Therefore, if L is an homogeneous Lagrangian, ξ_L is a spray and hence

$$\xi = \xi_L, \ \mathbf{H} = 0, \ T = 0.$$

Thus we have the following result due to Grifone [73].

Theorem 7.3.6 *If L is an homogeneous Lagrangian then there is a one and only one homogeneous connection Γ_L in TM whose geodesics are the solutions of the Euler–Lagrange equations.*

Proof: It follows directly from the Decomposition theorem (see Section 4.3).□

Next we consider non–homogeneous Lagrangians. Our purpose is to find a connection in TM such that its associated semispray is precisely ξ_L. From the Decomposition theorem, we need a semibasic vector 1–form T on TM such that $T^\circ + \xi_L^\star = 0$. Let $\bar{g}$ be the metric on $V(TM)$ defined in Proposition 7.3.3, i.e.,

$$\bar{g}(JX, JY) = \omega_L(JX, Y).$$

If we set

$$\Theta(X,Y) = \bar{g}(TX, JY) \tag{7.7}$$

then one has:
(1) $\Theta(JX, Y) = \bar{g}(TJX, JY) = 0$,
(2) $\Theta(X, JY) = \bar{g}(TX, JY) = 0$.
(We remark that Θ is semibasic). The problem now is to find a scalar semibasic 2–form Θ verifying (1) and (2) and, for such fixed Θ, we define T by (7.7). We have

$$\Theta^\circ(Y) = (i_{\xi_L}\Theta)(Y) = \Theta(\xi_L, Y) = \bar{g}(T\xi_L, JY)$$

$$= \omega_L(T^\circ, Y) = (i_{T^\circ}\omega_L)(Y),$$

i.e.,

$$\Theta^\circ = i_{T^\circ}\omega_L. \tag{7.8}$$

Now T must be such $T^\circ = -\xi_L^\star$. Thus, (7.8) becomes

$$\Theta^\circ = -i_{\xi_L^\star}\omega_L. \tag{7.9}$$

Let us consider the 2–form $\Theta = (i_C\omega_L) \odot \gamma$, where γ is a 1–form on TM such that $\gamma(JX) = 0$ and $\odot$ denotes the symmetric product. Then Θ verifies (1) and (2). If we take the potential of Θ, then

$$\Theta^\circ = (i_C\omega_L)^\circ\gamma + \gamma^\circ(i_C\omega_L),$$

i.e.,

$$-i_{\xi_L^\star}\omega_L = (i_C\omega_L)^\circ\gamma + \gamma^\circ(i_C\omega_L). \tag{7.10}$$

Therefore, we deduce that

$$-(i_{\xi_L^\star}\omega_L)^\circ = 2(i_C\omega_L)^\circ\gamma^\circ,$$

and so

$$\gamma^\circ = -\frac{1}{2}\frac{(i_{\xi_L^\star}\omega_L)^\circ}{(i_C\omega_L)^\circ}$$

$$= -\frac{1}{2}\frac{\omega_L(\xi_L^\star, \xi_L)}{\omega_L(C, \xi_L)}. \tag{7.11}$$

From (7.10) and (7.11) we have

$$\gamma = \frac{1}{\omega(C, \xi_L)} \left(-i_{\xi_L^*} \omega_L + \frac{1}{2} \frac{\omega_L(\xi_L^*, \xi_L)}{\omega_L(C, \xi_L)} (i_C \omega_L) \right), \tag{7.12}$$

from which one obtains $\Theta = i_C \omega_L \odot \gamma$ and a fortiori T. Hence we have proved the following.

Theorem 7.3.7 *(Grifone [73]). Let L be a regular Lagrangian on TM. Then there is a canonical connection Γ in TM such that its paths are the solutions of the Euler–Lagrange equations for L. This connection is given by*

$$\bar{\Gamma}_L = \Gamma_L + T = -L_{\xi_L} J + T,$$

where T is defined by (7.7) and γ by (7.12).

Now, let $\bar{g}$ be the metric in V defined in Proposition 7.3.3, Γ an arbitrary connection in TM and g_Γ the Riemannian prolongation of $\bar{g}$ along Γ.

Definition 7.3.8 Γ *is said to be* **simple** *if* $i_\Gamma \omega_L = 0$.

Proposition 7.3.9 Γ *is simple if and only if the Kähler form K_Γ coincides with ω_L.*

Proof: From Proposition 7.3.4 we have

$$K_\Gamma = i_v \omega_L,$$

where v is the vertical projector of Γ.

Hence

$$K_\Gamma = i_v \omega_L = i_{\frac{1}{2}(I-\Gamma)} \omega_L = \frac{1}{2} i_I \omega_L - \frac{1}{2} i_\Gamma \omega_L$$

$$= \omega_L - \frac{1}{2} i_\Gamma \omega_L.$$

Thus $K_\Gamma = \omega_L$ if and only if $i_\Gamma \omega_L = 0$.□

Let $E_L = CL - L$ be the energy function associated to L.

Definition 7.3.10 Γ *is said to be* **conservative** *if $d_h E_L = 0$, where h is the horizontal projector of* Γ.

Notice that $d_h E_L = 0$ is equivalent to the fact that E_L is constant along the horizontal curves with respect to Γ. In fact,

$$d_h E_L = i_h dE_L = h^*(dE_L).$$

Thus

$$(d_h L)(Z) = (dE_L)(hZ) = (hZ)E_L.$$

Proposition 7.3.11 *Suppose that Γ is simple. Then Γ is conservative if and only if it associated semispray ξ is precisely ξ_L.*

Proof: Since Γ is simple we have $K_\Gamma = \omega_L$. Hence we have

$$(i_\xi \omega_L)(X) = \omega_L(\xi, X) = K_\Gamma(\xi, X)$$

$$= g_\Gamma(J\xi, X) - g_\Gamma(\xi, JX) = g_\Gamma(C, X),$$

since ξ is horizontal. Thus

$$(i_\xi \omega_L)(X) = g_\Gamma(C, vX) = g_\Gamma(C, JFX) \text{ (since } v = JF)$$

$$= \bar{g}(C, JFX) = -\omega_L(C, FX) = -(i_C \omega_L)(FX)$$

$$= (-i_F i_C \omega_L)(X).$$

Therefore we obtain

$$i_\xi \omega_L = -i_F i_C \omega_L.$$

Since $i_C \omega_L = -d_J E_L$ (by Proposition 7.1.3), we have

$$i_\xi \omega_L = i_F d_J E_L = i_F i_J dE_L = i_{JF} dE_L = i_v dE_L.$$

$$= d_v E_L.$$

But $dE_L = d_h E_L + d_v E_L$ (since $Id = h + v$). Then $i_\xi \omega_L = dE_L$ if and only if $d_h E_L = 0$, or, equivalently

$$\xi = \xi_L \text{ if and only if } d_h E_L = 0. \square$$

Theorem 7.3.12 *Let L be an homogeneous Lagrangian. Then there exists a unique conservative connection with zero strong torsion.*

Proof: Since L is homogeneous then $E_L = L$, ξ_L is a spray and $\Gamma_L = -L_{\xi_L} J$ is a connection which associated semispray is precisely ξ_L. Moreover, Γ_L is simple (see Theorem 7.3.1). Hence Γ_L is conservative.

Now, let Γ be a conservative connection with zero strong torsion. We have

$$i_h \omega_L = -i_h dd_J L = i_h d_J dL$$

$$= d_J i_h dL + d_J dL \ \ (\text{since } [i_h, d_J] = i_h d_J + d_J i_h = d_J)$$

$$= d_J d_h L + d_J dL$$

$$= d_J dL \ \ (\text{since } d_h L = 0)$$

$$= -dd_J L$$

$$= \omega_L.$$

Thus

$$\frac{1}{2} i_\Gamma \omega_L = \frac{1}{2} i_{2h-Id} \omega_L = i_h \omega_L - \omega_L = 0,$$

which implies that Γ is simple. Hence, from Proposition 7.3.4, we see that the associated semispray to Γ is ξ_L. Then, from the Decomposition theorem one has $\Gamma = \Gamma_L$.$\square$

Remark 7.3.13 If L is the kinetic energy defined by a Riemannian metric g on M, then Theorem 7.3.12 is the fundamental theorem of Riemannian geometry (see Theorem 1.18.10).

Remark 7.3.14 In [27], [38] we have extended these results for higher order Lagrangians.

7.4 Semisprays and Lagrangian systems

From the above results we have seen that the Lagrangian formulation of Classical Mechanics may also be developed intrinsically through a symplectic structure. However this formulation depends directly on the choice of the Lagrangian function. This shows a particular difference from the Hamiltonian situation where an intrinsical symplectic structure is canonically defined on the cotangent bundle of a given configuration manifold, independent of the choice of the Hamiltonian function.

First-order differential equations (vector fields) are related with Hamiltonian functions via the symplectic structure. Let us now examine some results relating second-order differential equations (semisprays) with Lagrangian systems.

Let ξ be an arbitrary semispray on TM. We denote by $\Gamma = -L_\xi J$ the connection defined by ξ.

Then we have

$$\Gamma(\partial/\partial q^i) = \partial/\partial q^i + (\partial \xi^j/\partial q^i)\,\partial/\partial v^j,$$

$$\Gamma(\partial/\partial v^i) = -\partial/\partial v^i,$$

where $\xi = v^i\partial/\partial q^i + \xi^i\partial/\partial v^i$.

Hence

$$X^H = X^i(\partial/\partial q^i) - \frac{1}{2}X^i(\partial\xi^j/\partial q^i)\,(\partial/\partial v^j), \tag{7.13}$$

where $X = X^i\,(\partial/\partial q^i)$ is a vector field on M. By using (7.13) we have

$$[\xi, X^v] = [v^i(\partial/\partial q^i) + \xi^i(\partial/\partial v^i),\; X^j\partial/\partial v^j]$$

$$= -X^i(\partial/\partial q^i) + v^j(\partial X^i/\partial q^j)(\partial/\partial v^i) - X^j(\partial\xi^i/\partial v^j)(\partial/\partial v^i)$$

$$= -2X^i(\partial/\partial q^i) - X^j(\partial\xi^i/\partial v^j)(\partial/\partial v^i)$$

$$+X^i(\partial/\partial q^i) + v^j(\partial X^i/\partial q^j)(\partial/\partial v^i)$$

$$= -2X^H + X^C.$$

Therefore

$$[\xi, X^v] = -X^H - (X^H - X^C),$$

and, then, $X^H - X^C$ is the vertical component of $[\xi, X^v]$.

Proposition 7.4.1 *(Crampin [20]). Suppose that ω is a 2- form on TM such that*
(1) $L_\xi \omega = 0$,
(2) for every point $z \in TM$ the corresponding vertical subspace is Lagrangian with respect to ω and $i_Y d\omega$, where Y is an arbitrary horizontal vector field. Then ω is closed.

Proof: Let us first recall the following expression:

$$d\omega(X,Y,Z) = X\omega(Y,Z) - Y\omega(X,Z) + Z\omega(X,Y)$$

$$-\omega([X,Y],Z) + \omega([X,Z],Y) - \omega(X,[Y,Z]).$$

Therefore

$$d\omega(X^v, Y^v, Z^v) = 0,$$

since ω vanishes on pairs of vertical lifts (by (2)). Furthermore, for every vector field W on TM

$$d\omega(W, Y^v, Z^v) = d\omega(hW + vW, Y^v, Z^v)$$

$$= d\omega(hW, Y^v, Z^v) = (i_{hW}\, d\omega)(Y^v, Z^v) = 0.$$

Now,

$$(L_\xi d\omega)(X,Y,Z) = \xi d\omega(X,Y,Z) - d\omega([\xi, X], Y, Z)$$

$$-d\omega(X, [\xi, Y], Z) - d\omega(X, Y, [\xi, Z]).$$

Thus, from (1) and from the fact that

$$\xi d\omega(X^H, Y^v, Z^v) = 0,$$

one has

$$d\omega([\xi, X^H], Y^v, Z^v) + d\omega(X^H, [\xi, Y^v], Z^v)$$

$$+d\omega(X^H, Y^v, [\xi, Z^v]) = 0.$$

The first term vanishes because we have two vertical arguments in the expression. Now

$$[\xi, Y^v] = -Y^H - (Y^H - Y^C),$$

$$[\xi, Z^v] = -Z^H - (Z^H - Z^C)$$

and so only the horizontal components of $[\xi, Y^v]$ and $[\xi, Z^v]$ contribute to the remaining terms. Thus

$$d\omega(X^H, Y^H, Z^v) = d\omega(X^H, Z^H, Y^v).$$

But then

$$d\omega(X^H, Y^H, Z^v) = d\omega(X^H, Z^H, Y^v) = -d\omega(Z^H, X^H, Y^v)$$

$$= -d\omega(Z^H, Y^H, X^v) = d\omega(Y^H, Z^H, X^v)$$

$$= d\omega(Y^H, X^H, Z^v) = -d\omega(X^H, Y^H, Z^v)$$

and so

$$d\omega(X^H, Y^H, Z^v) = 0.$$

We conclude that $d\omega$ vanishes except when all its arguments are horizontal vector fields. As $Z^H - Z^C$ is vertical,

$$d\omega(X^H, Y^H, [\xi, Z^v]) = -d\omega(X^H, Y^H, Z^H).$$

Using again (1) for $d\omega(X^H, Y^H, Z^v)$ we obtain

$$d\omega(X^H, Y^H, [\xi, Z^v]) = 0. \square$$

Proposition 7.4.2 *(Crampin [20]). Let ω be as in Proposition 7.4.1. Then there is a C^∞ function K defined on an open subset of TM such that*

$$\omega = -dd_J K,$$

where J is the canonical almost tangent structure on TM.

Proof: From Proposition 7.4.1 we have that locally ω is exact, say $\omega = d\alpha$, for some 1–form α defined on some open subset U of TM. The restriction of $d\alpha$ to vertical subspaces is zero (they are Lagrangians for ω). Thus the restriction of α to each fibre is exact and there is a function F on an open set $V \subset U$ such that

$$\alpha(X^v) = X^v(F) = dF(X^v)$$

for any vector field X on $\tau_M(V) \subset M$. Set

$$\beta = \alpha - dF.$$

Then

$$d\beta = d\alpha = \omega \text{ and } \beta(X^v) = 0.$$

From (1) and (2) of Proposition 7.4.1, we have

$$\omega([\xi, X^v], Y^v) + \omega(X^v, [\xi, Y^v]) = 0$$

and as only the horizontal components of the brackets signify,

$$\omega(X^H, Y^v) = \omega(Y^H, X^v).$$

Thus, for vector fields X and Y on $\tau_M(V)$, we have

$$\omega(X^H, Y^v) = X^H(\beta(Y^v)) - Y^v(\beta(X^H)) - \beta([X^H, Y^v])$$

$$= -Y^v(\beta(X^H)) = \omega(Y^H, X^v) = -X^v(\beta(Y^H)),$$

since $[X^H, Y^v]$ is vertical. Put $\eta = S^*\beta$, where S is the almost complex structure defined by $\Gamma = -L_\xi J$. Let us recall that S is defined by

$$SX^H = -X^v, \; SX^v = X^H.$$

Hence

$$X^v(\eta(Y^v)) = X^v(\beta(Y^H)) = Y^v(\beta(X^H)) = Y^v(\eta(X^v)).$$

As before, the restriction of $d\eta$ to vertical subspaces is zero, and as before, there is a function $K : W \subset V \longrightarrow R$ such that $\eta(X^v) = -dK(X^v) = -X^v(K)$. Thus $\beta(X^v) = 0$ and $\beta(X^H) = \eta(X^v) = -dK(X^v)$, i.e.,

$$\beta(X^H) = (-d_J K)(X^H).$$

Therefore

$$\beta = -d_J K,$$

and, locally, up to a constant function along the fibres of TM, one has

$$\omega = d\beta = -dd_J K. \square$$

The following result is due to Crampin [20].

Theorem 7.4.3 *A necessary and sufficient condition for a semispray ξ on TM to be a (regular) Euler–Lagrange vector field is the existence of a 2-form ω on TM satisfying (1) and (2) of Proposition 7.4.1.*

Proof: If ξ is a Euler–Lagrange vector field then ξ is the unique solution of

$$i_\xi \omega_L = dE_L,$$

for some regular Lagrangian $L : TM \longrightarrow R$. The 2–form ω_L is symplectic and satisfies $L_\xi \omega_L = 0$. Also $d\omega_L = 0$ and then $i_Y d\omega_L = 0$. From Theorem 7.3.1 one has that the vertical subspaces are Lagrangian for ω_L.

Conversely, by using Propositions 7.4.1 and 7.4.2, one obtains a 2– form $\bar{\omega}$ of maximal rank such that $\bar{\omega} = -dd_J K$. The equation

$$i_{\xi_K} \bar{\omega} = dE_K, \; E_K = CK - K.$$

determines a semispray ξ_K. As $L_\xi \bar{\omega} = d(i_\xi \bar{\omega}) = 0$ there is at least locally some function E such that $i_\xi \bar{\omega} = dE$. Since ξ_K and ξ are both semisprays, then $\xi_K - \xi$ is vertical and since

$$i_{\xi_K - \xi} \bar{\omega} = d(E_K - E),$$

we deduce that $E_K - E$ is constant along the fibres ($\bar{\omega}$ vanishes on pairs of vertical vector fields). Thus

$$E_K - E = G,$$

where G is a function obtained by the pull–back of some function by $\tau_M : TM \longrightarrow M$, defined on some open set of M. If we set $L = K + G$ then

$$d_J L = d_J K = \beta,$$

$$CL - L = CK - K - G = E_K - G = E$$

and so ξ is (locally) a Euler–Lagrange vector field for such L.□

Remark 7.4.4 (Cantrijn et al. [11]). Let us suppose that X is a vector field on TM such that

$$(L_X J) \circ J = J, \; J \circ (L_X J) = -J. \tag{7.14}$$

(i.e., $-(L_X J)$ is a connection in TM). Then in a local coordinate system (q^i, v^i) X has the following expression

$$X = (v^i + \lambda^i(q)) \frac{\partial}{\partial q^i} + X^i(q, v) \frac{\partial}{\partial v^i},$$

i.e., X is a semispray modulo a τ_M–projectable vector field $\lambda^i(q)(\partial/\partial q^i)$.

Now, suppose that W is a $2m$–dimensional manifold endowed with an integrable almost tangent structure J and let γ be a vector field on W such that $L_\gamma J$ verifies (7.14).

Definition 7.4.5 *We say that γ is a* **regular Lagrangian dynamical system** *if there exists a symplectic form ω on W such that*
(1) the subspaces $(Im\, J)_z$ of $T_z W$ are Lagrangian with respect to ω_z for all $z \in W$;
(2) $L_\gamma \omega = 0$.

Using (1) and (2) it can be shown that the symplectic form ω and the almost tangent structure J are related by

$$i_{JX}\omega = -J^*(i_X \omega) = -i_J i_X \omega,$$

for all vector fields X on W. We prove this result by using the fact that γ defines "horizontal" and "vertical" distributions via the "connection" $-L_\gamma J$ and we proceed along the same lines as the ones given above in the case where $W = TM$. We also remark that the above equality implies that all subspaces $(Im\, J)_z$ are Lagrangian with respect to ω.

Now, using the same procedure as for the case TM and semisprays, we may show that if γ is a regular Lagrangian dynamical system on an integrable almost tangent manifold (W, J) then there exists (locally) a function L on W such that

$$\omega = -dd_J L,$$

$$i_\gamma \omega = dE_L, \; E_L = (J\gamma)L - L.$$

So, the Lagrangian formalism may be reobtained for the more general situation of integrable almost tangent manifolds.

7.5 A geometrical version of the inverse problem of Lagrangian dynamics

In this section, we will show that Theorem 7.4.2 gives a geometrical version of an old problem in Mechanics, attributed to Helmholtz (see for example Santilli [107]), known as the inverse problem of Lagrangian dynamics, which may be stated in the following form: given a semispray ξ on TM when is it possible to find a Lagrangian L such that ξ is precisely a Euler–Lagrange vector field for L?

The answer involves some conditions, called **Helmholtz conditions**, concerning a matrix $(\alpha_{ij}(q, \dot{q}))$. If we develop the Euler–Lagrange equations in local coordinates $(q, \dot{q})$ then we obtain an expression of the form

$$g_{ij}\ddot{q}^j + h_i = 0,$$

where g_{ij} and h_i are functions of $q^k, \dot{q}^k$ and

$$g_{ij} = \left(\partial^2 L / \partial \dot{q}^i \partial \dot{q}^j\right).$$

A system of second–order differential equations is derivable from a Lagrangian if it is self–adjoint. Let us see the conditions for self–adjointness. We will consider only systems of second–order differential equations of type

$$\ddot{q}^i = f^i(q, \dot{q}).$$

Then the necessary and sufficient conditions for a system of differential equations in this form to be derivable from a Lagrangian is that there exists a regular matrix (g_{ij}) depending on $(q, \dot{q})$ such that the equations

$$g_{ij}\ddot{q}^j - g_{ij}f^j = 0$$

are self–adjoint. In terms of the functions f^i these conditions are: there should exist functions g_{ij} such that

$$\det(g_{ij}) \neq 0, \tag{7.15}$$

$$g_{ij} = g_{ji}, \tag{7.16}$$

$$\frac{\partial g_{ij}}{\partial \dot{q}^k} = \frac{\partial g_{ik}}{\partial \dot{q}^j}, \tag{7.17}$$

$$\frac{d}{dt}(g_{ij}) + \frac{1}{2}\frac{\partial f^k}{\partial \dot{q}^j}g_{ik} + \frac{1}{2}\frac{\partial f^k}{\partial \dot{q}^i}g_{kj} = 0, \tag{7.18}$$

$$g_{ik}\left(\frac{d}{dt}\left(\frac{\partial f^k}{\partial \dot{q}^j}\right) - 2\frac{\partial f^k}{\partial q^j} - \frac{1}{2}\frac{\partial f^l}{\partial \dot{q}^j}\frac{\partial f^k}{\partial \dot{q}^l}\right)$$

$$= g_{jk}\left(\frac{d}{dt}\left(\frac{\partial f^k}{\partial \dot{q}^i}\right) - 2\frac{\partial f^k}{\partial q^i} - \frac{1}{2}\frac{\partial f^l}{\partial \dot{q}^i}\frac{\partial f^k}{\partial \dot{q}^l}\right). \tag{7.19}$$

Now, from the preceding geometrical study it is possible to show that (7.15)–(7.19) may be reobtained. In fact (we follow Crampin et al. [23]) let us first set the semispray ξ locally as

$$\xi = v^i\partial/\partial q^i + f^i\partial/\partial v^i.$$

Let $\Gamma = -L_\xi J$ be the connection determined by ξ. Then the horizontal distribution (with respect to Γ) is locally spanned by

$$D_i = \partial/\partial q_i + (1/2)(\partial f^j/\partial v^i)\partial/\partial v^j.$$

The dual basis of $\{D_i, V_i = \partial/\partial v^i\}$ is $\{\theta^i, \eta^i\}$, where

$$\theta^i = dq^i,\ \eta^i = -(1/2)(\partial f^i/\partial v^j)dq^j + dv^i.$$

The Lie derivatives of these 1-forms by ξ are

$$L_\xi \theta^i = (1/2)(\partial f^i/\partial v^j)\theta^j + \eta^i, \qquad (7.20)$$

$$L_\xi \eta^i = -\frac{1}{2}\gamma^i_j\theta^j + (1/2)(\partial f^i/\partial v^j)\eta^j, \qquad (7.21)$$

where

$$\gamma^i_j = \xi(\partial f^i/\partial v^j) - (1/2)(\partial f^i/\partial v^j) - 2(\partial f^i/\partial q^j).$$

Suppose that the form ω in Theorem 7.4.2 is locally expressed by

$$\omega = a_{ij}\theta^i \wedge \theta^j + g_{ij}\theta^i \wedge \eta^j,$$

where $a_{ij} + a_{ji} = 0$ (we may eliminate the terms in $\eta^i \wedge \eta^j$ from the expression of ω). If we use (7.20) and (7.21) we find that

$$L_\xi\omega = (\xi(a_{ij}) + a_{ik}(\partial f^k/\partial v^j) - (1/2)g_{ik}\gamma^k_j)\theta^i \wedge \theta^j$$

$$+(2a_{ij} + \xi(g_{ij}) + (1/2)g_{ik}(\partial f^k/\partial v^j) + (1/2)g_{kj}(\partial f^k/\partial v^i))\theta^i \wedge \eta^j$$

$$+g_{ij}\eta^i \wedge \eta^j.$$

As $L_\xi\omega = 0$, one finds that

$$g_{ij} = g_{ji}. \qquad ((7.16))$$

Therefore

$$a_{ij} = 0,$$

$$\xi(g_{ij}) + (1/2)g_{ik}(\partial f^k/\partial v^j) + (1/2)g_{kj}(\partial f^k/\partial v^i) = 0, \qquad ((7.18))$$

$$g_{ik}\gamma^k_j = g_{jk}\gamma^k_i. \qquad ((7.19))$$

Thus $\omega = g_{ij}\eta^i \wedge \eta^j$ and ω is of maximal rank if and only if (g_{ij}) is regular, i.e.,

$$\det(g_{ij}) \neq 0 \qquad ((7.15))$$

Finally, (7.17) is obtained from the condition $d\omega = 0$. As the only terms of interest in $d\omega$ are those involving $\theta^i \wedge \eta^j \wedge \eta^k$ and as the coefficients of such a term is $(\partial g_{ij}/\partial v^k)$ one has that

$$\partial g_{ij}/\partial v^k = \partial g_{ik}/\partial v^j.$$

and so

$$\partial g_{ji}/\partial v^k = \partial g_{jk}/\partial v^i = \partial g_{ij}/\partial v^k.$$

The self–adjointness conditions are thus equivalent to the conditions on the form ω stated in Theorem 7.4.2.

7.6 The Legendre transformation

Let $L : TM \longrightarrow R$ be a Lagrangian function and consider the 1–form

$$\alpha_L = d_J L.$$

Then α_L is locally expressed by

$$\alpha_L = \frac{\partial L}{\partial v^i} dq^i,$$

where (q^i, v^i) are the induced coordinates in TM.

Hence α_L is a semibasic 1–form on TM. If we take into account Theorem 4.2.16 one has the following

Theorem 7.6.1 *There exists a mapping $Leg : TM \longrightarrow T^*M$ such that the following diagram*

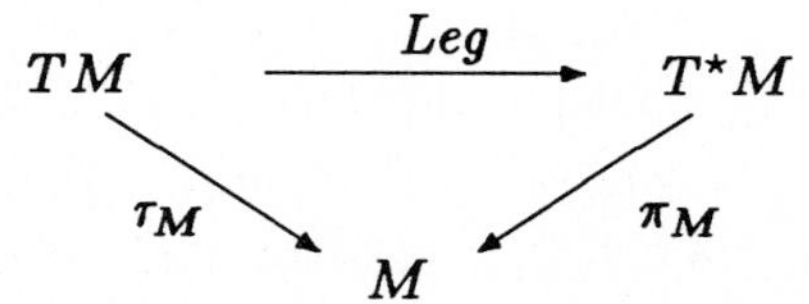

is commutative (τ_M and π_M are the canonical projections).

We locally have

$$Leg : (q^i, v^i) \longrightarrow (q^i, p_i),$$

where $p_i = \partial L / \partial v^i$.

Moreover, if λ_M denotes the Liouville form on T^*M, then

$$Leg^* \lambda_M = \alpha_L$$

(see Corollary 4.2.17). Then if $\omega_L = -d\lambda_M$ is the canonical symplectic form on T^*M, we have

$$Leg^* \omega_M = \omega_L,$$

where $\omega_L = -dd_J L$.

Theorem 7.6.2 *The following assertions are equivalent:*
(1) L is a regular Lagrangian.
(2) ω_L is a symplectic form on TM.
(3) $Leg : TM \longrightarrow T^*M$ is a local diffeomorphism.

In such a case, the mapping Leg is called the **Legendre transformation** *determined by L.*

Proof: The equivalence between (1) and (2) was proved in Section 7.1. On the other hand from $Leg^* \omega_M = \omega_L$ it follows that ω_L is symplectic if and only if $Leg : TM \longrightarrow T^*M$ is a symplectic map. □

Now, let $\sigma : R \longrightarrow M$ be a curve in M and $\dot{\sigma} : R \longrightarrow TM$ its natural prolongation to TM. Then along $\dot{\sigma}$, we have

$$p_i = \partial L / \partial \dot{q}^i,$$

since $v^i = dq^i/dt$, $\dot{\sigma}(t) = (q^i(t), \dot{q}^i(t))$. Hence

$$Leg : (q^i, \dot{q}^i) \longrightarrow (q^i, \partial L / \partial \dot{q}^i).$$

along $\dot{\sigma}$. We call $\partial L / \partial \dot{q}^i$ the **momentum** p_i, $1 \leq i \leq m$.

In general the Legendre transformation determined by a regular Lagrangian is not globally a diffeomorphism. If *Leg* is a global diffeomorphism then L is said to be **hyperregular**. As $\pi_M \circ Leg = \tau_M$ we see that *Leg* is a global diffeomorphism if and only if its restriction to each fibre of $\tau_M : TM \longrightarrow M$ is one–to–one.

Suppose that L is regular. Let $E_L = CL - L$ be the energy of L and consider the equation

$$i_{\xi_L}\omega_L = dE_L.$$

Then the vector field $\bar{\xi} = T\,Leg \circ \xi_L \circ Leg^{-1}$ on T^*M is given by

$$i_{\bar{\xi}}\omega_M = d(E_L \circ Leg^{-1}).$$

Indeed,

$$i_{\bar{\xi}}\omega_M = i_{\bar{\xi}}((Leg^{-1})^*\omega_L) = (Leg^{-1})^*(i_{\xi_L}\omega_L)$$

$$= (Leg^{-1})^*(dE_L) = d(E_L \circ Leg^{-1}).$$

Hence,

Proposition 7.6.3 *If γ is an integral curve of ξ_L then $\gamma = Leg \circ \sigma$ is an integral curve of $\bar{\xi}$ and we have*

$$\pi_M \circ \gamma = \tau_M \circ \sigma.$$

The energy function E_L of a regular Lagrangian L (and the function $E_L \circ Leg^{-L}$ on T^*M if L is hyperregular) will be called **Hamiltonian energy**, corresponding to L and will be denoted by H. It is now clear that Leg is the map which permits to pass from a regular Lagrangian formalism to a Hamiltonian formalism. Thus the above vector field $\bar{\xi}$ is the Hamiltonian vector field of energy $H = E_L \circ Leg^{-1}$.

We have seen above the relation between the Lagrangian and Hamiltonian formulation of Classical Mechanics which are equivalent in the hyperregular case. Our exposition was presented in terms of the tangent bundle geometry on TM. Let us examine this theory without using the almost tangent procedure (see Sternberg [114], Mac Lane [92], Abraham and Marsden [1]).

Let $L : TM \longrightarrow R$ be a Lagrangian function. For each tangent vector $v \in T_xM$, $x \in M$, we consider the natural identifications

$$\varphi_v : T_xM \longrightarrow T_v(T_xM)$$

and

$$\varphi_v^* : T_x^*M \longrightarrow T_v^*(T_xM).$$

Let $L_x : T_xM \longrightarrow R$ be the restriction of L to T_xM. We define a mapping $Leg_L : TM \longrightarrow T^*M$ by

$$Leg_L(v) = (\varphi_v^*)^{-1}(dL_x(v)).$$

Let (q^i, v^i) and (q^i, p_i) be the induced coordinates in TM and T^*M, respectively. Then the isomorphism φ_v is given by

$$\varphi_v(\partial/\partial q^i)_x = (\partial/\partial q^i)_v. \tag{7.22}$$

Hence

$$\varphi_v^*((dq^i)_x) = (dq^i)_v \tag{7.23}$$

From (7.22) and (7.23), one has

$$Leg_L(q^i, v^i) = (q^i, \frac{\partial L}{\partial v^i}),$$

i.e., Leg_L coincides with Leg. Sometimes it is found in the litterature the name of **fiber derivative** of L for Leg_L.

Example.- Let g be a Riemannian metric on M and L the **kinetic energy** defined by g, i.e.,

$$L(v) = (1/2)g_x(v, v), \; v \in T_xM, \; x \in M.$$

In local coordinates (q^i, v^i) we have

$$L(q^i, v^i) = (1/2)g_{ij}v^iv^j.$$

Hence $E_L = L$ and L is regular:

$$\frac{\partial^2 L}{\partial v^i \partial v^j} = g_{ij}.$$

Thus the corresponding Legendre transformation is given by

$$Leg(q^i, v^i) = (q^i, g_{ij}v^j),$$

i.e., $Leg(v) \in T_x^*M$ is the cotangent vector of M at x defined by

$$< u, Leg(v) >= g_x(u, v),$$

for all $u \in T_xM$. Hence $Leg : TM \longrightarrow T^*M$ is a global diffeomorphism. In fact, Leg is injective since $Leg(v) = Leg(\bar{v})$ implies that $g_x(u, v) = g_x(u, \bar{v})$, for all $u \in T_xM$ and so $v = \bar{v}$. Moreover, Leg is surjective, since, if $\alpha \in T_x^*M$, then the tangent vector $v \in T_xM$ such that

$$< u, \alpha >= g_x(u, v), \text{ for all } u \in T_xM,$$

verifies $Leg(v) = \alpha$.

The same is true for a Lagrangian $\bar{L} : TM \longrightarrow R$ defined by

$$\bar{L} = L + V \circ \tau_M,$$

where L is the kinetic energy of a Riemann metric g on M and $V : M \longrightarrow R$ is the potential energy.

7.7 Non–autonomous Lagrangians

In this section and in the followings we will use jet manifolds. More details about jets are given in Appendix B.

Let $(E = M \times R, p, R)$ be a trivial bundle, where M is a m– dimensional manifold. As (E, p, R) is trivial we may identify mappings from R to M with sections of (E, p, R) as well as their k–jets. The velocity space associated to M is the manifold $J_0^1(R, M)$ of all 1–jets of mappings $\sigma : R \longrightarrow M$ with source at the origin $0 \in R$. Thus we may identify $J_0^1(R, M)$ with the tangent bundle TM of M. The **evolution space** of M is the fibred manifold $J^1(R, M)$ of all 1–jets of mappings from R to M.

A C^∞ function $L : J^1(R, M) \longrightarrow R$ defined on the evolution space of M is said to be **non–autonomous** (or **time–dependent**).

It is our purpose to stablish an intrinsical description of non–autonomous Lagrangian systems.

For this, let us first notice that $J^1(R, M)$ can be canonically identified with $R \times TM$ by the map

$$j_t^1\sigma \longrightarrow (t, \dot{\sigma}(t)),$$

where $j_t^1\sigma$ denotes the 1–jet with source at t of a curve $\sigma : R \longrightarrow M$ and $\dot{\sigma}(t)$ is the tangent vector of σ at the point $\sigma(t)$.

If (q^1) are local coordinates for M then (t, q^i, v^i) are the induced coordinates for $J^1(R, M)$.

Therefore we transport all geometric structures defined on TM to $J^1(R, M)$, via this identification. Indeed J and C may be considered on $J^1(R, M)$ (and $J(\partial/\partial t) = 0$). Furthermore, we define a new tensor field $\tilde{J}$ of type (1,1) on $J^1(R, M)$ by

$$\tilde{J} = J - C \otimes dt \tag{7.24}$$

From (7.24), we deduce that $\tilde{J}$ is locally characterized by

$$\tilde{J}(\partial/\partial z) = -C, \; \tilde{J}(\partial/\partial q^i) = \partial/\partial v^i, \; \tilde{J}(\partial/\partial v^i) = 0. \tag{7.25}$$

Hence $\tilde{J}$ has rank m and satisfies $(\tilde{J})^2 = 0$. We define the adjoint of $\tilde{J}$, denoted by $\tilde{J}^*$, as the endomorphism of the exterior algebra $\wedge(J^1(R, M))$ of $J^1(R, M)$ locally given by

$$\tilde{J}^*(dt) = 0, \; \tilde{J}^*(dq^i) = 0, \; \tilde{J}^*(dv^i) = dq^i - v^i dt. \tag{7.26}$$

Definition 7.7.1 *A p-form α on $J^1(R, M)$ is said to be π_2-***semibasic** *(resp. π-***semibasic***) if α belongs to $Im\, J^*$ (resp. $Im\, \tilde{J}^*$). (Here $\pi_2 : J^1(R, M) \longrightarrow M$ and $\pi : J^1(R, M) \longrightarrow R \times M$ are the canonical projections).*

From (7.26) we easily deduce that a 1-form α on $J^1(R, M)$ is π_2-semibasic (resp. π-semibasic) if and only if α is locally expressed by

$$\alpha = \alpha_i(t, q, v) dq^i \tag{7.27}$$

(resp.

$$\alpha = \alpha_i(t, q, v) dq^i - \bar{\alpha}_i(t, q, v) v^i dt) \tag{7.28}$$

Let α be a π_2-semibasic 1-form on $J^1(R, M)$. Then we may define a differentiable mapping $D : R \times TM \longrightarrow R \times T^*M$ as follows:

$$D(t, v) = (t, p),$$

where $v \in T_x M$ and $p \in T_x^* M$ is given by

$$p(X) = \alpha(t, v)(\bar{X}), \; X \in T_x M,$$

with $\bar{X} \in T_{(t,v)}(R \times TM)$ projecting onto X, i.e., $(\pi_2)_* \bar{X} = X$. Hence if α is locally given by (7.27), we have

$$D(t, q^i, v^i) = (t, q^i, \alpha_i)$$

Like in the autonomous situation we associate to $\tilde{J}$ operators $i_{\tilde{J}}$ and $d_{\tilde{J}}$ on the algebra $\wedge(J^1(R, M))$ by

$$(i_{\tilde{J}}\omega)(X_1, \ldots, X_p) = \sum_{i=1}^{p} \omega(X_1, \ldots, \tilde{J}X_i, \ldots, X_p),$$

$$d_{\tilde{J}} = [i_{\tilde{J}}, d] = i_{\tilde{J}}d - di_{\tilde{J}},$$

and so from (7.25) we have

$$\begin{array}{rcl} i_{\tilde{J}}(df) & = & \tilde{J}^*(df), \\ i_{\tilde{J}}(dt) & = & i_{\tilde{J}}(dq^i) = 0, \\ i_{\tilde{J}}(dv^i) & = & dq^i - v^i dt, \\ d_{\tilde{J}}f & = & (\partial f / \partial v^i)(dq^i - v^i dt), \\ d_{\tilde{J}}(dt) & = & d_{\tilde{J}}(dq^i) = 0, \\ d_{\tilde{J}}(dv^i) & = & -d(dq^i - v^i dt) = dv^i \wedge dt, \end{array} \tag{7.29}$$

for all $f \in C^\infty(J^1(R, M))$.

Remark 7.7.2 According to the theory of Frölicher–Nijenhuis $i_{\tilde{J}}$ (resp. $d_{\tilde{J}}$) is the derivation of type i_* (resp. d_*) associated to the tensor field $\tilde{J}$.

Remark 7.7.3 In the following we will put

$$\theta^i = dq^i - v^i dt, \; 1 \leq i \leq m, \tag{7.30}$$

which are 1–forms on $J^1(R, M)$.

Let us characterize a semispray on $J^1(R, M)$ by means of the tensor fields J and $\tilde{J}$. We denote by $\pi_1 : J^1(R, M) = R \times TM \longrightarrow R$ the canonical projection defined by $\pi_1(j_t^1\sigma) = t$. Then the **1–jet prolongation** of a curve $\sigma : R \longrightarrow M$ in M is a section denoted by $j^1\sigma$ of the fibred manifold $(J^1(R, M), \pi_1, R)$ defined by

$$j^1\sigma : t \in R \longrightarrow (j^1\sigma)(t) = j^1_t\sigma \in J^1(R, M).$$

Alternatively, $j^1\sigma$ may also be regarded as a curve in $J^1(R, M)$ and for simplicity we will call it the **canonical prolongation** of σ to $J^1(R, M)$. However it is clear that not every section of $(J^1(R, M), \pi_1, R)$ has this particular form. A (local) section μ of $\pi_1 : J^1(R, M) \longrightarrow R$ will be the canonical prolongation of a curve in M if and only if

$$\mu^* \theta^i = 0, \ 1 \leq i \leq m \tag{7.31}$$

In such a case we say that μ is **holonomic**.

Definition 7.7.4 *A vector field ξ on $J^1(R, M)$ whose integral curves are all holonomics is called a* **semispray** *(or* **second-order differential equation***).*

It follows that a vector field ξ on $J^1(R, M)$ is a semispray if and only if

$$dt(\xi) = 1, \ \theta^i(\xi) = 0, \ 1 \leq i \leq m.$$

Then a semispray ξ on $J^1(R, M)$ is locally given by

$$\xi = \partial/\partial t + v^i \partial/\partial q^i + \xi^i \partial/\partial v^i, \tag{7.32}$$

where $\xi^i = \xi^i(t, q, v)$ is a C^∞ function on $J^1(R, M)$, $1 \leq i \leq m$.

Proposition 7.7.5 *A vector field ξ on $J^1(R, M)$ is a semispray if and only if $J\xi = C$ and $\tilde{J}\xi = 0$.*

Proof: In fact, if ξ is a semispray we easily deduce from (7.25) and (7.32) that $J\xi = C$ and $\tilde{J}\xi = 0$.

Conversely, suppose that ξ is locally expressed by

$$\xi = \tau \partial/\partial t + X^i \partial/\partial q^i + Y^i \partial/\partial v^i.$$

As $J\xi = C$ we obtain that $X^i = v^i$. Thus

$$\tilde{J}\xi = (1 - \tau) v^i \partial/\partial v^i = 0$$

and so $(1 - \tau)v^i = 0$, $1 \leq i \leq m$, where $\tau = \tau(t, q, v)$. If some $v^i \neq o$, we deduce that $\tau = 1$ and by continuity $\tau = 1$ on all $J^1(R, M)$, showing that ξ is a semispray. □

Definition 7.7.6 *Let ξ be a semispray on $J^1(R, M)$. A curve σ in M is called a* **path** *or* **solution** *of ξ if its canonical prolongation is an integral curve of ξ.*

Let σ be a curve in M, locally given by $(q^i(t))$. Then the 1–jet $(j^1\sigma)(t) = (t, q^i(t), (dq^i/dt)(t))$ and so σ is a path of ξ if and only if it satisfies the following non–autonomous system of second–order differential equations

$$\frac{d^2q^i}{dt^2} = \xi^i(t, q, \frac{dq}{dt}),\ 1 \leq i \leq m,$$

where ξ is given by (7.32).

Suppose that a non–autonomous regular Lagrangian L is given on M, i.e., L is a function on $J^1(R, M)$ such that the matrix

$$(\partial^2 L/\partial v^i \partial v^j)$$

is non–singular. Then we define a 2–form Ω_L on $J^1(R, M)$ by

$$\Omega_L = dd_{\tilde{J}}L + dL \wedge dt;$$

Ω_L is called the **Poincaré–Cartan 2–form** and $d_{\tilde{J}}L + Ldt$ is called the **Poincaré–Cartan 1–form.**

A straightforward computation in local coordinates shows that

$$\Omega_L = ((\partial^2 L/\partial v^i \partial t) + v^j(\partial^2 L/\partial v^j \partial q^i) - (\partial L/\partial q^i))dt \wedge dq^i$$

$$-v^j(\partial^2 L/\partial v^j \partial v^i)dv^i \wedge dt + (\partial^2 L/\partial v^i \partial q^j)dq^j \wedge dq^i$$

$$+ (\partial^2 L/\partial v^i \partial v^j)dv^i \wedge dq^j. \tag{7.33}$$

Therefore

$$\Omega_L^m = \pm \det(\partial^2 L/\partial v^i \partial v^j)dq^1 \wedge \ldots \wedge dq^m \wedge dv^1 \wedge \ldots \wedge dv^m.$$

Since L is regular we have

$$\Omega_L^m \wedge dt \neq 0.$$

Then (Ω_L, dt) determines an almost cosymplectic structure on $J^1(R, M)$. Moreover, Ω_L is exact, since

$$\Omega_L = d(d_{\tilde{J}} L + L dt).$$

Let

$$R_{\Omega_L} = \{X \in T(J^1(R, M)) / i_X \Omega_L = 0\}$$

be the characteristic bundle of Ω_L. Then R_{Ω_L} is an orientable line bundle and there exists a unique vector field ξ_L on $J^1(R, M)$ such that

$$i_{\xi_L} \Omega_L = 0, \quad i_{\xi_L} dt = 1. \tag{7.34}$$

Proposition 7.7.7 *Let L be a non–autonomous regular Lagrangian on $J^1(R, M)$ and ξ_L the vector field given by (7.34). Then ξ_L is a semispray on $J^1(R, M)$ whose paths are the solutions of the Euler–Lagrange equations*

$$\frac{d}{dt}\left(\frac{\partial L}{\partial \dot{q}^i}\right) - \frac{\partial L}{\partial q^i} = 0, \quad 1 \leq i \leq m.$$

Proof: Since $i_{\xi_L} dt = 1$, then ξ_L is locally given by

$$\xi_L = \partial/\partial t + X^i \partial/\partial q^i + \xi^i \partial/\partial v^i$$

Furthermore, ξ_L satisfies $i_{\xi_L} \Omega_L = 0$. Therefore, from (7.33), we have

$$X^i((\partial^2 L/\partial v^i \partial t) + v^j(\partial^2 L/\partial v^j \partial q^i) - (\partial L/\partial q^i))$$

$$+ \xi^i v^j (\partial^2 L/\partial v^j \partial v^i) = 0 \tag{7.35}$$

$$(\partial^2 L/\partial v^i \partial t) + v^j(\partial^2 L/\partial v^j \partial q^i) + X^j(\partial^2 L/\partial v^i \partial q^j)$$

$$- X^j(\partial^2 L/\partial v^j \partial q^i) - (\partial L/\partial q^i) + \xi^j(\partial^2 L/\partial v^i \partial v^j) = 0 \tag{7.36}$$

$$v^j(\partial^2 L/\partial v^j \partial v^i) = X^j(\partial^2 L/\partial v^j \partial v^i). \tag{7.37}$$

As L is regular, from (7.37), we have $X^i = v^i$, $1 \leq i \leq m$. Hence (7.35) and (7.36) become

$$v^i((\partial^2 L/\partial v^i \partial t) + v^j(\partial^2 L/\partial v^i \partial q^j) + \xi^j(\partial^2 L/\partial v^i \partial v^j)$$

$$- (\partial L/\partial q^i)) = 0, \quad (7.38)$$

$$(\partial^2 L/\partial v^i \partial t) + v^j(\partial^2 L/\partial v^i \partial q^j) + \xi^j(\partial^2 L/\partial v^i \partial v^j) - \partial L/\partial q^i = 0. \quad (7.39)$$

Now, let σ be a path of ξ_L. Then from (7.39) we have

$$(\partial^2 L/\partial v^i \partial t) + (dq^j/dt)(\partial^2 L/\partial v^i \partial q^j) + (d^2 q^j/dt^2)(\partial^2 L/\partial v^i \partial q^j)$$

$$- (\partial L/\partial q^i) = 0 \quad (7.40)$$

along $j^1\sigma$. But (7.40) are the Euler–Lagrange equations for L. □

We call ξ_L the **Euler–Lagrange vector field** for L.

7.8 Dynamical connections

The tensor fields J and $\tilde{J}$ on $J^1(R, M)$ permit us to give a characterization of a kind of connections in the fibred bundle $\pi : J^1(R, M) \longrightarrow R \times M$.

Definition 7.8.1 *By a* **dynamical connection** *on* $J^1(R, M)$ *we mean a tensor field* Γ *of type (1,1) on* $J^1(R, M)$ *satisfying*

$$J\Gamma = \tilde{J}\Gamma = \tilde{J}, \; \Gamma\tilde{J} = -\tilde{J}, \; \Gamma J = -J. \quad (7.41)$$

From a straightforward computation from (7.41) we deduce that the local expressions of Γ are

$$\left.\begin{array}{rcl} \Gamma(\partial/\partial t) & = & -v^i(\partial/\partial q^i) + \Gamma^i(\partial/\partial v^i), \\ \Gamma(\partial/\partial q^i) & = & \partial/\partial q^i + \Gamma_i^j(\partial/\partial v^j), \\ \Gamma(\partial/\partial v^i) & = & -\partial/\partial v^i \end{array}\right\} \quad (7.42)$$

The functions $\Gamma^i = \Gamma^i(t, q, v)$ and $\Gamma_i^j = \Gamma_i^j(t, q, v)$ will be called the **components** of Γ. From (7.42) we easily deduce that

$$\Gamma^3 - \Gamma = 0 \text{ and rank } \Gamma = 2m.$$

Then Γ defines an $f(3,-1)$–structure of rank $2m$ on $J^1(R,M)$. Now, we can associate to Γ two canonical operators ℓ and m given by

$$\ell = \Gamma^2, \; m = -\Gamma^2 + Id.$$

Then we have

$$\ell^2 = \ell, \; m^2 = m, \; \ell m = m\ell = 0, \; \ell + m = Id, \tag{7.43}$$

and ℓ and m are complementary projectors. From (7.43) we deduce that ℓ and m are locally given by

$$\begin{aligned}
\ell(\partial/\partial t) &= -v^i(\partial/\partial q^i) - (\Gamma^i + v^j\Gamma^i_j)(\partial/\partial v^i),\\
\ell(\partial/\partial q^i) &= \partial/\partial q^i, \; \ell(\partial/\partial v^i) = \partial/\partial v^i,\\
m(\partial/\partial t) &= \partial/\partial t + v^i(\partial/\partial q^i) + (\Gamma^i + v^j\Gamma^i_j)(\partial/\partial v^i),\\
m(\partial/\partial q^i) &= m(\partial/\partial v^i) = 0.
\end{aligned} \tag{7.44}$$

If we set $L = Im\,\ell$, $\mathcal{M} = Im\,m$ then we have that L and $\mathcal{M}$ are complementary distributions on $J^1(R,M)$, i.e.,

$$T(J^1(R,M)) = \mathcal{M} \oplus L.$$

From (7.44) we deduce that L is $2m$–dimensional and is locally spanned by $\{\partial/\partial q^i, \; \partial/\partial v^i\}$. $\mathcal{M}$ is 1–dimensional and globally spanned by the vector field $\xi = m(\partial/\partial t)$. Taking into account the local expression of ξ, we deduce that ξ is a semispray which will called the **canonical semispray associated to the dynamical connection** Γ.

Furthermore, we have $\Gamma^2\ell = \ell$ and $\Gamma m = 0$. Thus Γ acts on L as an almost product structure and trivially on $\mathcal{M}$. Since $\mathcal{M} = Ker\ \Gamma$, Γ is an $f(3,-1)$–structure of rank $2m$ and parallelizable kernel. Moreover, Γ/L has eigenvalues +1 and -1. From (7.42), the eigenspaces corresponding to the eigenvalue -1 are the π–vertical subspaces V_z, $z \in J^1(R,M)$. Recall that for each $z \in J^1(R,M)$, V_z is the set of all tangent vectors to $J^1(R,M)$ at z which are projected to 0 by $T\pi$. Thus V is a distribution given by $z \longrightarrow V_z$. The eigenspace at $z \in J^1(R,M)$ corresponding to the eigenvalue +1 will be denoted by H_z and called the **strong horizontal subspace at** z. We have a canonical decomposition

$$T_z(J^1(R,M)) = \mathcal{M}_z \oplus H_z \oplus V_z,$$

and then

$$T(J^1(R,M)) = \mathcal{M} \oplus H \oplus V, \tag{7.45}$$

where H is the distribution $z \longrightarrow H_z$.

Remark 7.8.2 We note that a dynamical connection Γ in $J^1(R,M)$ induces an almost product structure on $J^1(R,M)$ given by three complementary distributions for the three eigenvalues 0, +1 and -1. However, for a connection Γ in TM, the corresponding almost product structure on TM was given by two complementary distributions corresponding to the eigenvalues +1 and -1 of Γ.

Let us put $H'_z = \mathcal{M}_z \oplus H_z$; H'_z will be called the **weak horizontal subspace** at z. Then we have the following decompositions:

$$T_z(J^1(R,M)) = H'_z \oplus V_z, \; z \in J^1(R,M),$$

and

$$T(J^1(R,M)) = H' \oplus V, \tag{7.46}$$

where $z \longrightarrow H'_z$ is the corresponding distribution.

We notice that $L, \mathcal{M}, H$ and H' may be considered as vector bundles over $J^1(R,M)$; the bundles H and H' will be called **strong** and **weak horizontal bundles**, respectively. Thus, from (7.46), Γ defines a connection in the fibred manifold $\pi : J^1(R,M) \longrightarrow R \times M$, with horizontal bundle H'.

A vector field X on $J^1(R,M)$ which belongs to H (resp. H') will be called a **strong** (resp. **weak**) **horizontal** vector field. From (7.46), we have that the canonical projection $\pi : J^1(R,M) \longrightarrow R \times M$ induces an isomorphism $\pi_* : H'_z \longrightarrow T_{\pi(z)}(R \times M)$, $z \in J^1(R,M)$.

Then, if X is a vector field on $R \times M$, there exists a unique vector field $X^{H'}$ on $J^1(R,M)$ which is weak horizontal and projects to X; $X^{H'}$ will be called the **weak horizontal lift** of X to $J^1(R,M)$. The projection of $X^{H'}$ to H will be denoted by X^H and called the **strong horizontal lift** of X to $J^1(R,M)$.

From (7.42) and by a straightforward computation, we obtain

$$\begin{aligned}
(\partial/\partial t)^{H'} &= \partial/\partial t + (\Gamma^j + (1/2)v^i\Gamma_i^j)(\partial/\partial v^j), \\
(\partial/\partial q^i)^{H'} &= \partial/\partial q^i + (1/2)\Gamma_i^j(\partial/\partial v^j).
\end{aligned} \tag{7.47}$$

Then, if we put $D_i = (\partial/\partial q^i)^{H'}$ and $V_i = \partial/\partial v^i$, one deduces that $\{\xi, D_i, V_i\}$ is a local basis of vector fields on $J^1(R, M)$. In fact, $\mathcal{M} =< \xi >$, $H =< D_i >$, and $V =< V_i >$. Then $\{\xi, D_i, V_i\}$ is an adapted basis to the $f(3,-1)$–structure Γ. In terms of $\{\xi, D_i, V_i\}$. (7.47) becomes

$$(\partial/\partial t)^{H'} = \xi - v^i D_i,\ (\partial/\partial q^i)^{H'} = D_i.$$

Therefore we obtain

$$(\partial/\partial t)^H = -v^i D_i, (\partial/\partial q^i)^H = D_i.$$

If $X = \tau(\partial/\partial t) + X^i(\partial/\partial q^i)$ is a vector field on $R \times M$ we have

$$X^H = (X^i - \tau v^i)D_i. \tag{7.48}$$

We notice that the dual local basis of 1–forms of the adapted basis $\{\xi, D_i, V_i\}$ is given by $\{dt, \theta^i, \eta^i\}$, where

$$\eta^i = -(\Gamma^i + (1/2)v^j\Gamma^i_j)dt - (1/2)\Gamma^i_j dq^j + dv^i.$$

Remark 7.8.3 If we set

$$h = (1/2)(I + \Gamma)\ell,\ v = (1/2)(I - \Gamma)\ell$$

then we have

$$H = Im\, h,\ V = Im\, v,$$

$$D_i = h(\partial/\partial q^i),$$

$$\Gamma\xi = 0, \Gamma D_i = D_i,\ \Gamma V_i = -V_i,$$

$$h\xi = 0,\ hD_i = D_i,\ hV_i = 0,$$

$$v\xi = 0,\ vD_i = 0,\ vV_i = V_i.$$

h and v will be called the **strong horizontal** and **vertical projectors** associated to Γ. If we set $h' = m + h$, then $Im\, h' = H'$ and h' is called the **weak horizontal projector**.

Definition 7.8.4 *Let Γ be a dynamical connection in $J^1(R,M)$. A curve $\sigma : R \longrightarrow M$ is said to be a* **path** *of Γ if and only if $j^1\sigma$ is a weak horizontal curve in $J^1(R,M)$, i.e., the tangent vector $\overline{\dot{j^1\sigma}(t)}$ belong to $H'_{j^1\sigma(t)}$, for every $t \in R$.*

From (7.44) we easily deduce that σ is a path of Γ if and only if σ satisfies the following system of non–autonomous second–order differential equations:

$$\frac{d^2q^i}{dt^2} = \Gamma^i(t,q,\dot{q}) + \Gamma^i_j(t,q,\dot{q})\frac{dq^j}{dt} \tag{7.49}$$

From (7.32), (7.44) and (7.49) we obtain the following.

Proposition 7.8.5 *A dynamical connection Γ and its associated semispray ξ have the same paths.*

Now, we introduce the curvature and torsion of a dynamical connection.

Before proceeding further, we stablish the following lemma (the proof is obtained from a long by straightforward computation in local coordinates).

Lemma 7.8.6 *Let Γ be a dynamical connection in $J^1(R,M)$ and $\{\xi, D_i, V_i\}$ an adapted basis to Γ. Then we have*

$$[\xi,\xi] = 0\,,$$

$$[\xi, D_i] = (1/2)\Gamma^r_i D_r + A^r_i V_r\,,$$

$$[\xi, V_i] = -D_i + B^r_i V_r\,,$$

$$[D_i, D_j] = C^r_{ij} V_r\,,$$

$$[D_i, V_j] = (1/2)(\partial\Gamma^r_i/\partial v^j)V_r\,,$$

$$[V_i, V_j] = 0. \tag{7.50}$$

where

$$A^r_i = (3/4)\Gamma^s_i\Gamma^r_s - (1/2)(\partial\Gamma^r_i/\partial t) - (1/2)v^s(\partial\Gamma^r_i/\partial q^s)$$

$$+(\partial\Gamma^r/\partial q^i) + v^s(\partial\Gamma^r_s/\partial q^i) + (1/2)\Gamma^s(\partial\Gamma^r_i/\partial v^s)$$

$$+(1/2)v^s\Gamma^a_s(\partial\Gamma^r_i/\partial v^a) + (1/2)\Gamma^s_i(\partial\Gamma^r/\partial v^s)$$

$$+(1/2)v^a\Gamma^s_i(\partial\Gamma^r_a/\partial v^s),$$

$$B^r_i = (\partial\Gamma^r/\partial v^i) + (1/2)\Gamma^r_i + v^s(\partial\Gamma^r_s/\partial v^i),$$

$$C^r_{ij} = (1/2)(\partial\Gamma^v_i/\partial q^j) - (1/2)(\partial\Gamma^r_j/\partial q^i) + (1/4)\Gamma^s_i(\partial\Gamma^v_j/\partial v^s)$$

$$-(1/4)\Gamma^s_j(\partial\Gamma^r/\partial v^s).$$

Definition 7.8.7 *The tensor fields R and R' of type (1,2) on $J^1(R,M)$ given by*

$$R = \frac{1}{2}[h,h],\ R' = \frac{1}{2}[h',h']. \tag{7.51}$$

will be called, respectively the **strong** *and* **weak curvatures** *of* Γ.

From (7.50) and (7.51) we easily deduce the following.

Proposition 7.8.8 *We have*
$R(\xi,\xi) = R(\xi,D_i) = R(\xi,V_i) = R(D_i,V_j) = R(V_i,V_j) = 0,$
$R(H_i,H_j) = C^r_{ij}V_r,$
$R'(\xi,\xi) = R'(D_i,V_j) = R'(V_i,V_j) = 0,$
$R'(\xi,D_i) = A^r_iV_r,$
$R'(\xi,V_i) = D_i,$
$R'(D_i,D_j) = C^r_{ij}V_r,$
$\ell^* \circ R' = R.$

Definition 7.8.9 *The tensor fields $t, t', \tilde{t}$ and $\tilde{t}'$ of type (1,2) on $J^1(R,M)$ given by*

$$t = [J,h],\ t' = [J,h'],\ \tilde{t} = [\tilde{J},h],\ \tilde{t}' = [\tilde{J},h'] \tag{7.52}$$

will be called, respectively, the **strong J–torsion, weak J–torsion, strong $\tilde{J}$–torsion** *and* **weak $\tilde{J}$–torsion of** Γ.

Proposition 7.8.10 *We have*
$t(\xi,\xi) = t(D_i, V_j) = t(V_i, V_j) = 0,$
$t(\xi, D_i) = D_i + (1/2)(\Gamma^r_i - v^s(\partial\Gamma^r_i/\partial v^s)V_r,$
$t(\xi, V_i) = -V_i,$
$t(D_i, D_j) = (1/2)((\partial\Gamma^r_i/\partial v^j) - (\partial\Gamma^r_j/\partial v^i))V_r,$
$t'(\xi,\xi) = t'(\xi, V_i) = t'(D_i, V_j) = t'(V_i, V_j) = 0,$
$t'(\xi, D_i) = (1/2)((\Gamma^r_i - v^s(\partial\Gamma^r_i/\partial v^s)) + B^r_i - (1/2)\Gamma^r_i))V_r,$
$t'(D_i, D_j) = (1/2)((\partial\Gamma^r_i/\partial v^j) - (\partial\Gamma^r_j/\partial v^i))V_r,$
$\tilde{t}(\xi,\xi) = \tilde{t}(D_i, V_j) = \tilde{t}(V_i, V_j) = 0,$
$\tilde{t}(\xi, D_i) = D_i,$
$\tilde{t}(\xi, V_i) = -V_i,$
$\tilde{t}(D_i, D_j) = (1/2)((\partial\Gamma^r_i/\partial v^j) - (\partial\Gamma^r_j/\partial v^i))V_r,$
$\tilde{t'}(\xi,\xi) = \tilde{t'}(D_i, V_j) = \tilde{t'}(V_i, V_j) = 0,$
$\tilde{t'}(\xi, D_i) = (B^r_i - (1/2)\Gamma^r_i)V_r,$
$\tilde{t'}(\xi, V_i) = D_i - V_i,$
$\tilde{t'}(D_i, D_j) = (1/2)((\partial\Gamma^r_i/\partial v^j) - (\partial\Gamma^r_j/\partial v^i))V_r,$
$\ell^* \circ t = \ell^* \circ \tilde{t} = \ell^* \circ t' = \ell^* \circ \tilde{t'}.$

Next, we associate to each dynamical connection Γ in $J^1(R,M)$ an almost contact structure on $J^1(R,M)$.

Let Γ be a dynamical connection in $J^1(R,M)$ with canonical projectors m and ℓ and let h and v be the strong horizontal and vertical projectors of Γ, respectively. Let us now define a tensor field ϕ of type (1,1) on $J^1(R,M)$ by

$$\phi J = h,\ \Phi h = -J,\ \Phi m = 0.$$

Then, if $\{\xi, D_i, V_i\}$ is an adapted basis to Γ, we have

$$\phi V_i = D_i,\ \phi D_i = -V_i,\ \phi\xi = 0. \tag{7.53}$$

Hence we deduce that

$$\phi^2 = -Id + \xi \otimes \eta,$$

where $\eta = dt$ and ξ is the associated semispray to Γ. Then (ϕ,ξ,η) is an almost contact structure on $J^1(R,M)$ which will called the **canonical almost contact structure associated to** Γ.

Next, we characterize the normality of (ϕ,ξ,η) in terms of the connection Γ. As we have seen in Section 3.6, an almost contact structure (ϕ,ξ,η) is normal if and only if

$$N_\phi + 2\xi \otimes (d\eta) = 0,$$

where $N_\phi = (1/2)[\phi, \phi]$ is the Nijenhuis tensor of ϕ. In our case, $\eta = dt$ implies $d\eta = 0$. Hence the canonical almost contact structure (ϕ, ξ, η) associated to Γ is normal if and only if $N_\phi = 0$.

Now, from Lemma 7.8.6 and (7.53), we have the following.

Proposition 7.8.11 *The Nijenhuis tensor N_ϕ of ϕ is given by*

$$\begin{aligned}
N_\phi(\xi, \xi) &= 0, \\
N_\phi(\xi, D_i) &= (B^r - (1/2)\Gamma_i^r)D_r + (\delta_i^r - A_i^r)V_r, \\
N_\phi(\xi, V_i) &= (\delta_i^r - A_i^r)D_r + ((1/2)\Gamma_i^r - B_i^r)V_r, \\
N_\phi(D_i, D_j) &= (1/2)((\partial\Gamma_i^r/\partial v^j) - (\partial\Gamma_j^r/\partial v^i))H_r - C_{ij}^r V_r, \\
N_\phi(D_i, V_j) &= -C_{ij}^r D_r - (1/2)((\partial\Gamma_i^r/\partial v^j) - (\partial\Gamma_j^r/\partial v^i))V_r, \\
N_\phi(V_i, V_j) &= (C_{ij}^r - (1/2)((\partial\Gamma_i^r/\partial v^j) - (\partial\Gamma_j^r/\partial v^i))D_r.
\end{aligned}$$

Moreover, the Lie derivative of ϕ with respect to ξ is given by

$$(L_\xi\phi)(\xi) = 0,$$

$$(L_\xi\phi)(D_i) = (\delta_i^r - A_i^r)D_r + ((1/2)\Gamma_i^r - B_i^r)V_r,$$

$$(L_\xi\phi)(V_i) = ((1/2)\Gamma_i^r - B_i^r)D_r + (A_i^r - \delta_i^r)V_r.$$

From Propositions 7.8.8, 7.8.10 and 7.8.11, we easily deduce the following.

Theorem 7.8.12 *(ϕ, ξ, η) is normal if and only if $L_\xi\phi = 0$, $R = 0$ and $\ell^* \circ t = 0$.*

Next, we shall construct an adapted metric to (ϕ, ξ, η). Let $\bar{g}$ be a metric on the vertical bundle V over $J^1(R, M)$. Then we define a Riemannian metric g_Γ on $J^1(R, M)$ by

$$g_\Gamma(X, Y) = \bar{g}(JX, JY) + \bar{g}(vX, vY) + \eta(X)\eta(Y). \tag{7.54}$$

A simple computation from (7.54) shows that

$$g_\Gamma(\xi,\xi)=1,\ g_\Gamma(\xi,D_i)=g_\Gamma(\xi,V_i)=g_\Gamma(D_i,V_j)=0. \tag{7.55}$$

From (7.55) one easily deduces that $\mathcal{M}, H$ and V are orthogonal distributions, $\eta = i_\xi g_\Gamma$ and g_Γ is an adapted metric to (ϕ,ξ,η). Let us now consider the fundamental 2–form K_Γ associated to the almost contact metric structure (ϕ,ξ,η,g_Γ) defined by

$$K_\Gamma(X,Y)=g_\Gamma(\phi X,Y).$$

In terms of an adapted basis $\{\xi, D_i, V_i\}$, we have

$$K_\Gamma(\xi,\xi)=K_\Gamma(\xi,D_i)=K_\Gamma(\xi,V_i)=K_\Gamma(D_i,D_j)=K_\Gamma(V_i,V_j)=0,$$

$$K_\Gamma(D_i,V_j)=-\bar{g}(V_i,V_j). \tag{7.56}$$

7.9 Dynamical connections and non–autonomous Lagrangians

In this section we construct a dynamical connection canonically associated to a non–autonomous Lagrangian whose paths are the solutions of the Euler–Lagrange equations.

First, we examine the relation between semisprays and dynamical connections.

Let ξ be an arbitrary semispray on $J^1(R,M)$ and suppose that ξ is locally given by

$$\xi=\partial/\partial t+v^i\partial/\partial q^i+\xi^i\partial/\partial v^i.$$

Then a simple computation shows that

$$\left.\begin{aligned}
[\xi,\partial/\partial t] &= -(\partial\xi^j/\partial t)(\partial/\partial v^j),\\
\left[\xi,\partial/\partial q^i\right] &= -(\partial\xi^j/\partial q^i)(\partial/\partial v^j),\\
\left[\xi,\partial/\partial v^i\right] &= -\partial/\partial q^i-(\partial\xi^j/\partial v^i)(\partial/\partial v^j)
\end{aligned}\right\} \tag{7.57}$$

Now, let $\Gamma=-L_\xi\tilde{J}$. Then from (7.57) we have

$$\left.\begin{aligned}\Gamma(\partial/\partial t) &= -v^i(\partial/\partial q^i) - (v^j(\partial\xi^i/\partial v^j) - \xi^i)(\partial/\partial v^i),\\ \Gamma(\partial/\partial q^i) &= \partial/\partial q^i + (\partial\xi^j/\partial v^i)(\partial/\partial v^j),\\ \Gamma(\partial/\partial v^i) &= -\partial/\partial v^i\end{aligned}\right\} \tag{7.58}$$

From (7.58) we easily deduce the following.

Proposition 7.9.1 *Let ξ be a semispray on $J^1(R,M)$. Then $\Gamma = -L_\xi \tilde{J}$ is a dynamical connection in $J^1(R,M)$ whose associated semispray is ξ.*

Now, let L be a non–autonomous regular Lagrangian on $J^1(R,M)$. We set

$$\Gamma_L = -L_{\xi_L}\tilde{J},$$

where ξ_L is the Euler–Lagrange vector field for L. From Proposition 7.8.5, we deduce that **the paths of Γ_L are the solutions of the Euler–Lagrange equations for L.**

A direct computation shows that

$$\Omega_L(\xi_L,\xi_L) = \Omega_L(\xi_i,D_i) = \Omega_L(\xi_L,V_i) = \Omega_L(D_i,D_j)$$

$$= \Omega_L(V_i,V_j) = 0,$$

$$\Omega_L(D_i,V_j) = -(\partial^2 L/\partial v^i\partial v^j), \tag{7.59}$$

where $\{\xi_L, D_i, V_i\}$ is an adapted basis for L.

Next we shall construct an adapted metric on $J^1(R,M)$ to the canonical almost contact structure (ϕ_L,ξ_L,η) associated to Γ_L. First, we define a metric $\bar{g}_{\Gamma_L}$ on the vertical bundle V by

$$\bar{g}_{\Gamma_L}(JX,\tilde{J}Y) = \Omega_L(JX,Y) - \eta(Y). \tag{7.60}$$

In terms of an adapted basis $\{\xi_L, D_i, V_i\}$ for Γ_L we have

$$\bar{g}_{\Gamma_L}(V_i,V_j) = \Omega_L(V_i,D_j) = (\partial^2 L/\partial v^i\partial v^j).$$

Hence we can construct a Riemannian metric g_{Γ_L} defined by (7.54), i.e.,

$$g_{\Gamma_L}(X,Y) = \bar{g}_{\Gamma_L}(JX,JY) + \bar{g}_{\Gamma_L}(vX,vY) + \eta(X)\eta(Y).$$

Thus g_{Γ_L} is an adapted metric to (ϕ_L, ξ_L, η) and $(\phi_L, \xi_L, \eta, g_{\Gamma_L})$ is an almost contact metric structure on $J^1(R, M)$. Now, let K_{Γ_L} be the fundamental 2–form associated to $(\phi_L, \xi_L, \eta, g_{\Gamma_L})$. Then we have

$$K_{\Gamma_L}(\xi_L, \xi_L) = K_{\Gamma_L}(\xi_L, D_i) = K_{\Gamma_L}(\xi_L, V_i) = K_{\Gamma}(D_i, D_j)$$

$$= K_{\Gamma_L}(V_i, V_j) = 0,$$

$$K_{\Gamma_L}(D_i, V_j) = -\bar{g}_{\Gamma_L}(V_i, V_j) = -(\partial^2 L/\partial v^i \partial v^j). \tag{7.61}$$

From (7.59) and (7.61) we easily deduce the following.

Theorem 7.9.2 *Let L be a non–autonomous regular Lagrangian on $J^1(R, M)$ and ξ_L the Euler–Lagrange vector field for L. Let $(\phi_L, \xi_L, \eta, g_{\Gamma_L})$ be the almost contact metric structure canonically associated to the dynamical connection $\Gamma_L = -L_{\xi_L}\tilde{J}$. Then the Poincaré–Cartan form Ω_L is the fundamental 2–form K_{Γ_L} defined from $(\phi_L, \xi_L, \eta, g_{\Gamma_L})$, i.e.,*

$$\Omega_L = K_{\Gamma_L}.$$

To end this section, we reobtain some results of Crampin et al. [23]. In fact, we have an adapted basis $\{\xi_L, D_i, V_i\}$ to Γ_L, where

$$D_i = \partial/\partial q^i + (1/2)(\partial \xi^j/\partial v^i)(\partial/\partial v^j).$$

Thus the corresponding dual basis is $\{dt, \theta^i, \eta^i\}$, where

$$\eta^i = -(\xi^i - (1/2)v^j(\partial \xi^i/\partial v^j))dt$$

$$-(1/2)(\partial \xi^i/\partial v^j)dq^j + dv^i.$$

The significance of this dual basis is that the form Ω_L can be re–written as follows:

$$\Omega_L = (\partial^2 L/\partial v^i \partial v^j)\ \theta^i \wedge \eta^j$$

and so the semispray ξ_L is uniquely determined by the equations

$$i_{\xi_L}\theta^i = i_{\xi_L}\eta^i = 0,\ i_{\xi_L}dt = 1.$$

Remark 7.9.3 In [39] we have extended the results of this section for higher order Lagrangians.

7.10 The variational approach

Let M be a manifold of dimension m. Consider the evolution space $J^1(R, M)$ which may be canonically identified with $R \times TM$ (see Section 7.7). If we consider the trivial fibred manifold $(R \times M, p, R)$ then it is easy to see that we may identify the corresponding jet manifold of 1–jets of sections of $(R \times M, p, R)$ with $J^1(R, M)$. In the following we set $N = R \times M$, $J^1N = J^1(R, M) = R \times TM$. We have the following canonical projections:

$$\pi : J^1N \longrightarrow N, \ \rho : J^1N \longrightarrow R.$$

which are locally defined by

$$\pi(t, q^i, v^i) = (t, q^i), \ \rho(t, q^i, v^i) = t$$

Let $z \in J^1N$, $y = \pi(z) \in N$ and $s : R \longrightarrow N$ a section such that $s(t) = y$, where $t = \rho(z)$. We define the following linear difference from $T_z(J^1N)$ to T_yN:

$$T\pi - (Ts) \circ (T\rho) : T_z(J^1N) \longrightarrow T_yN \tag{7.62}$$

using the tangent prolongations of π, s and ρ. The following diagram illustrates this definition:

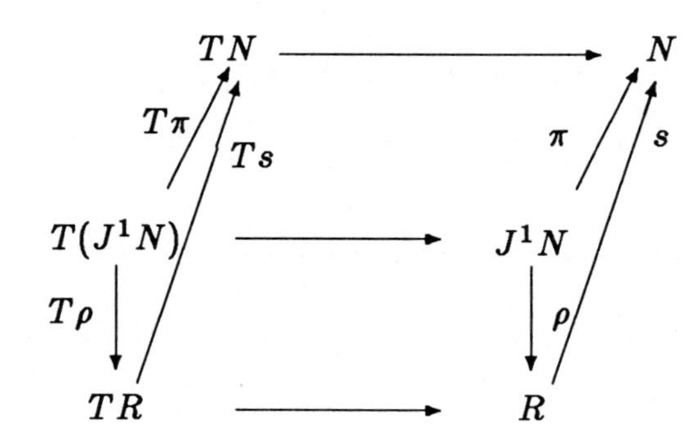

Suppose that $X \in T_z(J^1N)$ is locally given by

$$X = \tau\partial/\partial t + X^i\partial/\partial q^i + Y^i\partial/\partial v^i$$

Then

$$T\pi(z)(X) - ((Ts)(t) \circ T\rho(z))(X) = \tau\partial/\partial t + X^i\partial/\partial q^i$$

$$-(Ts)(t)(\tau\partial/\partial t)$$

$$= \tau\partial/\partial t + X^i\partial/\partial q^i - \tau\partial/\partial t - v^i\tau\partial/\partial q^i$$

$$= (X^i - \tau v^i)\partial/\partial q^i, \tag{7.63}$$

i.e., $(T\pi - (Ts)\circ(T\rho))(X)$ is a vertical tangent vector with respect to the fibration $p : R \times M \longrightarrow R$.

Definition 7.10.1 *The TN–valued 1–form given by (7.62) is called* **structure** *(or* **canonical***)* **1–form** *and it is represented by* θ.

Now, let $s : R \longrightarrow N$ be a section of $p : N \longrightarrow R$ and consider the 1–jet prolongation j^1s. Then $T(j^1s) : TR \longrightarrow T(J^1N)$ gives

$$T(j^1s)(\tau\partial/\partial t) = \tau\partial/\partial t + v^i\partial/\partial q^i + z^i\partial/\partial v^i,$$

where $s(t) = (t, q^i(t))$, $v^i(t) = (dq^i/dt)(t)$ and $z^i(t) = (d^2q^i/dt^2)(t)$.

Therefore

$$((Ts)\circ(T\rho))(T(j^1s)(\tau\partial/\partial t)) = \tau\partial/\partial t + v^i\partial/\partial q^i,$$

$$(T\pi)(T(j^1s)(\tau\partial/\partial t)) = \tau\partial/\partial t + v^i\partial/\partial q^i,$$

and so

$$\theta(T(j^1s)(\tau\partial/\partial t)) = 0. \tag{7.64}$$

Also it is easy to verify that if $Z : N \longrightarrow T(J^1N)$ is vertical over N then

$$\theta(Z) = T\pi(Z). \tag{7.65}$$

From (7.63) we may define 1–forms θ^i, i.e., setting $i_X\theta^i = X^i - \tau v^i$, we have

$$\theta^i = dq^i - v^i dt. \tag{7.66}$$

Proposition 7.10.2 *For every section* $s : R \longrightarrow N$ *the structure form* θ *is the unique 1–form verifying (7.64) and (7.65).*

Proof: Suppose that ω is a $V(N)$–valued 1–form verifying (7.64) and (7.65). Then

$$\omega(z) : T_z(J^1N) \longrightarrow V_y(N) \subset T_yN,$$

$z \in J^1N$, $y = \pi(z)$. Suppose that $z = j^1s(t)$, where s is a section of $p : N \longrightarrow R$ and $t \in R$. If $X \in T_z(J^1N)$ then

$$X - (T(j^1s) \circ (T\rho))(X) = Y \in V_y(N) \subset T_yN.$$

As for vertical tangent vectors Z, $(\omega - \theta)(Z) = 0$, one obtains

$$\begin{aligned} (\omega - \theta)(X) &= (\omega - \theta)(Y + (T(j^1s) \circ (T\rho))(X)) \\ &= (\omega - \theta)(T(j^1s) \circ (T\rho))(X)) \\ &= (\omega - \theta)(T(j^1s)((T\rho)(X))) \\ &= 0, \end{aligned}$$

where the last equality come from (7.64).□

Definition 7.10.3 *Let W be a submanifold of N. We say that W is a* **cross–section submanifold** *of N if p/W is a bijection of W onto a submanifold of R and for every $y \in W$, $Tp : T_yW \longrightarrow T_{p(y)}R$ is an isomorphism.*

If we consider a section $s : R \longrightarrow N$ then $s(R)$ defines a cross–section submanifold of N. Sometimes we set $s(R) = R_s$. Also, the 1–jet prolongation $j^1s : R \longrightarrow J^1N$ defines a cross–section submanifold of J^1N and we set $(j^1s)(R) = R_{j^1s}$. In the following we identify j^1s with R_{j^1s} and we will say that j^1s **is a cross–section submanifold of** J^1N.

Proposition 7.10.4 *The only cross–section submanifolds of J^1N for which θ vanishes ($\theta \equiv 0$ along such submanifolds) are the 1–jet prolongations of sections of (N, p, R).*

This proposition is a consequence of the following:

Proposition 7.10.5 *Let $u : R \longrightarrow J^1N$ be a section and R_u the corresponding cross–section submanifold. Suppose that $N_u \subset \pi^{-1}(y)$, $y \in N$. Then there is a section $s : R \longrightarrow N$ such that $u = j^1s$ (i.e., u is a 1–jet prolongation) if and only if*

$$\theta^i / N_u = 0.$$

Proof: Suppose that there is a section $s : R \longrightarrow N$ such that $u = j^1 s$. Then from Proposition 7.10.2

$$i_{T(j^1 s)(X)}\theta = 0,$$

and so θ^i vanishes alons N_u, since

$$(i_Y\theta)(u(t)) = (i_Y\theta^i)(u(t))(\partial/\partial q^i).$$

Conversely, let be $u : R \longrightarrow J^1 N$ such that $N_u \subset (\pi^{-1})(t, q^i)$, where (t, q^i) are the local coordinates of $y \in N$. We have

$$u(t) = (t, q^i, v^i).$$

But

$$\theta^i = dq^i - v^i dt / N_u \equiv 0$$

and so

$$v^i = \frac{dq^i}{dt} = \dot{q}^i.$$

Therefore taking $s(t) = (t, q^i(t)) = (\pi \circ u)(t)$ one has $u = j^1 s$.$\square$

Now, let (E, p, Q) be a fibred manifold and α a k–form on E. We suppose that S is a compact k–dimensional submanifold with smooth boundary ∂S. We denote by $Sec(E)$ the sections of (E, p, Q).

Definition 7.10.6 *A* **variational problem** *with respect to α on a domain S is given by a mapping $J : Sec(E) \longrightarrow R$ defined by*

$$J(s) = \int_S s^\star \alpha \tag{7.67}$$

called **Hamiltonian functional** *of α for the sections $s : S \longrightarrow E$ of $Sec(E)$. A* **variation** *of (7.67) is given by a C^∞ 1-parameter family of sections $s_t \in Sec(E)$, with $s_0 = s$ such that*

$$J(s_t) = \int_S s_t^\star \alpha.$$

$$\int_{\bar{\epsilon}} (j^1 s)^* L_{\tilde{X}^1}(L\,dt) = 0,$$

where $\tilde{X}^1$ are such that $\tilde{X}^1/_{(j^1 s)(\partial\bar{\epsilon})} \equiv 0$. As the above integral may be re–written as

$$\int_{(j^1 s)(\bar{\epsilon})} L_{\tilde{X}^1}(L\,dt) = 0 \tag{7.74}$$

and taking into account Stoke's theorem and the Cartan formula, (7.74) takes the form

$$\int L_{\tilde{X}^1}(L\,dt) = \int i_{\tilde{X}^1}(dL \wedge dt) = 0 \tag{7.75}$$

along $(j^1 s)(\bar{\epsilon})$.

Let us examine (7.75) in local coordinates. For X vertical we have

$$\tilde{X}^1 = X^i(\partial/\partial q^i) + ((\partial X^i/\partial t) + v^j(\partial X^i/\partial q^j))(\partial/\partial v^i)$$

(see (7.72)). Thus

$$L_{\tilde{X}^1}(Ldt) = X^i(\partial L/\partial q^i)dt + ((\partial X^i/\partial t) + v^j(\partial X^i/\partial q^j))(\partial L/\partial v^i)dt.$$

Taking into account that

$$(\partial L/\partial v^i)d(X^i) = d((\partial L/\partial v^i)X^i) - X^i d(\partial L/\partial v^i)$$

and

$$(\partial X^i/\partial q^j)(\partial L/\partial v^i)v^j dt = (\partial X^i/\partial q^j)(\partial L/\partial v^i)(dq^j - \theta^j)$$

we see that

$$L_{\tilde{X}^1}(Ldt) = X^i(\partial L/\partial q^i)dt + (\partial X^i/\partial t)(\partial L/\partial v^i)dt$$

$$+(\partial X^i/\partial q^j)(\partial L/\partial v^i)dq^j - (\partial X^i/\partial q^j)(\partial L/\partial v^i)\theta^j$$

$$= X^i(\partial L/\partial q^i)dt + (\partial L/\partial v^i)((\partial X^i/\partial t)dt + (\partial X^i/\partial q^j)dq^j)$$

$$-(\partial X^i/\partial q^j)(\partial L/\partial v^i)\theta^j$$

$$= X^i(\partial L/\partial q^i)dt + (\partial L/\partial v^i)dX^i - (\partial X^i/\partial q^j)(\partial L/\partial v^i)\theta^j$$

$$= X^i(\partial L/\partial q^i)dt + d((\partial L/\partial v^i)X^i) - X^i d(\partial L/\partial v^i) - (\partial X^i/\partial q^j)(\partial L/\partial v^i)\theta^j.$$

Therefore

$$L_{\tilde{X}^1}(Ldt) = X^i((\partial L/\partial q^i)dt - d(\partial L/\partial v^i))$$

$$+d((\partial L/\partial v^i)X^i) - (\partial X^i/\partial q^j)(\partial L/\partial v^i)\theta^j.$$

Now, the second term in the right side vanishes when we apply Stoke's theorem and as $\theta^j = 0$ along $(j^1 s)(\bar{\epsilon})$ (see Proposition 7.10.4) one has

$$X^i((\partial L/\partial q^i)dt - d(\partial L/\partial v^i)) = 0,$$

and thus (as $X^i \neq 0$)

$$\frac{\partial L}{\partial q^i} - \frac{d}{dt}\left(\frac{\partial L}{\partial v^i}\right) = 0, \quad 1 \leq i \leq m. \tag{7.76}$$

Then if $s : [0, \epsilon] \longrightarrow N$ is a section, s is an extremal of $J(s)$ if and only if $s(t) = (t, q^i(t), \dot{q}^i(t))$ is a solution of (7.76), i.e., one has

$$\frac{d}{dt}\left(\frac{\partial L}{\partial \dot{q}^i}\right) - \frac{\partial L}{\partial q^i} = 0, \quad 1 \leq i \leq m. \tag{7.77}$$

Let us now examine the **modified Hamilton's principle** given by a functional I defined on $Sec\,(J^1 N)$. For this we consider the 1–form

$$\theta_L = d_{\tilde{J}} L + Ldt \tag{7.78}$$

where $\tilde{J}$ is the tensor field of type (1,1) on $R \times TM$ given by (7.24). In local coordinates

$$\theta_L = \frac{\partial L}{\partial v^i}\theta^i + Ldt.$$

Note that if $j^1 s$ is the 1–jet prolongation of a secction s, then

$$\theta_L/_{j^1 s} = (Ldt)/_{j^1 s}.$$

The modified Hamilton's principle is defined by

$$I(\mu) = \int_{\bar{\epsilon}^1} \theta_L, \tag{7.79}$$

where now $\mu : [0, \epsilon] \longrightarrow J^1 N$ is a section (not necessarily a 1–jet prolongation) and $\bar{\epsilon}^1 = \mu([0, \epsilon])$. The section μ is an extremal of (7.79) if for all vertical vector field Y on $J^1 N$ over R one has

$$\int_{\bar{\epsilon}^1} L_Y \theta_L = 0$$

which leads to the equation

$$i_Y d\theta_L = 0 \text{ along } \bar{\epsilon}^1 = \mu([0, \epsilon]). \tag{7.80}$$

In local coordinates θ_L is written as

$$\theta_L = f_i dq^i + (L - f_i v^i)dt,$$

where, for simplicity, we have set $f_i = (\partial L/\partial v^i)$. Thus

$$d\theta_L = \left(-\frac{\partial f_i}{\partial t} + \frac{\partial L}{\partial t} - v^j \frac{\partial f_j}{\partial q^i}\right) dq^i \wedge dt$$

$$+\frac{\partial f_i}{\partial q^j} dq^j \wedge dq^i + \frac{\partial f_i}{\partial v^j} dv^j \wedge dq^i - v^i \frac{\partial f_i}{\partial v^j} dv^j \wedge dt.$$

For a vertical vector field Y on $J^1 N$ along R we have

$$Y = Y^i(\partial/\partial q^i) + \bar{Y}^i(\partial/\partial v^i)$$

and (7.80) takes the form

$$i_Y d\theta_L = \left(-\frac{\partial f_i}{\partial t} + \frac{\partial L}{\partial t}\right) Y^i dt - \left(v^j \frac{\partial f_j}{\partial q^i}\right) Y^i dt$$

$$+\frac{\partial f_i}{\partial q^j} Y^j dq^i - \frac{\partial f_i}{\partial q^j} Y^i dq^j$$

$$+\frac{\partial f_i}{\partial v^j}\bar{Y}^j dq^i - \frac{\partial f_i}{\partial v^j} Y^i dv^j$$

$$-\left(v^i \frac{\partial f_i}{\partial v^j}\right)\bar{Y}^j dt. \tag{7.81}$$

Now, (7.80) is considered along $\bar{\epsilon}^1$, thus the coordinates are taken along the curve $\mu(t)$, i.e., $\mu(t) = (t,\ q^i(t),\ v^i(t))$. Then

$$dq^i = \frac{dq^i}{dt}dt,\quad dv^i = \frac{dv^i}{dt}dt,$$

and so (7.81) is

$$i_Y d\theta_L = \left(-\frac{\partial f_i}{\partial t} + \frac{\partial L}{\partial t}\right) Y^i dt - (v^j \frac{\partial f_j}{\partial q^i}) Y^i dt$$

$$+\left(\frac{\partial f_i}{\partial q^j}\frac{dq^i}{dt}\right) Y^j dt - \left(\frac{\partial f_i}{\partial q^j}\frac{dq^j}{dt}\right) Y^i dt$$

$$+\left(\frac{\partial f_i}{\partial v^j}\frac{dq^i}{dt}\right) \bar{Y}^j dt - \left(\frac{\partial f_i}{\partial v^j}\frac{dv^j}{dt}\right) Y^i dt$$

$$-\left(v^i \frac{\partial f_i}{\partial v^j}\right)\bar{Y}^j dt,$$

i.e.,

$$i_Y d\theta_L = \left(-\frac{\partial f_i}{\partial t} - \frac{\partial f_i}{\partial q^j}\frac{dq^j}{dt} - \frac{\partial f_i}{\partial v^j}\frac{dv^j}{dt} + \frac{\partial L}{\partial t}\right) Y^i dt$$

$$+\left(\left(\frac{dq^i}{dt} - v^i\right)\frac{\partial f_i}{\partial q^j}\right) Y^j dt$$

$$+\left(\left(\frac{dq^i}{dt} - v^i\right)\frac{\partial f_i}{\partial v^j}\right) \bar{Y}^j dt = 0,$$

which leads to the following set of equations (from the coefficients of Y^i and $\bar{Y}^i$):

$$\frac{\partial L}{\partial q^i} - \frac{d}{dt}\left(\frac{\partial L}{\partial v^i}\right) = 0, \tag{7.82}$$

$$\left.\begin{array}{rcl} \frac{\partial^2 L}{\partial q^j \partial v^i}\left(\frac{dq^i}{dt} - v^i\right) & = & 0 \\ \frac{\partial^2 L}{\partial v^j \partial v^i}\left(\frac{dq^i}{dt} - v^i\right) & = & 0 \end{array}\right\} \tag{7.83}$$

In general the set of solutions of such system is different from the set of solutions of (7.76). However, if we suppose that $L : J^1N \longrightarrow R$ is a regular function then one has

$$\frac{dq^i}{dt} = \dot{q}^i = v^i. \tag{7.84}$$

In other words, if L is regular, the expression $\mu^*(i_Y d\theta_L) = 0$ is equivalent to (7.77). Thus the regularity of L gives the 1–jet prolongation condition (7.84) and the extremals are the same for both variational problems. We have shown the following:

Theorem 7.10.10 *Let $L : J^1N \longrightarrow R$ be a regular Lagrangian function. Then the Hamilton's principle and the modified Hamilton's principle are equivalent in the sense that we may stablish an injective–surjective mapping between the set of extremals of both problem.*

7.11 Special symplectic manifolds

In Hamiltonian Classical Mechanics we have considered the cotangent bundle of the configuration manifold. However, sometimes it is convenient to consider symplectic manifolds diffeomorphic to cotangent bundles. In this section we introduce the notion of special symplectic manifolds due to Tulczyjew [116], [117].

Definition 7.11.1 *A* **special symplectic manifold** *is a quintuple (X, M, π, θ, A) where $\pi : X \longrightarrow M$ is a fibred manifold, θ is a 1–form on X and $A : X \longrightarrow T^*M$ is a diffeomorphism such that $\pi = \pi_M \circ A$ and $\theta = A^*\lambda_M$.*

Now, let K be a submanifold of M and $F : K \longrightarrow R$ a function. Set

$$N = \{p \in X/\pi(p) \in K, < u,\theta > = < T\pi(u), dL >,$$

$$\text{for any } u \in TX \text{ such that } \tau_X(u) = p \text{ and } T\pi(u) \in TK \subset TM\}. \quad (7.85)$$

Suppose that dim $K = k \leq$ dim $M = m$. Then we can choose local coordinates (q^a, q^r), $1 \leq a \leq k$, $k+1 \leq r \leq m$ for M such that K is locally given by $q^{k+1} = \ldots = q^m = 0$. Let (q^i, p_i), $1 \leq i \leq m$ be the induced coordinates for T^*M. Since A is a diffeomorphism then (q^i, p_i) may be taken as local coordinates for X in such a way that A is locally written as the identity map. Hence

$$\theta = p_i dq^i$$

and, from (7.85) we deduce that a point (q^i, p_i) of X is in N if and only if

$$q^r = 0, \; p_a = \partial F/\partial q^a, \; 1 \leq a \leq k, \; k+1 \leq r \leq m.$$

Thus (q^a, p_r) are local coordinates for N and dim $N = m$. One can easily check that N is a Lagrangian submanifold in the symplectic manifold $(X, d\theta)$; we call it the **Lagrangian submanifold generated by** L.

Let Q be an arbitrary m–dimensional manifold and T^*Q its cotangent bundle. Let $\beta_Q : TT^*Q \longrightarrow T^*T^*Q$ be the vector bundle isomorphism defined by the canonical symplectic structure ω_Q on T^*Q, i.e.,

$$\beta_Q(X) = i_X \omega_Q.$$

Then we have

$$\tau_{T^*Q} = \pi_{T^*Q} \circ \beta_Q$$

If $(q^i, p_i, \dot{q}^i, \dot{p}_i)$ are local coordinates for TT^*Q one has

$$\beta_Q(q^i, p_i, \dot{q}^i, \dot{p}_i) = (q^i, p_i, -\dot{p}_i, \dot{q}^i)$$

We can easily check that $(TT^*Q, T^*Q, \tau_{T^*Q}, \chi_Q, \beta_Q)$ is a special symplectic manifold, where $\chi_Q = \beta_Q^* \lambda_{T^*Q}$. Hence

$$\chi_Q = -\dot{p}_i dq^i + \dot{q}^i dp_i. \quad (7.86)$$

Let $H : T^*Q \longrightarrow R$ be a Hamiltonian function. Hence $dH = (\partial H/\partial q^i)dq^i + (\partial H/\partial p_i)dp_i$. Choose $u = (q^i, p_i, \dot{q}^i, \dot{p}_i;\; \delta q^i, \delta p_i, \delta \dot{q}^i, \delta \dot{p}_i) \in TTT^*Q$. Then

$$T\tau_{T^*Q}(u) = (q^i, p_i, \delta q^i, \delta p_i),$$

and $< u, \chi_Q > = < T\tau_{T^*Q}(u), dH >$ is equivalent to

$$-\dot{p}_i(\delta q^i) + \dot{q}_i(\delta p_i) = (\partial H/\partial q^i)(\delta q^i) + (\partial H/\partial p_i)(\delta p_i) \quad (7.87)$$

From (7.64) we have

$$\dot{q}_i = \partial H/\partial p_i,\ \dot{p}_i = -(\partial H/\partial q^i),\ 1 \leq i \leq m,$$

which are the Hamilton equations for H.

Moreover, if X_H is the Hamiltonian vector field on T^*Q corresponding to H then N is the image of X_H an $X_H = (\beta_Q)^{-1} \circ (dH)$. The following diagram illustrates the above situation:

$$\begin{array}{ccccc} TT^*Q & \xrightarrow{\beta_Q} & T^*T^*Q & & \\ & \nwarrow^{X_H} & \pi_{T^*Q}\downarrow \uparrow dH & & \\ & & T^*Q & \xrightarrow{H} & R \end{array}$$

Now, we can define a canonical diffeomorphism $A_Q : TT^*Q \longrightarrow T^*TQ$. Indeed, given $v \in TT^*Q$ we must define $A_Q(v) \in T^*TQ$ by means of its pairing on any element $w \in TTQ$, where $\tau_{TQ}(w) = T\pi_Q(v)$. Given two curves $\gamma : R \longrightarrow TQ$ and $\eta : R \longrightarrow T^*Q$ such that $\dot{\gamma}(o) = \sigma_Q(w)$, $\dot{\eta}(o) = v$ and $\tau_Q \circ \gamma = \pi_Q \circ \eta$, we define

$$< w, A_Q(v) > = d/dt < \gamma, \eta > /_{t=0}$$

(here $\sigma_Q : TTQ \longrightarrow TTQ$ is the **canonical involution** of TTQ defined in Exercise 7.16.3).

A simple computation shows that A_Q is locally given by

$$A_Q(q^i, p_i, \dot{q}^i, \dot{p}_i) = (q^i, \dot{q}^i, \dot{p}_i, p_i)$$

Set $\alpha_Q = A_Q^*(\lambda_{TQ})$. Then we locally have

$$\alpha_Q = \dot{p}_i dq^i + \dot{p}_i d\dot{q}^i \quad (7.88)$$

Proposition 7.11.2 *The quintuple $(TT^*Q, TQ, T\pi_Q, \alpha_Q, A_Q)$ is a special symplectic manifold.*

The proof is left to the reader as an exercise.

Remark 7.11.3 We notice that $\chi_Q \neq \alpha_Q$ but their sum is an exact 1–form. To show this, we define a canonical function $\Omega_Q : TT^*Q \longrightarrow R$ as follows:

$$\Omega_Q(v) = < T\pi_Q(v),\ \tau_{T^*Q}(v) >,\ v \in TT^*Q.$$

A simple computation in local coordinates shows that

$$\Omega_Q(q^i, p_i, \dot{q}^i, \dot{p}_i) = p_i \dot{q}^i$$

From (7.86) and (7.88) we obtain

$$\chi_Q + \alpha_Q = \dot{q}^i dp_i + p_i d\dot{q}^i = d(\dot{q}^i p_i) = d\Omega_Q.$$

Hence χ_Q and $-\alpha_Q$ define the same symplectic structure on TT^*Q (see de León and Lacomba [31]).

Now, we assume that $L : TQ \longrightarrow R$ is a Lagrangian function. Consider the above construction for $K = M = TQ$, $X = TT^*Q$. Hence

$$dL = (\partial L/\partial q^i)dq^i + (\partial L/\partial \dot{q}^i)d\dot{q}^i.$$

Choose

$$U = (q^i, p_i, \dot{q}^i, \dot{p}_i, \delta q^i, \delta p_i, \delta\dot{q}^i, \delta\dot{p}_i) \in TTT^*Q.$$

Then

$$TT\pi_Q(u) = (q^i, \dot{q}^i, \delta q^i, \delta\dot{q}^i) \in TTQ$$

and

$$< u, \alpha_Q > = < TT\pi_Q(u),\ dL >$$

is equivalent to

$$\dot{p}^i(\delta q^i) + p_i(\delta\dot{q}^i) = (\partial L/\partial q^i)(\delta q^i) + (\partial L/\partial \dot{q}^i)(\delta\dot{q}^i)$$

which implies

$$p_i = \partial L/\partial \dot{q}^i, \tag{7.89}$$

$$\dot{p}^i = \partial L / \partial q^i, \tag{7.90}$$

$1 \le i \le m$. Thus (7.66) give the momentum p_i and (7.67) are the Euler–Lagrange equations for L.

Moreover we can check that the Lagrangian submanifold N given by (7.85) is

$$N = (A_Q^{-1} \circ dL)(TQ).$$

Then we have the following diagram:

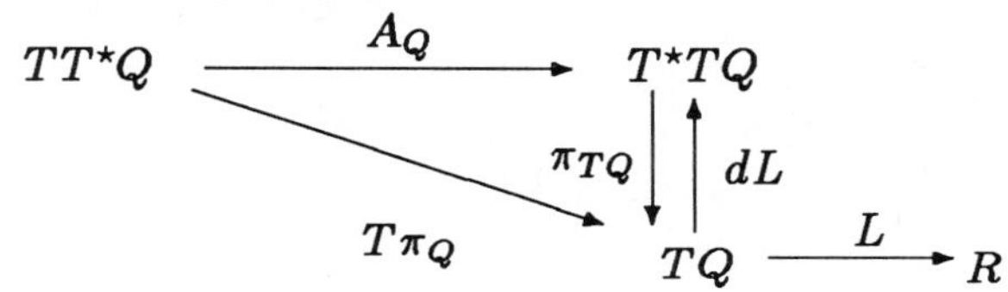

Remark 7.11.4 We notice that the result holds even if L is a degenerate Lagrangian (see Chapter 8 for further information about degenerate Lagrangians). Also the Lagrangian L may be defined on a submanifold K of TQ.

In order to connect the Lagrangian with the Hamiltonian diagrams, which exist independently, we need the Lagrangian L to be regular. In such a case there exists a unique vector field ξ_L on TQ such that

$$i_{\xi_L} \omega_L = dE_L$$

(ξ_L is the Lagrange vector field for L). Let $Leg : TQ \longrightarrow T^*Q$ be the Legendre transformation defined by L. Hence the Lagrange vector field $\xi_L : TQ \longrightarrow TTQ$ satisfies the relation

$$A_Q \circ T\, Leg \circ \xi_L = dL$$

and we can construct the following diagram:

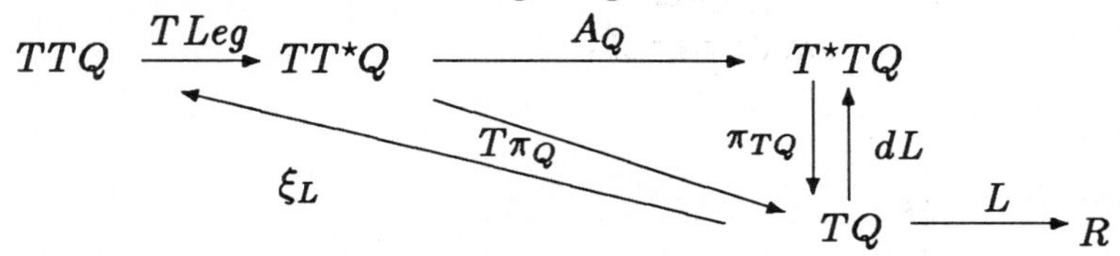

Since *Leg* is at least locally inversible, then we can define the Hamiltonian function $H : T^*Q \longrightarrow R$ as $H = E_L \circ Leg^{-1}$. Also the equality $N = X_H(T^*Q)$ holds at least locally. Then we can construct the following diagram:

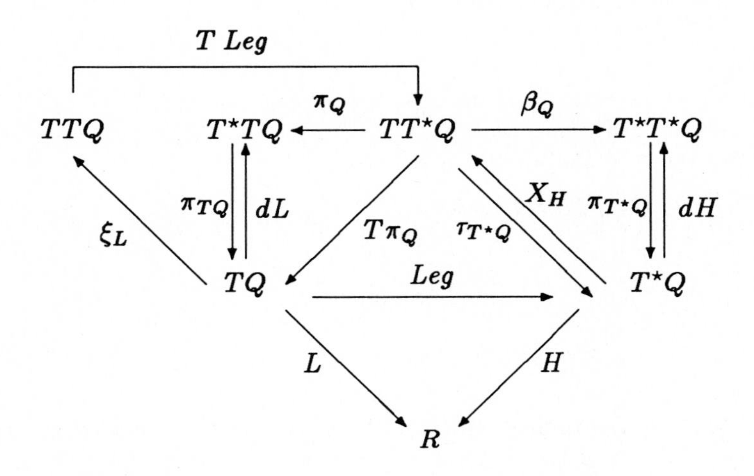

Remark 7.11.5 In de León and Lacomba [30] this construction has been extended for higher order mechanical systems (see de León and Rodrigues [38]).

7.12 Noether's theorem. Symmetries

In this section we first discuss Noether's theorem for mechanical Lagrangian systems and then symmetries of Euler–Lagrange vector fields. We follow Arnold [4] and Crampin [21].

Definition 7.12.1 *(1) Let $L : TM \longrightarrow R$ be a regular Lagrangian function and $F : M \longrightarrow M$ a map. We say that L* **admits** *the map F if $L \circ TF = L$, i.e., the following diagram*

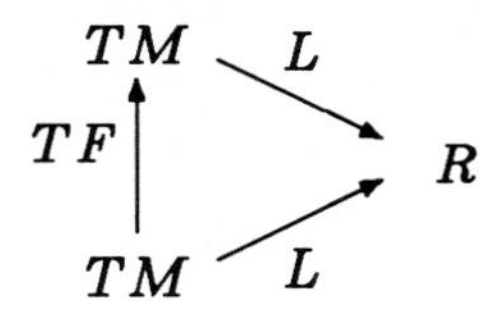

is commutative.
(2) Let X be a vector field on M and ϕ_t the local 1–parameter group of local transformations generated by X. We say that L **admits** *X if L admits ϕ_t, for all t.*

Proposition 7.12.2 *L admits X if and only if $X^c L = 0$, where X^c is the complete lift of X to TM.*

Proof: In fact, X^c is the infinitesimal generator of the local 1–parameter group of local transformations $T\phi_t$. Then $L_{X^c}L = X^c L = 0$ if and only if $L \circ T\phi_t = L$, for all t.□
Example: Let $L : TR^3 \longrightarrow R$ be a Lagrangian function defined by

$$L(x^i, v^i) = 1/2 \sum_{i=1}^{3} (v^i)^2 - U(x^1, x^2),$$

where (x^1, x^2, x^3) are the canonical coordinates on R^3. Then L admits the vector field $X = \partial/\partial x^3$ on R^3.

Theorem 7.12.3 (Noether's Theorem) *If L admits a vector field X on M, then $X^v L$ is a first integral of L, i.e., $\xi_L(X^v L) = 0$, where ξ_L is the Euler–Lagrange vector field for L.*

Proof: In fact, ξ_L is given by the equation

$$i_{\xi_L}\omega_L = dE_L.$$

Hence we have

$$(i_{\xi_L}\omega_L)(X^c) = \omega_L(\xi_L, X^c) = -dd_J L(\xi_L, X^c)$$

$$= -\xi_L((d_J L)X^c) + X^c((d_J L)\xi_L) + d_J L[\xi_L, X^c]$$

$$= -\xi_L((JX^c, L) + X^c(CL) + J[\xi_L, X^c]L$$

$$= -\xi_L(X^v L) + X^c(CL),$$

since $JX^c = X^v$, $J\xi_L = C$ and $J[\xi_L, X^c] = 0$. On the other hand, we have

$$dE_L(X^c) = X^c E_L = X^c(CL - L) = X^c(CL) - X^c L.$$

Thus, we obtain

$$\xi_L(X^v L) = X^c L$$

and, consequently, $X^c L = 0$ implies $\xi_L(X^v L) = 0$.□

Example. Consider the Lagrangian

$$L = TR^3 \longrightarrow R$$

given by

$$L(x^1, x^2, x^3, v^1, v^2, v^3) = \sum_{i=1}^{3} m_i(v_i^2/2) - U(x^1, x^2, x^3)$$

where the m_i's are positive real numbers and $U : R^3 \longrightarrow R$. The Lagrangian L corresponds to a system of point masses with masses m_i on R^3. Suppose that L admits translations along the x^1–axis. This is equivalent to say that L admits the vector field $X = \partial/\partial x^1$ on R^3. (In fact, the 1–parameter group of transformations of R^3 generated by X is $\phi_t(x^1, x^2, x^3) = (x^1 + t, x^2, x^3)$). Then $X^c L = 0$ and, according to Noether's theorem $X^v L$ is a first integral of ξ_L (i.e., a constant of motion). But

$$X^v L = \frac{\partial L}{\partial v^1}$$

is exactly the momentum p_1. Thus p_1 is conserved.

Next we shall discuss symmetries.

Definition 7.12.4 *Let ξ be a semispray on TM. A* **Lie symmetry** *of ξ is a vector field X on M such that $[X^c, \xi] = 0$.*

From Exercise 1.22.4, we deduce that X is a Lie symmetry of ξ if and only if every $T\phi_t$ commutes with every ψ_s, where ϕ_t (resp. ψ_s) is the local 1–parameter group of local transformations of M (resp. TM) generated by X (resp. ξ).

Proposition 7.12.5 *Let L be a regular Lagrangian on TM and ξ_L the corresponding Euler–Lagrange vector field. If X is a vector field on M such that $L_{X^c} d_J L$ is closed and $d(X^c E_L) = 0$, then X is a Lie symmetry of ξ_L.*

Proof: In fact,

$$i_{[X^c, \xi_L]} \omega_L = L_{X^c}(i_{\xi_L} \omega_L) - i_{\xi_L}(L_{X^c} \omega_L)$$

$$= L_{X^c} dE_L + i_{\xi_L} d(L_{X^c} d_J L)$$

$$= d(X^c E_L) + i_{\xi_L} d(L_{X^c} d_J L)$$

$$= 0. \square$$

If, in addition, $L_{X^c} d_J L$ is exact, i.e., $L_{X^c} d_J L = df$, and $X^c E_L = 0$, then such a symmetry X is called a **Noether symmetry**. Clearly, a Noether symmetry of ξ_L is also a Lie symmetry, since Proposition 7.12.5.

Proposition 7.12.6 *Suppose that X is a Noether symmetry of ξ_L. Then $f - X^v L$ is a first integral of ξ_L.*

Proof: In fact, we have

$$df = L_{X^c} d_J L = i_{X^c} dd_J L + di_{X^c} d_J L$$

$$= -i_{X^c} \omega_L + d(X^v L) \text{ (since } i_{X^c} d_J L = (JX^c) L = X^v L)$$

$$= -dE_L + d(X^v L) = d(X^v L - E_L)$$

Thus

$$\xi_L(f - X^v L) = d(f - X^v L)(\xi_L) = -(dE_L)(\xi_L)$$

$$= -\xi_L E_L = 0. \square$$

Remark 7.12.7 Proposition 7.12.6 is a generalization of Noether's theorem. In fact, let us suppose that f is a constant function (for instance $f = 0$). Then

$$L_{X^c} d_J L = 0.$$

Therefore we have

$$0 = (L_{X^c} d_J L)(\xi_L) = X^c((d_J L)(\xi_L)) - d_J L[X^c, \xi_L]$$

$$= X^c(CL) - J[X^c, \xi_L]L = X^c(CL).$$

Definition 7.12.8 *(1) A* **dynamical symmetry** *of a semispray ξ on TM is a vector field $\tilde{X}$ on TM such that $[\tilde{X}, \xi] = 0$.*
(2) A **Cartan symmetry** *of an Euler–Lagrange vector field ξ_L is a vector field $\tilde{X}$ on TM such that $L_{\tilde{X}}(d_J L)$ is exact, namely, $L_{\tilde{X}}(d_J L) = df$, and $\tilde{X} E_L = 0$.*

Remark 7.12.9 The argument used in Proposition 7.12.5 applies to show that a Cartan symmetry of ξ_L is also a dynamical symmetry.

Proposition 7.12.10 *If $\tilde{X}$ is a Cartan symmetry of ξ_L then $f - (J\tilde{X})L$ is a first integral of ξ_L.*

Proof: In fact, we have

$$df = L_{\tilde{X}} d_J L = i_{\tilde{X}} d d_J L + d i_{\tilde{X}} d_J L$$

$$= -i_{\tilde{X}} \omega_L + d((J\tilde{X})L)$$

$$(\text{since } i_{\tilde{X}} d_J L = (J\tilde{X})L)$$

$$= -dE_L + d((J\tilde{X})L) = d((J\tilde{X})L - E_L).$$

Hence

$$\xi_L(f - (J\tilde{X})L) = d(f - (J\tilde{X})L)\xi_L = -dE_L(\xi_L)$$

$$= -\xi_L E_L = 0. \square$$

Thus, Cartan symmetries give rise to constants of motion.

Remark 7.12.11 We remit to Prince [105] for a discussion of symmetries when L is a time–dependent Lagrangian.

7.13 Lagrangian and Hamiltonian mechanical systems with constraints

In this section we reformulate some results of Weber [124] for Lagrangian and Hamiltonian mechanical systems with constraints.

Let $L : TM \longrightarrow R$ be a regular Lagrangian. Then (TM, ω_L) is a symplectic manifold.

Definition 7.13.1 *Let $C = \{\theta_1, \ldots, \theta_r\}$ be a system of constraints on TM. We call (TM, ω_L, E_L, C) a* **regular Lagrangian system with constraints**.

The constraints C are said to be **classical constraints** if the 1–forms θ_a are basic. Then holonomic classical constraints define foliations on the configuration manifold M, but holonomic constraints also admit foliations on the phase space of velocities TM.

Next we seek conditions such as to make a Lagrangian system with constraints defines a mechanical system.

Consider the equation

$$i_\xi \omega_L = dE_L + \wedge^a \theta_a, \; \theta_a(\xi) = 0 \tag{7.91}$$

Definition 7.13.2 *We say that (TM, ω_L, E_L, C)* **defines a mechanical system with constraints** *if the vector field ξ given by (7.91) is a semispray.*

Theorem 7.13.3 *(TM, ω_L, E_C, C) defines a mechanical system with constraints if and only if the 1–forms θ_a are semibasic.*

Proof: Let X_a be the vector fields on TM given by

$$i_{X_a} \omega_L = \theta_a, \;\; 1 \leq a \leq r.$$

Then we have

$$\xi = \xi_L + \wedge^a \theta_a,$$

where ξ_L is given by $i_{\xi_L} \omega_L = dE_L$.

Thus ξ is a semispray if and only if $JX_a = 0$ for all a, $1 \leq a \leq r$. Since $i_J \omega_L = 0$, we have

$$i_J i_{X_a} \omega_L = i_{X_a} i_J \omega_L - i_{JX_a} \omega_L = -i_{JX_a} \omega_L.$$

But $i_J \theta_a = J^* \theta_a$. Hence ξ is a semispray if and only if $J^* \theta_a = 0$. Thus, since a 1–form θ on TM is semibasic if and only if $J^* \theta = 0$, we obtain the required result. □

Remark 7.13.4 Let (TM, ω_L, E_L, C) be a mechanical system with constraints. From Theorem 7.13.3 the 1–forms θ_a are semibasic. Since $m+1$ semibasic 1–forms on TM, $m = dim\ M$, are always linearly dependent, then we deduce that the number of constraints is $r \leq m$.

Let (TM, ω_L, E_L, C) be a mechanical system with constraints. Then the vector field ξ given by (7.91) is a semispray. One can easily check that the paths of ξ satisfy the **Euler–Lagrange equations with constraints**:

$$\frac{\partial L}{\partial q^i} - \frac{d}{dt}\left(\frac{\partial L}{\partial \dot{q}^i}\right) = \wedge^a (\theta_a)_i, \quad 1 \leq i \leq m,$$

where $\theta_a = (\theta_a)_i(q, \dot{q}) dq^i$.

As we have seen, in Classical Mechanics, the phase spacce of momenta is the cotangent bundle T^*M of the configuration manifold M. The bundle structure of T^*M allows to define distinguished 1–forms on T^*M.

Definition 7.13.5 *A 1–form θ on T^*M is said to be* **semibasic** *if $\theta(X) = 0$ for all π_M–vertical vector field X on T^*M. A 1–form θ on T^*M is called* **basic** *if $\theta = \pi_M^* \eta$, where η is a 1–form on M.*

Hence a 1–form θ on T^*M is semibasic (resp. basic) if and only if it is locally expressed by

$$\theta = \theta_i(q, p) dq^i,$$

(resp.

$$\theta = \theta_i(q) dq^i)$$

where (q^i, p_i) are the induced coordinates for T^*M.

Next, we consider a Hamiltonian system (T^*M, ω_M, H, C) with constraints C, where $C = \{\theta_a;\ 1 \leq a \leq r\}$ is a set of r linearly independent

1–forms on T^*M. As above, we may distinguish classical and non–classical constraints if the forms θ_a are basic or semibasic.

A first question is to seek conditions such that (T^*M, ω_M, H, C) defines a mechanical system with constraints.

For this we re–examine the relationship between Lagrangian and Hamiltonian formalisms.

Let $H : T^*M \longrightarrow R$ be a Hamiltonian and suppose that H is **regular**, i.e., the Hessian matrix

$$(\partial^2 H/\partial p_i \partial p_j)$$

is non–singular. Then we define a map $Ham : T^*M \longrightarrow TM$ as follows. Let x be a point of M and denote by $H_x : T_x^*M \longrightarrow R$ the restriction of H to the fibre T_x^*M. If $\gamma \in T_x^*M$, then we have

$$dH_x(\gamma) : T_\gamma(T_x^*M) \longrightarrow R.$$

Since $T_\gamma(T_x^*M)$ may be canonically identified with T_x^*M, then $dH_x(\gamma)$ is a linear map from T_x^*M into R, i.e., an element of the dual space $(T_x^*M)^*$ which may be canonically identified with T_xM. We define

$$Ham(\gamma) = dH_x(\gamma).$$

In local coordinates, we have

$$Ham(q^i, p_i) = (q^i, \partial H/\partial p_i).$$

Since H is regular then Ham is a local diffeomorphism. Thus we can define a (local) function L on TM by

$$L(q^i, v^i) = (\partial H/\partial p^i)p^i - H,$$

where $v^i = \partial H/\partial p_i$ holds. It is clear that since H is regular, then L is regular also. Actually, the Legendre transformation corresponding to L is the (local) inverse of Ham.

Now, we can transport the regular Hamiltonian system with constraints (T^*M, ω_M, H, C) into the Lagrangian system with constraints (TM, ω_L, E_L, C^*), where

$$\omega_L = Leg^*\omega_M, \; E_L = CL - L, \text{ and } C^* = \{\theta_a^* = Leg^*\theta_a\}.$$

Definition 7.13.6 *We say that* (T^*M, ω_M, H, C) **defines** *a mechanical system with constraints if* (TM, ω_L, E_L, C^*) *defines a mechanical system with constraints (as defined in Definition 7.13.2).*

From Theorem 7.13.3 the Lagrangian system (TM, ω_L, E_L, C^*) defines a mechanical system with constraints if and only if the 1–forms θ_a^* are semibasic. Since $Leg : TM \longrightarrow T^*M$ is fibre preserving then θ_a^* is semibasic on TM if and only if θ_a is semibasic on T^*M. Thus (T^*M, ω_M, H, C) defines a mechanical system with constraints if and only if the 1–forms θ_a are semibasic. This shows that the essential condition to obtain mechanical systems for constrained (Lagrangian or Hamiltonian) systems is the semibasic character of the 1–forms defining C. In both cases the number of constraints must be $r \leq m$.

Remark 7.13.7 Let $C = \{\theta_a;\ 1 \leq a \leq r\}$ be a system of constraints on the symplectic manifold (T^*M, ω_M). For the holonomic case it is possible to show that there exists a local canonical transformation $F : U \subset T^*M \longrightarrow T^*M$ such that the original 1–forms θ_a can be transformated via F in semibasic forms if and only if the 1–forms $\{\theta_a\}$ are in involution (for a proof see Jacobi's theorem in Duistermaat [50], p. 100 and the Lie's corollary in Abraham and Marsden [1], p. 419). This result is valid, under certain circumstances, for the non–holonomic case (see Weber [124]).

7.14 Euler–Lagrange equations on $T^*M \oplus TM$

In this section we will adopt the following concise notation. Suppose that M is an m–dimensional manifold and $(q) = (q^i)$, $1 \leq i \leq m$, are local coordinates for M; then:

$$(q, v) = (q^i, v^i) \text{ are local coordinates for } TM;$$

$$(q, p) = (q^i, p_i) \text{ are local coordinates for } T^*M;$$

$$(q, v, \delta q, \delta v) = (q^i, v^i, \delta q^i, \delta v^i) \text{ are local coordinates for } TTM;$$

$$(q, p, \dot{q}, \dot{p}) = (q^i, p_i, \dot{q}^i, \dot{p}_i) \text{ are local coordinates for } TT^*M.$$

In the following W represents the Whitney sum of T^*M and TM; $W = T^*M \oplus TM$. A point of W is locally represented by (q, v, p) and $\pi_1 : W \longrightarrow$

T^*M, $\pi_2 : W \longrightarrow TM$ are the obvious projections on the first and second factors of W. The tangent prolongations of π_1 and π_2 are locally given by

$$T\pi_1 : TW \longrightarrow TT^*M,\ (q,v,p,\delta q,\delta v,\delta p) \longrightarrow (q,p,\delta q,\delta p),$$

$$T\pi_2 : TW \longrightarrow TTM,\ (q,v,p,\delta q,\delta v,\delta p) \longrightarrow (q,v,\delta q,\delta v).$$

Let X_2 be a semispray (or second–order differential equation) on M.

Definition 7.14.1 *We will say that X_2 is* **related** *to a vector field X_1 on T^*M if for every solution (or path) $\sigma(t)$ of X_2 there is a unique solution (or integral curve) $\tau(t)$ of X_1 such that*

$$\tau_M(\dot{\sigma}(t)) = \pi_M(\tau(t)).$$

In other words, X_2 is related to a vector field X_1 on T^*M if and only if there is a C^∞ map $\varphi : TM \longrightarrow T^*M$ such that $T\varphi \circ X_2 = X_1 \circ \varphi$ and $\pi_M \circ \varphi = \tau_M$ (φ is fiber–preserving). The following diagram illustrates this situation:

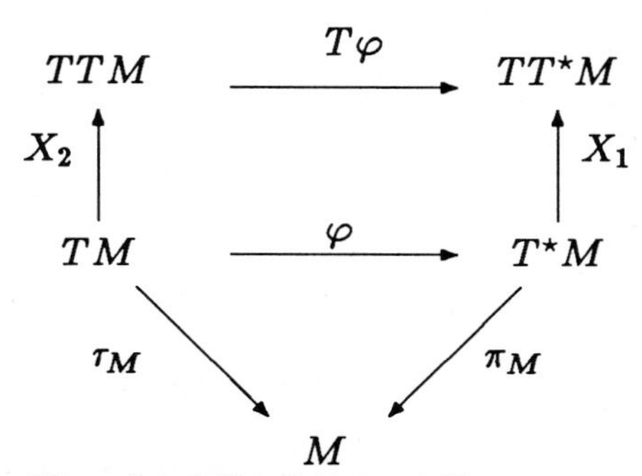

In such a case X_1 verifies the following equality:

$$T\pi_M \circ X_1 \circ \varphi = Id_{TM} \tag{7.92}$$

In fact, as $\pi_M \circ \varphi = \tau_M$ then $T\pi_M \circ T\varphi = T\tau_M$ and so $T\pi_M \circ T\varphi \circ X_2 = T\tau_M \circ X_2 = Id_{TM}$, since X_2 is a semispray. But $T\varphi \circ X_2 = X_1 \circ \varphi$.

Suppose that X_2 is related to X_1 and consider the graph of φ, Graph $\varphi \subset T^*M \times TM$. Since the diagram

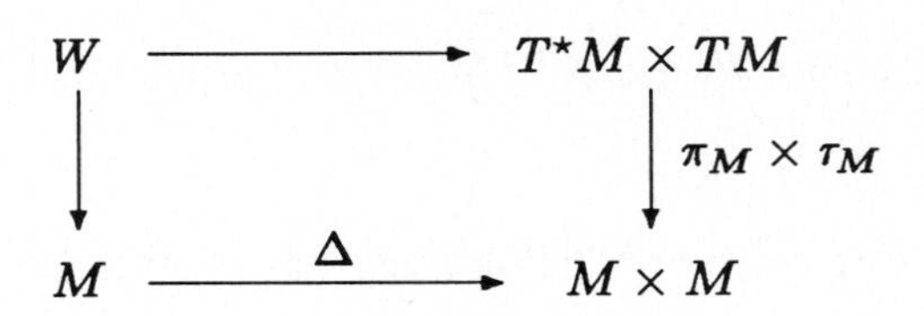

where Δ is the diagonal mapping, is commutative, we shall consider Graph $\varphi \subset W$. Then the related vector fields $X_2 : TM \longrightarrow TTM$ and $X_1 : T^*M \longrightarrow TT^*M$ determine a vector field X : Graph $\varphi \subset W \longrightarrow TW$ such that X projects by π_1 and π_2 to X_1 and X_2, respectively, i.e.,

$$T\pi_1 \circ X = X_1 \circ \pi_1, \quad T\pi_2 \circ X = X_2 \circ \pi_2. \tag{7.93}$$

Thus we have the following diagram:

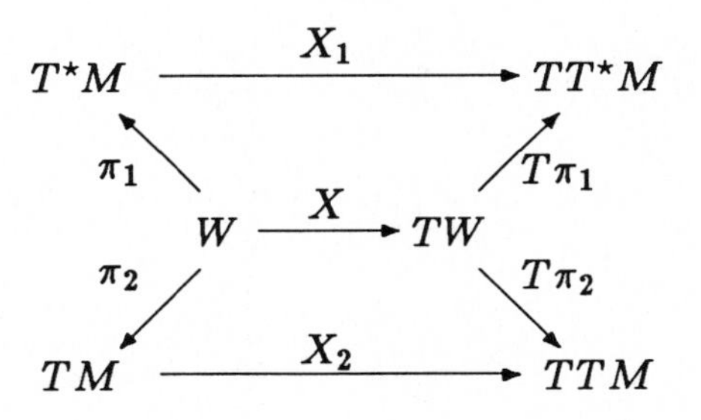

We set Graph $\varphi = W_1$ and we shall suppose that $X = W_1 \longrightarrow TW_1$, i.e., X does not generate solutions curves leading off the (constraint) submanifold W_1 (see Section 8.1.1).

Suppose now that $X : W \longrightarrow TW$ is an arbitrary vector field such that X projects by π_1 to X_1 and by π_2 to a vector field ξ on TM, i.e.,

$$T\pi_1 \circ X = X_1 \circ \pi_1, \quad T\pi_2 \circ X = \xi \circ \pi_2. \tag{7.94}$$

Proposition 7.14.2 *If the above X verifies*

$$T\pi_M \circ T\pi_1 \circ X = \pi_2, \tag{7.95}$$

then ξ *in (7.94) is a semispray.*

Proof: In fact, we have

$$\pi_2 = T\pi_M \circ T\pi_1 \circ X = T(\pi_M \circ \pi_1) \circ X$$

$$= T(\tau_M \circ \pi_2) \circ X \quad (\text{since } \pi_M \circ \pi_1 = \tau_M \circ \pi_2)$$

$$= T\tau_M \circ T\pi_2 \circ X = T\tau_M \circ \xi \circ \pi_2,$$

which implies

$$T\tau_M \circ \xi = Id_{TM}$$

Hence ξ is a semispray. □

Equation (7.95) gives a sufficient condition for a vector field X on the Whitney sum W be projected by π_2 to a semispray. If X is locally given by $(q, p, v, \delta q, \delta p, \delta v)$ and verifies (7.95) then a simple computation shows that $\delta q = v$ and so $T\pi_2 \circ X = (q, v, v, \delta v)$ is a semispray.

Let λ_M be the Liouville form on T^*M and set

$$\lambda_\oplus = \pi_1^*(\lambda_M).$$

Then $\lambda_\oplus$ is locally given by

$$\lambda_\oplus = p_i dq^i.$$

Consider the function $< \pi_2, \pi_1 >$ on W, i.e.,

$$< \pi_2, \pi_1 > (q, p, v) =< (q, v), (q, p) >= pv.$$

A direct computation shows that

$$< X, \lambda_\oplus >=< \pi_2, \pi_1 > . \tag{7.96}$$

Let $Z \in Ker\ T\pi_2$ and consider the action of the Lie derivative L_Z on both sides of (7.96):

$$L_Z < X, \lambda_\oplus >= L_Z < \pi_2, \pi_1 > \tag{7.97}$$

We have

$$L_Z < X, \lambda_\oplus >= L_Z(\lambda_\oplus(X)) = Z(\lambda_\oplus(X))$$

On the other hand,

$$< X, L_Z\lambda_\oplus >= (L_Z\lambda_\oplus)(X) = Z(\lambda_\oplus(X)) - \lambda_\oplus([Z, X])$$

$$= Z(\lambda_\oplus(X)),$$

since $\lambda_\oplus([Z, X]) = 0$.

Thus

$$L_Z < X, \lambda_\oplus >=< X, L_Z\lambda_\oplus >,$$

for all $Z \in Ker\ T\pi_2$, and (7.97) becomes

$$< X, L_Z\lambda_\oplus >= L_Z < \pi_2, \pi_1 >,$$

from which one obtains

$$i_X i_Z d\lambda_\oplus = i_Z d < \pi_2, \pi_1 >,$$

since $i_Z\lambda_\oplus = 0$ and $i_Z < \pi_2, \pi_1 >= 0$. Setting

$$\omega_\oplus = -d\lambda_\oplus,$$

we have

$$i_Z i_X \omega_\oplus = i_Z d < \pi_2, \pi_1 >,$$

i.e.,

$$i_X\omega_\oplus = d < \pi_2, \pi_1 > +\lambda,$$

with $i_Z\lambda = 0$, for all $Z \in Ker\ T\pi_2$.

Theorem 7.14.3 *(Skinner [110]) If we suppose that $\lambda = -d(\pi_2^\star L) = -d(L \circ \pi_2)$, where $L : TM \longrightarrow R$ is a Lagrangian function, then*

$$i_X\omega_\oplus = d < \pi_2, \pi_1 > -d(\pi_2^\star L) = d(< \pi_2, \pi_1 > -\pi_2^\star L >,$$

and the projections on M of the integral curves of X are the Euler–Lagrange equations for L.

Proof: It remains to prove the second assertion. As X verifies (7.95), then X is locally given by

$$X = (q, p, v, v, \delta p, \delta v).$$

Hence

$$i_X\omega_\oplus = i_X(dq \wedge dp) = (i_X dq)dp - (i_X dp)dq = -(\delta p)dq + vdp.$$

As $< \pi_2, \pi_1 >= pv$, we have

$$\begin{aligned} d(< \pi_2, \pi_1 > -\pi_2^\star L) &= d(pv) - \pi_2^\star(dL) \\ &= pdv + vdp - \frac{\partial L}{\partial q}dq - \frac{\partial L}{\partial v}dv \\ &= -\frac{\partial L}{\partial q}dq + vdp + (p - \frac{\partial L}{\partial v})dv. \end{aligned}$$

So,

$$\delta p = \frac{\partial L}{\partial q}, \quad p = \frac{\partial L}{\partial v}. \tag{7.98}$$

Now, if $(q(t),\ p(t),\ v(t))$ is an integral curve of X, then $\delta q = v = \dot{q} = (d/d)(q(t))$ and $\delta p = \dot{p} = (d/dt)(p(t))$. Thus (7.98) implies

$$\frac{d}{dt}\left(\frac{\partial L}{\partial \dot{q}}\right) - \frac{\partial L}{\partial q} = 0,$$

which are the Euler–Lagrange equations for L.□

It is clear that the converse is also true, i.e., the solutions of the Euler–Lagrange equations for L are the solutions of the vector field X. So the intrinsical expression $i_{X_2}\omega_L = dE_L$ of the Euler–Lagrange equations is equivalent to the above $i_X\omega_\oplus = dD$, where $D : W \longrightarrow R$ is the function defined by $D =< \pi_2, \pi_1 > -\pi_2^\star L$.

We finish this section by remarking that no hypothesis on the regularity of L was made, i.e., the above procedure may be adopted for regular (or hyperregular) Lagrangians as well as degenerate Lagrangians. We will return to this topic in Section 8.1.1.

7.15 More about semisprays

One of the most interesting results about the Hamiltonian formulation of Classical Mechanics is that vector fields and one–forms on the phase space are related by an isomorphism induced by the canonical symplectic form on the cotangent bundle of the configuration manifold. For the Lagrangian approach however this correspondence depends directly on the Lagrangian function, that is, for each regular Lagrangian we may define a symplectic form and so the induced isomorphism depends on this choice. Sarlet et al. [109] proposed a geometric study relating semisprays and one–forms for which it is possible to generate Lagrangian systems. The method is inspired on the definition of gradient vector fields and associates to every semispray on the tangent bundle a set of one–forms defined on such bundle. We start Sarlet et al. study by considering the **conservative case**.

We consider a special set of one–forms $\wedge^1_\xi$ defined by

$$\wedge^1_\xi = \{\alpha \in \wedge^1(TM)/(L_\xi \circ J^\star)\alpha = \alpha\},$$

where ξ is a semispray on TM, $\wedge^1(TM)$ is the space of all 1–forms on TM and $J^\star$ is the adjoint endomorphism on $\wedge^1(TM)$ induced by the canonical almost tangent structure J on TM. We will see that for an appropiate choice of $\alpha \in \wedge^1_\xi$ we may obtain the Euler–Lagrange equations of motion in its intrinsical form.

Let X be a vector field on TM. Then E_X is a R–linear operator on 1–forms on TM defined by

$$E_X = Id - L_X \circ J^\star.$$

Thus

$$\wedge^1_X = Ker\ E_X.$$

This set is in fact a vector space over R, by the linearity of E_X. Moreover, we have

$$E_X(f\alpha) = (Xf)\alpha + f(E_X\alpha).$$

Therefore $\wedge^1_X$ is not a module over the ring of real functions but it is a module over the ring of real functions satisfying $Xf = 0$ (the constants of motion).

Next we shall compute the local expressions of E_X and $\wedge^1_X$. Suppose that X is locally expressed by

$$X = X^i \partial/\partial q^i + \bar{X}^i \partial/\partial v^i,$$

where (q^i, v^i) are induced coordinates for TM. If we set

$$\alpha = \alpha_i dq^i + \bar{\alpha}_i dv^i$$

one easily deduces

$$E_X \alpha = (\alpha_i - X(\bar{\alpha}_i) - \bar{\alpha}_j \frac{\partial X^j}{\partial q^i}) dq^i + \bar{\alpha}_j (\delta_i^j - \frac{\partial X^j}{\partial v^i}) dv^i$$

(δ_i^j = Kronecker's delta). Hence, the elements of $\wedge^1_X$ are locally of the form

$$\alpha = (X(\bar{\alpha}_i) - \bar{\alpha}_j \frac{\partial X^j}{\partial q^i}) dq^i + \bar{\alpha}_j \frac{\partial X^j}{\partial v^i} dv^i \tag{7.99}$$

Proposition 7.15.1 *Let R be a tensor field of type (1,1) on TM such that $RJ = JR$ and $J(L_X R) = 0$. Then*

$$R^\star \circ E_X = E_X \circ R^\star$$

Proof: Let $\alpha \in \wedge^1(TM)$ and $Y \in \chi(TM)$. Then

$$(R^\star(E_X \alpha))(Y) = (E_X \alpha)(RY) = \alpha(RY) - (L_X(J^\star \alpha))(RY)$$

$$= \alpha(RY) - X(\alpha(JRY)) + \alpha(J[X, RY])$$

$$= \alpha(RY) - X(\alpha(RJY)) + \alpha(RJ[X, Y]),$$

since $RJ = JR$ and $J(L_X R) = 0$. Thus

$$(R^\star(E_X \alpha))(Y) = (E_X(R^\star \alpha))(Y). \square$$

Corollary 7.15.2 *In order that a tensor field R of type (1,1) on TM preserves $\wedge^1_X$ it is sufficient that R commutes with J and satisfies $J(L_X R) = 0$.*

An example of a tensor field of type $(1,1)$ on TM satisfying the hypothesis of Proposition 7.15.1 is obtained by using the theory of lifts of tensor fields studied in Chapter 2. If we take $R = R_0^c$ as the complete lift of a tensor field R_0 on M then we have

$$R_0^c J = J R_0^c = R_0^v,$$

where R_0^v is the vertical lift of R_0. Also, for a semispray ξ one has

$$J(L_\xi R_0^c)(Y^c) = J([\xi, R_0^c Y^c] - R_0^c[\xi, Y^c])$$

$$= J[\xi, (R_0 Y)^c] - R_0^v[\xi, Y^c] = 0,$$

since $[\xi, Y^c]$ is a vertical vector field for all $Y \in \chi(M)$.

Proposition 7.15.3 *Let Y be a vector field on TM. Then the tensor field*

$$R_Y = (L_Y J)(L_\xi J) + (L_{[\xi,Y]}J)J$$

verifies the hypothesis of Proposition 7.13.1.

The proof is obtained from a straightforward computation. We note that $R_{Y^v} = R_{Y^c} = 0$ and $R_\xi = Id$ for any semispray ξ on TM.

Now, suppose that $X = \xi$ is a semispray on TM. We call E_ξ by **Lagrange operator**. Using the local expression of ξ,

$$\xi = v^i \partial/\partial q^i + \xi^i \partial/\partial v^i,$$

we see that

$$E_\xi \alpha = (\alpha_i - \xi(\bar{\alpha}_i)) dq^i.$$

Therefore the elements of $\wedge_\xi^1$ are locally characterized by

$$\alpha = \xi(\bar{\alpha}_i) dq^i + \bar{\alpha}_i dv^i.$$

Moreover, from the above expression for E_ξ we deduce that $E_\xi \alpha$ **is a semibasic form**, that is, $J^*(E_\xi \alpha) = 0$.

Definition 7.15.4 *Let $\alpha \in \wedge^1_\xi$. Then α is said to be* **regular** *if the 2-form*

$$\omega_\alpha = -d(J^\star \alpha)$$

is symplectic. A semispray ξ is called a **regular Lagrangian vector field** *if there is a regular form $\alpha \in \wedge^1_\xi$ (called the associated form of ξ) which is exact, i.e., $\alpha = dL$ for some C^∞ function $L : TM \longrightarrow R$.*

Proposition 7.15.5 *Let ξ be a regular Lagrangian vector field. Then*

$$E_\xi(\alpha) = 0$$

is the intrinsical form of the Euler–Lagrange equations of motions, where α is the associated regular form of ξ.

Proof: $E_\xi \alpha = 0$ implies

$$\alpha = L_\xi(J^\star \alpha),$$

that is,

$$dL = L_\xi(J^\star dL) = L_\xi d_J L$$

$$= i_\xi dd_J L + di_\xi d_J L.$$

Thus

$$i_\xi dd_J L = d(L - i_\xi d_J L) = -dE_L,$$

where $E_L = i_\xi d_J L - L = CL - L$ (since $J\xi = C$). Then the intrinsical form of the Euler–Lagrange equation is

$$i_\xi \omega_L = dE_L,$$

where ω_L is defined by $\omega_L = -dd_J L.\square$

Since ξ is supposed to be regular Lagrangian the form ω_L is symplectic and so L is a regular Lagrangian (of course the converse is also true).

Now, we shall introduce a kind of dual of $\wedge^1_\xi$. We define a subset of $\chi(TM)$ by

$$\chi_\xi = \{Y \in \chi(TM) / J[\xi, Y] = 0\}.$$

Then χ_ξ is a real vector space. As $J[\xi, fY] = (\xi f)JY + fJ[\xi, Y]$ we have that χ_ξ is not a $C^\infty(TM)$–module. However it is a module over the ring of constants of motion. Locally the elements of χ_ξ have the form

$$Y = Y^i \partial/\partial q^i + \xi(Y^i)\partial/\partial v^i.$$

Let us give a geometrical interpretation of χ_ξ. If $Y \in \chi_\xi$ then $J[Y, \xi] = 0$, i.e., the Lie derivative $L_Y \xi$ of ξ in the "direction" Y is a vertical vector field. Now, for every point x one has

$$(L_Y \xi)_x = \lim(1/t)[\xi(x) - (T\varphi_t(\xi))(x)],$$

where φ_t is the 1–parameter group generated by Y. Thus we may say that, at least to first order in t, $(T\varphi_t)\xi$ differs from ξ by a vertical vector field. As ξ is a semispray, $J\xi = C$, then, again at least to first order, $(T\varphi_t)\xi$ is a semispray.

Definition 7.15.6 *χ_ξ is called the set of all* **variation vector fields** *of the semispray ξ.*

This definition is suggested by the fact that the 1–parameter group φ_t is a variation from a point x to a point $\varphi_t(x)$.

A complete lift of any vector field X on M belongs to χ_ξ for any semispray ξ on TM. In fact, there is a unique Y such that $JY = X^v$ which is X^c. As $[X^c, \xi]$ is vertical for any ξ, $J[X^c, \xi] = 0$. Moreover

Proposition 7.15.7 *(Crampin [21]). Let ξ be a semispray on TM. If V is a vertical vector field on TM then there is a unique vector field in χ_ξ, denoted by V_ξ, such that $JV_\xi = V$.*

Proof: It is sufficient to show that the linear mapping $J : Y \in \chi_\xi \longrightarrow JY \in V(TM)$ is bijective.

It is injective. Suppose that $JY = 0$. Then Y is vertical. Consider the connection $\Gamma = -L_\xi J$ defined by ξ. Then, if h is the horizontal projector of Γ, we have

$$0 = hY = \frac{1}{2}(Y - (L_\xi J)Y) = \frac{1}{2}(Y - [\xi, JY] + J[\xi, Y])$$

which implies

$$Y = [\xi, JY] - J[\xi, Y] = 0.$$

It is surjective. Let Y be a vertical vector field and define

$$Y_\xi = -\Gamma([\xi, Y])$$

Then

$$JY_\xi = -J\Gamma[\xi, Y] = -J[\xi, Y] = Y.$$

Now

$$J[\xi, Y_\xi] = -J([\xi, [\xi, Y] + J[\xi, [\xi, Y]]])$$

$$= -J[\xi, [\xi, Y]] - J[\xi, J[\xi, [\xi, Y]]]$$

$$= -J[\xi, [\xi, Y]] + J[\xi, [\xi, Y]] = 0$$

and so $Y_\xi \in \chi_\xi$.□

Let us consider the pairing $< Y, J^\star\alpha >$. Then

$$L_\xi < Y, J^\star\alpha >=< L_\xi Y, J^\star\alpha > + < Y, L_\xi(J^\star\alpha) >$$

$$=< J(L_\xi Y), \alpha > + < Y, L_\xi(J^\star\alpha) >$$

Thus, if $Y \in \chi_\xi$ and $\alpha \in \wedge^1_\xi$ we have

$$L_\xi < Y, J^\star\alpha >=< Y, \alpha > .\square \tag{7.100}$$

The following proposition shows a kind of duality between χ_ξ and $\wedge^1_\xi$.

Proposition 7.15.8 *A vector field Y (resp. 1-form α) on TM belongs to χ_ξ (resp. $\wedge^1$ ξ) if and only if (7.100) holds for all $\alpha \in \wedge^1_\xi$ (resp. $Y \in \chi_\xi$).*

Proof: In fact, if $Y \in \chi_\xi$ and $\alpha \in \wedge^1_\xi$, then (7.100) holds. Now, let $Y \in \chi(TM)$ (resp. $\alpha \in \wedge^1(TM)$) such that (7.100) holds for all $\alpha \in \wedge^1_\xi$ (resp. for all $Y \in \chi_\xi$). Then we have

$$< Y, \alpha >=< J(L_\xi Y), \alpha > + < Y, \alpha >$$

(resp.

$$< Y, \alpha >=< Y, L_\xi(J^\star \alpha) >)$$

which implies

$$J(L_\xi Y) = J[\xi, Y] = 0$$

(resp.

$$L_\xi(J^\star \alpha) = \alpha). \square$$

Proposition 7.15.9 *Let R be a tensor field of type (1,1) satisfying the hypothesis of Proposition 7.15.1. Then for all $\alpha \in \wedge_\xi^1$ and $Y \in \chi_\xi$ we have*

$$< RY, \alpha >= L_\xi(< RY, J^\star \alpha >).$$

Also,

$$R_Y = (L_Y J) \circ (L_\xi J).$$

Proof:

$$< RY, \alpha >=< Y, R^\star \alpha >= L_\xi < Y, J^\star R^\star \alpha > \ \text{(since } R^\star \alpha \in \wedge_\xi^1)$$

$$= L_\xi < Y, R^\star J^\star \alpha >= L_\xi < RY, J^\star \alpha > .$$

If $Y \in \chi_\xi$ then $J[\xi, Y] = 0$ and so

$$(L_{[\xi,Y]} J) J = 0.$$

Thus

$$R_Y = (L_{YJ}) \circ (L_\xi J). \square$$

Proposition 7.15.10 *("Conservation of energy"). Let ξ be an arbitrary semispray on TM. For each $\alpha \in \wedge_\xi^1$ we have*

$$i_\xi(\alpha - d < C, \alpha >) = 0.$$

Therefore, if $\alpha = dL$, we have $i_\xi dE_L = 0$.

Proof:

$$L_\xi(J^\star\alpha) = \alpha \Longleftrightarrow i_\xi dJ^\star\alpha + d < \xi, J^\star\alpha >= \alpha$$

$$\Longleftrightarrow i_\xi dJ^\star\alpha = \alpha - d < J\xi, \alpha >$$

$$\Longleftrightarrow i_\xi dJ^\star\alpha = \alpha - d < C, \alpha >$$

Thus

$$i_\xi(\alpha - d < C, \alpha >) = 0.$$

If $\alpha = dL$ then $\alpha - d < C, \alpha >= -dE_L$, since

$$E_L = CL - L = dL(C) - L =< C, \alpha > -L. \square$$

Proposition 7.15.11 *("gauge freedom"). Let f be a C^∞ function on M and ξ a semispray on TM. If $\alpha \in \wedge^1_\xi$ then*

$$\bar{\alpha} = \alpha + df^c \in \wedge^1_\xi,$$

where f^c is the complete lift of f to TM.

Proof: In fact, $J^\star(df^c) = df^v$. Therefore

$$E_\xi(\alpha + df^c) = E_\xi\alpha + E_\xi(df^c)$$

$$= E_\xi(df^c) = df^c - L_\xi(J^\star(df^c))$$

$$= df^c - L_\xi df^v$$

$$= 0,$$

since $L_\xi(df^v) = d(L_\xi f^v) = df^c. \square$

If $\alpha = dL$ in Proposition 7.15.11, then one has the well-known gauge-freedom in Lagrangian mechanics (see Exercise 7.17.1).

Now, suppose that Y is a Lie symmetry of ξ, i.e., $[Y, \xi] = 0$ and $Y = Z^c$, for some vector field Z on M. Then

$$L_Y J = 0, \quad L_Y \circ J = J \circ L_Y.$$

Proposition 7.15.12 *Let Y be a Lie symmetry of ξ. Then*

$$L_Y \circ E_\xi = E_\xi \circ L_Y$$

Proof:

$$L_Y(E_\xi \alpha) = L_Y \alpha - L_Y L_\xi(J^\star \alpha)$$

$$= L_Y \alpha - L_\xi L_Y(J^\star \alpha) \ \ (\text{since } [Y, \xi] = 0)$$

$$= L_Y \alpha - L_\xi J^\star(L_Y \alpha)$$

$$= E_\xi(L_Y \alpha). \square$$

Corollary 7.15.13 *The Lie derivative with respect to a Lie symmetry Y of ξ preserves $\wedge^1_\xi$, i.e.,*

$$L_Y(\wedge^1_\xi) \subset \wedge^1_\xi.$$

Now, let Y be a dynamical symmetry of ξ, i.e., $[Y, \xi] = 0$.

Proposition 7.15.14 *If Y is a dynamical symmetry of ξ and $\alpha \in \wedge^1_\xi$ satisfies*

$$J^\star R^\star_Y \alpha = J^\star df,$$

for some function f on TM, then $\alpha' \in \wedge^1_\xi$, where

$$\alpha' = L_Y \alpha - d(\xi f).$$

Proof: We have

$$J^\star \alpha' = J^\star(L_Y \alpha) - J^\star d(\xi f)$$

$$= L_Y(J^\star \alpha) - (L_Y J)^\star \alpha - J^\star d(\xi f),$$

since

$$(L_Y J)^\star = L_Y \circ J^\star - J^\star \circ L_Y .$$

Therefore

$$L_\xi(J^\star \alpha') = L_\xi L_Y(J^\star \alpha) - L_\xi (L_Y J)^\star \alpha - L_\xi J^\star(d(\xi f))$$

$$= L_Y L_\xi(J^\star \alpha) - L_\xi((L_Y J)^\star \alpha) - L_\xi(J^\star d(\xi f))$$

(since $L_Y \circ L_\xi = L_\xi \circ L_Y$).

$$= L_Y \alpha - L_\xi((L_Y J)^\star \alpha) - L_\xi(J^\star d(\xi f)) \tag{7.101}$$

On the other hand, $R_Y^\star \alpha \in \wedge_\xi^1$ and so

$$R_Y^\star \alpha = L_\xi(J^\star R_Y^\star \alpha) = L_\xi(J^\star df).$$

As $R_Y = (L_Y J) \circ (L_\xi J)$ we deduce

$$(L_\xi J)^\star (L_Y J)^\star \alpha = (L_\xi J)^\star df + J^\star L_\xi(df).$$

Operating on both sides with $(L_\xi J)^\star$ we see that

$$(L_Y J)^\star \alpha = df - J^\star d(\xi f).$$

Therefore, taking again the Lie derivative of the above expression with respect to ξ, we obtain

$$L_\xi((L_Y J)^\star \alpha) = d(\xi f) - L_\xi(J^\star d(\xi f)).$$

From this we see that (7.101) reduces to

$$L_\xi(J^\star \alpha') = L_Y \alpha - d(\xi f) = \alpha'.$$

Thus $\alpha' \in \wedge_\xi^1$.□

Remark 7.15.15 We have a converse of Proposition 7.15.14. In fact, if $Y \in \chi_\xi$ and if there is a regular element $\alpha \in \wedge_\xi^1$ such that $J^\star R_Y^\star \alpha = J^\star df$, for some $f : TM \longrightarrow R$ with $\alpha' \in \wedge_\xi^1$ then Y is a dynamical symmetry of ξ.

Finally, the following theorem gives a generalization of Noether's theorem.

Theorem 7.15.16 *Let α be a regular element of $\wedge^1_\xi$. If a vector field Y on TM satisfies*
(1) $L_Y(J^\star\alpha) = df$, for some $f : TM \longrightarrow R$
(2) $i_Y(\alpha - d < C, \alpha >) = 0$,
then $g = f - < Y, J^\star\alpha >$ is a first integral of ξ. Conversely, to each first integral g of ξ there corresponds a vector field Y on TM satisfying (1) and (2). In particular, if $\alpha = dL$ then $[Y, \xi] = 0$ and so we have Noether's theorem for Lagrangian mechanics.

Proof: From (1) we obtain

$$i_Y d(J^\star\alpha) = df - di_Y(J^\star\alpha) = d(f - < Y, J^\star\alpha >) = dg.$$

But $\alpha \in \wedge^1_\xi$ and so

$$i_\xi d(J^\star\alpha) = \alpha - d < C, \alpha > .$$

Consequently

$$L_\xi g = i_\xi dg = i_\xi i_Y d(J^\star\alpha) = -i_Y i_\xi d(J^\star\alpha)$$

$$= -i_Y(\alpha - d < C, \alpha >) = 0,$$

i.e.,

$$\xi g = 0.$$

Conversely, as $d(J^\star\alpha)$ is symplectic, each first integral g of ξ defines a unique vector field Y on TM satisfying

$$i_Y d(J^\star\alpha) = dg.$$

If we put $f = g + < Y, J^\star\alpha >$, then

$$L_Y(J^\star\alpha) = i_Y d(J^\star\alpha) + di_Y(J^\star\alpha)$$

$$= dg + d < Y, J^\star\alpha > = df.$$

Furthermore, since $L_\xi g = 0$, we obtain

$$0 = L_\xi g = i_\xi dg = i_\xi i_Y d(J^\star \alpha)$$

$$= -i_Y i_\xi d(J^\star \alpha) = -i_Y(\alpha - d < C, \alpha >).$$

Finally,

$$i_{[Y,\xi]} d(J^\star \alpha) = L_Y i_\xi d(J^\star \alpha) = i_Y d\alpha.$$

If $\alpha = dL$, then

$$i_{[Y,\xi]} dJ^\star(dL) = i_{[Y,\xi]} dd_J L = 0$$

and, as $dd_J L$ is symplectic, we obtain $[Y, \xi] = 0$. Then, if $Y = X^c$ for some vector field X on M, we deduce that X is a Noether symmetry of ξ and we reobtain Proposition 7.12.6.□

We examine now the **non–conservative** case, i.e., mechanical systems (M, F, ρ) where ρ is not a closed semibasic form. As the results obtained are similar to the conservative case we will only examine the main modifications.

Let ξ be a semispray on TM. We associate to ξ a R–bilinear operator

$$\wedge^1(TM) \times \wedge^1(TM) \xrightarrow{E_{\xi,J}} \wedge^1(TM)$$

defined by

$$E_{\xi,J}(\alpha, \beta) = E_\xi \alpha + J^\star \beta.$$

Now, let us put

$$\wedge^1_{\xi,J} = Ker\ E_{\xi,J},$$

i.e.,

$$\wedge^1_{\xi,J} = \{(\alpha, \beta) \in \wedge^1(TM) \times \wedge^1(TM) / E_\xi \alpha = -J^\star \beta\}.$$

The elements of $\wedge^1_{\xi,J}$ are locally of the form

$$\alpha = (\xi(\bar{\alpha}_i) + \bar{\beta}_i) dq^i + \bar{\alpha}_i dv^i,$$

$$\beta = \beta_i dq^i + \bar{\beta}_i dv^i.$$

We note that

$$E_{\xi,J}(f\alpha, f\beta) = (\xi f)(J^\star\alpha) + fE_{\xi,J}(\alpha,\beta).$$

Hence $\wedge^1_{\xi,J}$ is a real vector space but is not a module over the ring of the constants of motion. Next we shall relate $\wedge^1_\xi$ and $\wedge^1_{\xi,J}$. Let $j : \wedge^1(TM) \longrightarrow \wedge^1(TM) \times \wedge^1(TM)$, $j(\alpha) = (\alpha, 0)$ be the canonical injection. Then $E_{\xi,J} \circ j = E_\xi$ and so $j(\wedge^1_\xi) \subset \wedge^1_{\xi,J}$.

Definition 7.15.17 *A pair of 1–forms (α,β) is said to be* **regular** *if $(\alpha,\beta) \in \wedge^1_{\xi,J}$ and the 2–form $\omega_\alpha = -d(J^\star\alpha)$ is symplectic. The form β is called the* **force field**. *A semispray ξ is called a* **non–conservative regular Lagrangian vector field** *if there is a regular pair $(\alpha,\beta) \in \wedge^1_{\xi,J}$ such that α is exact, say $\alpha = dF$.*

The condition $E_\xi(dF) = -J^\star\beta$ which characterizes ρ may be rewritten in the form

$$i_\xi\omega_F = dE_F + \rho \ \ (\omega_F \stackrel{def}{=} \omega_{dF}),$$

where $\rho = -J^\star\beta$ is a semibasic form on TM. We deduce

$$\xi\left(\frac{\partial F}{\partial v^i}\right) - \frac{\partial F}{\partial q^i} = -\bar{\beta}_i, \ \ 1 \leq i \leq m = dim\, M,$$

which yields a system of Euler–Lagrange equations with non–conservative forces. Similar results presented above may be re–obtained.

For instance, it is easy to see that

"If R is a tensor field of type (1,1) on TM satisfying $JR = RJ$ and $J(L_\xi R) = 0$, then

$$R^\star \circ E_{\xi,J} = E_{\xi,J} \circ (R^\star \times R^\star)$$

and so $R^\star \times R^\star$ preserves $\wedge^1_{\xi,J}$".

We may also introduce a kind of dual of $\wedge^1_{\xi,J}$. Let χ_ξ be the subset of $\chi(TM)$ defined as above.

If we define $< Y, (\alpha,\beta) >= (< Y, \alpha >, < Y, \beta >)$ then we have $< Y, (\alpha,\beta) >= ((\xi(\bar{\alpha}_i) + \bar{\beta}_i)Y^i + \bar{\alpha}_i\xi(Y^i), \beta_i Y^i + \bar{\beta}_i\xi(Y^i))$, for all $(\alpha,\beta) \in \wedge^1_{\xi,J}$ and $Y \in \chi_\xi$. As in the conservative situation, we deduce that if the above local expression holds for all $(\alpha,\beta) \in \wedge^1_{\xi,J}$ then Y must be in χ_ξ. Next, let us examine symmetries in the non–conservative case.

Proposition 7.15.18 *If Y is a dynamical symmetry of ξ and $(\alpha, \beta) \in \wedge^1_{\xi,J}$ satisfies*

$$J^\star R_Y^\star \alpha = J^\star(df), \quad R_Y^\star \beta = dg,$$

for some $f, g \in C^\infty(TM)$ then

$$(\alpha', \beta') \in \wedge^1_{\xi,J},$$

where $\alpha' = L_Y\alpha - d(\xi f)$ and $\beta' = L_Y\beta - d(\xi g)$.

Proof: From the above definition of α', β' we have

$$J^\star \alpha' = J^\star(L_Y\alpha) - J^\star d(\xi f)$$

$$= L_Y(J^\star\alpha) - (L_Y J)^\star\alpha - J^\star d(\xi f)$$

since $(L_Y J)^\star = L_Y \circ J^\star - J^\star \circ L_Y$. Therfore we have

$$L_\xi(J^\star\alpha') = L_\xi(L_Y(J^\star\alpha)) - L_\xi((L_Y J)^\star\alpha) - L_\xi(J^\star d(\xi f))$$

$$= L_Y(L_\xi(J^\star\alpha)) - L_\xi((L_Y J)^\star\alpha) - L_\xi(J^\star d(\xi f))$$

(since $L_Y \circ L_\xi = L_\xi \circ L_Y$)

$$= L_Y\alpha + L_Y(J^\star\beta) - L_\xi((L_Y J)^\star\alpha) - L_\xi(J^\star d(\xi f)). \tag{7.102}$$

On the other hand, $(R_Y^\star\alpha, R_Y^\star\beta) \in \wedge^1_{\xi,J}$; then we have

$$R_Y^\star\alpha = L_\xi(J^\star R_Y^\star\alpha) - J^\star R_Y^\star\beta = L_\xi(J^\star df) - J^\star(dg).$$

As the vector field Y is a dynamical symmetry of ξ, then $R_Y = (L_Y J) \circ (L_\xi J)$. Thus

$$R_Y^\star\alpha = (L_\xi J)^\star df + J^\star(L_\xi(df)) - J^\star(dg).$$

Operating on both sides with $(L_\xi J)^\star$, we see that

$$(L_Y J)^\star\alpha = df - J^\star d(\xi f) + J^\star(dg). \tag{7.103}$$

Now, taking the Lie derivative of (7.103) with respecto to ξ, we obtain

$$L_\xi((L_Y J)^\star \alpha) = d(\xi f) + L_\xi(J^\star d(\xi f)) + L_\xi J^\star(dg) \tag{7.104}$$

With (7.104), the relation (7.102) reduces to

$$L_\xi(J^\star \alpha') = L_Y \alpha + L_Y(J^\star \beta) - d(\xi f) + L_\xi(J^\star dg)$$

Thus

$$\alpha' - L_\xi(J^\star \alpha') = -L_Y(J^\star \beta) + L_\xi(J^\star dg)$$

$$= -(L_Y J)^\star \beta - J^\star(L_Y \beta) + (L_\xi J)^\star(dg) + J^\star d(\xi g).$$

But, since

$$(L_\xi J)^\star (L_Y J)^\star \beta = dg.$$

Now, operating again on both sides with $(L_\xi J)^\star$, we deduce

$$(L_Y J)^\star \beta = (L_\xi J)^\star dg.$$

Hence

$$\alpha' - L_\xi(J^\star \alpha') = -J^\star(L_Y \beta - d(\xi g))$$

or, equivalently,

$$E_\xi \alpha' = -J^\star \beta',$$

and so $(\alpha', \beta') \in \wedge^1_{\xi,J}$.□

Proposition 7.15.19 *(Noether's Theorem). Let (α, β) be a regular element of $\wedge^1_{\xi,J}$. If $Y \in \chi(TM)$ satisfies*
(1) $L_Y(J^\star \alpha) = df$, for some $f \in C^\infty(TM)$,
(2) $i_Y(\alpha - d < C, \alpha > + J^\star \beta) = 0$,
then $F = f - < Y, J^\star \alpha >$ is a first integral of ξ. Conversely: to each first integral F of ξ there corresponds a vector field Y satisfying (1) and (2).

Proof: The same as for the conservative case.□

7.16 Generalized Caplygin systems

Let $P(M,G)$ be a principal bundle with projection $\pi : P \longrightarrow M$ and suppose that $L : TP \longrightarrow R$ is a G–invariant Lagrangian, i.e.,

$$L \circ TR_a = L \text{ for all } a \in G.$$

We impose as constraints on TP the horizontal subspaces determined by a connection Γ in P. This means that only horizontal paths are allowed. We call $(P(M,G), L, \Gamma)$ a **generalized Caplygin system** (see Koiller [82]).

Given a generalized Caplygin system $(P(M,G), L, \Gamma)$ we can define a Lagrangian $L^\star : TM \longrightarrow R$ as follows:

$$L^\star(X_q) = L((X_q)_p^H) \text{ for all } X_q \in T_qM,$$

where $(X_q)_p^H$ is the horizontal lift of X_q at p, $\pi(p) = q$, with respect to the connection Γ. We notice that $L^\star$ is well–defined since L is G–invariant and $TR_a(X^H) = X^H$, for any vector field X on M and for all $a \in G$.

Next, we shall obtain the Euler–Lagrange equations for $L^\star$.

First, we choose local coordinates for P. For each point x of M there exist a coordinate neighborhood (U, q^i), $1 \leq i \leq m = dim\ M$, of x and a diffeomorphism $\psi : \pi^{-1}(U) \longrightarrow U \times G$. We need to introduce local coordinates for G. Let $\exp : g \longrightarrow G$ be the exponential map from the Lie algebra g of G into G. As we know exp gives a diffeomorphism of a neighborhood $\tilde{W}$ of O in g onto a neighborhood W of e in G. Thus, if we choose a basis $\{e_r; 1 \leq r \leq n = dim\ G\}$ of g we define local coordinates in W as follows: the point $\exp(\sum_{r=1}^n \pi^r e_r)$ has local coordinates (π^r). For an arbitrary point $a \in G$, we consider the coordinate neighborhood $\ell_a(W)$. These coordinates (π^r) are called **normal coordinates**. So we obtain local coordinates (q^i, π^a) in P.

We denote by $\sigma : U \longrightarrow P$ the local section over U given by $\sigma(q) = \psi^{-1}(q, e)$. In local coordinates we have

$$\sigma(q^i) = (q^i, 0).$$

Now, let ω be the connection form of ω. Then $\omega = \omega^r e_r$, where ω^r are 1–forms on P. We have

$$\sigma^\star\omega = (\sigma^\star\omega^r)e_r,$$

where $\sigma^\star\omega^r$ are 1–forms on U. Thus

$$\sigma^\star \omega^r = \omega_i^r dq^i, \ 1 \leq r \leq n. \tag{7.105}$$

Let X be a vector field on M and suppose that X is locally given by $X = \dot{q}^i \partial/\partial q^i$ on U. Then the horizontal lift X^H of X to P is given by

$$(X^H)_{\sigma(q)} = \dot{q}^i (\partial/\partial q^i)_{\sigma(q)} + A^r (\partial/\partial \pi^r)_{\sigma(q)},$$

since $T\pi(X^H) = X$. Now, from (7.105), we obtain

$$0 = \omega(X^H)_{\sigma(q)} = \dot{q}^i \omega(\partial/\partial q^i)_{\sigma(q)} + A^r \omega(\partial/\partial \pi^r)_{\sigma(q)}$$

$$= [\dot{q}^i \omega_i^r + A^r] e_r, \ (\text{since } (\partial/\partial \pi^r)_{\sigma(q)} = (\lambda e_r)_{\sigma(q)})$$

which implies

$$A^r = -\dot{q}^i \omega_i^r.$$

Thus

$$(X^H)_{\sigma(q)} = \dot{q}^i (\partial/\partial q^i)_{\sigma(q)} - \dot{q}^i \omega_i^r (\partial/\partial \pi^r)_{\sigma(q)}. \tag{7.106}$$

But since $(TR_a)(X^H) = X^H$, we obtain from (7.106) that

$$(X^H)_{\sigma(q)a} = TR_a[\dot{q}^i (\partial/\partial q^i)_{\sigma(q)} - \dot{q}^i \omega_i^r (\partial/\partial \pi^r)_{\sigma(q)}]$$

$$= \dot{q}^i (\partial/\partial q^i)_{\sigma(q)a} - \dot{q}^i \omega_i^r (TR_a)(\partial/\partial \pi^r)_{\sigma(q)}.$$

Now, in local coordinates we have

$$R_a(q^i, \pi^r) = (q^i, \bar{\pi}^r),$$

where, if $b = \exp(\pi^r e_r)$, then $\bar{\pi}^r = \pi^r(ab)$. Hence we obtain

$$(X^H)_{\sigma(q)a} = \dot{q}^i (\partial/\partial q^i)_{\sigma(q)a} - \dot{q}^i \omega_i^r \frac{\partial \bar{\pi}^s}{\partial \pi^r} \left(\frac{\partial}{\partial \pi^s} \right)_{\sigma(q)a},$$

or, for sake of simplicity

$$X^H = \dot{q}^i \frac{\partial}{\partial q^i} - \dot{q}^i \omega_i^r (\partial \bar{\pi}^s / \partial \pi^r) \partial / \partial \pi^s. \tag{7.107}$$

From the Campbell–Haussdorff formula (see Exercise 1.17.8) we have

$$\bar{\pi}^s = \pi^s + \tau^s + \frac{1}{2} C^s_{rv} \pi^r \tau^v + \text{ terms of order } \geq 3,$$

where $a = \exp(\tau^r e_r)$ and C^s_{rv} are the structure constants of G with respect to the basis $\{e_r\}$. Hence

$$(\partial \bar{\pi}^s / \partial \pi^r)_{(q^i,0)} = \delta^s_r + \frac{1}{2} C^s_{rv} \tau^v$$

and then

$$X^H = \dot{q}^i \partial/\partial q^i - \dot{q}^i [\omega^s_i + \frac{1}{2} \omega^r_i C^s_{rv} \tau^v] \partial/\partial \pi^s$$

Thus we have

$$L^\star(q^i, \dot{q}^i) = L(q^i, \tau^s, \dot{q}^i, -\dot{q}^j(\omega^s_j + \frac{1}{2}\omega^r_j C^s_{rv} \tau^v))$$

$$= L(q^i, \dot{q}^i, -\dot{q}^j(\omega^s_j + \frac{1}{2}\omega^r_j C^s_{rv} \tau^v)),$$

since L does not depends on τ^s.

An easy computation shows that

$$\frac{\partial L^\star}{\partial q^i} = \frac{\partial L}{\partial q^i} - \dot{q}^j \left(\frac{\partial \omega^s_j}{\partial q^i} + \frac{1}{2} \frac{\partial \omega^r_j}{\partial q^i} C^s_{rv} \tau^v \right) \frac{\partial L}{\partial \dot{\pi}^s},$$

$$\frac{\partial L^\star}{\partial \dot{q}^i} = \frac{\partial L}{\partial \dot{q}^i} - (\omega^s_i + \frac{1}{2}\omega^r_i C^s_{rv} \tau^v) \frac{\partial L}{\partial \dot{\pi}^s}$$

Thus

$$\frac{d}{dt}\left(\frac{\partial L^\star}{\partial \dot{q}^i}\right) - \frac{\partial L^\star}{\partial q^i} = \frac{d}{dt}\left(\frac{\partial L}{\partial \dot{q}^i}\right) - \frac{\partial L}{\partial q^i} - \frac{d}{dt}\left(\frac{\partial L}{\partial \dot{\pi}^s}\right)(\omega^s_i + \frac{1}{2}\omega^r_i C^s_{rv} \tau^v)$$

$$-\dot{q}^j \frac{\partial L}{\partial \dot{\pi}^s} \left[\frac{\partial \omega^s_i}{\partial q^j} - \frac{\partial \omega^s_j}{\partial q^i} + \frac{1}{2}\left(\frac{\partial \omega^r_i}{\partial q^j} - \frac{\partial \omega^r_j}{\partial q^i} \right) C^s_{rv} \tau^v \right.$$

$$\left. + \frac{1}{2} \omega^r_i C^s_{rv} \dot{\tau}^v \right] \tag{7.108}$$

If we evaluate the right–hand of (7.108) at the point $\sigma(q)$, i.e.,

$$\tau^v = 0,\ \dot{\tau}^v = -\dot{q}^j \omega_j^v,$$

then we obtain

$$\frac{d}{dt}\left(\frac{\partial L^\star}{\partial \dot{q}^i}\right) - \frac{\partial L^\star}{\partial q^i} = -\dot{q}^j \frac{\partial L}{\partial \dot{\pi}^s}\left[\frac{\partial \omega_i^s}{\partial q^j} - \frac{\partial \omega_j^s}{\partial q^i} + \frac{1}{2} C_{rv}^s \omega_i^r \omega_j^s\right], \tag{7.109}$$

since the Euler–Lagrange equations for L are

$$\frac{d}{dt}\left(\frac{\partial L}{\partial \dot{q}^i}\right) - \frac{\partial L}{\partial q^i} = 0,\ \frac{d}{dt}\left(\frac{\partial L}{\partial \dot{\pi}^s}\right) = 0.$$

If a solution $q(t)$ is known, then we recover the dynamics in P by considering the horizontal lift of the curve $q(t)$ in M to P.

We remark that (7.109) generalizes Caplygin's formula by the appearance of terms involving the structure constants of g. In fact, if G is Abelian then (7.109) becomes

$$\frac{d}{dt}\left(\frac{\partial L^\star}{\partial \dot{q}^i}\right) - \frac{\partial L^\star}{\partial q^i} = -\dot{q}^j \frac{\partial L}{\partial \dot{\pi}^s}\left[\frac{\partial \omega_i^s}{\partial q^j} - \frac{\partial \omega_j^s}{\partial q^i}\right]$$

(see Neimark and Fufaev [99], and Koiller [82]).

Next we examine the geometrical meaning of the right–hand of (7.109).

Let us recall the structure equation for Γ:

$$d\omega = -\frac{1}{2}[\omega, \omega] + \Omega, \tag{7.110}$$

where Ω is the curvature form of Γ. If

$$\sigma^\star \Omega = \sigma^\star(\Omega^r e_r) = \Omega_{ij}^r dq^i \wedge dq^j,$$

then (7.110) becomes

$$\frac{\partial \omega_i^s}{\partial q^j} - \frac{\partial \omega_j^s}{\partial q^i} = -\frac{1}{2} C_{rv}^s \omega_i^r \omega_j^v + \Omega_{ij}^s$$

and then

$$\frac{d}{dt}\left(\frac{\partial L^\star}{\partial \dot{q}^i}\right) - \frac{\partial L^\star}{\partial q^i} = -\dot{q}^j \frac{\partial L}{\partial \dot{\pi}^s} \Omega_{ij}^s. \tag{7.111}$$

A simple computation shows that the right–hand of (7.111) does not depends on the choice of the coordinate neighborhood (U, q^i) and the section σ. Thus there exists a 1–form α on TM locally given by

$$\alpha = [-\dot{q}^j \frac{\partial L}{\partial \dot{\pi}^s} \Omega^s_{ij}] dq^i$$

Then (7.111) **are the Euler–Lagrange equations for a mechanical system** (M, L^*, α).

We notice that this result was obtained by Koiller [82] by using Hamel's formalism of "quasi–coordinates". We remit to the paper of Koiler for more details and examples.

7.17 Exercises

7.17.1 Let L_1 and L_2 be two regular Lagrangians on TM and ξ_{L_1}, ξ_{L_2} the corresponding Euler–Lagrange vector fields. Prove that the following two assertions are equivalent:
(1) $L_1 = L_2 + \hat{\alpha} +$ constant, where α is a closed 1–form on M and $\hat{\alpha} : TM \longrightarrow R$ is the map defined by $\hat{\alpha}(X_x) = < X_x, \alpha(x) >$, $X_x \in T_xM$.
(2) $\xi_{L_1} = \xi_{L_2}$ and $\omega_{L_1} = \omega_{L_2}$.
7.17.2 Let $L : TM \longrightarrow R$ be a regular Lagrangian. Prove that $Im\ J$ is a Lagrangian subbundle of the symplectic vector bundle (TM, ω_L).
7.17.3 Let Q be an m–dimensional differentiable manifold and TQ its tangent bundle. If $(q^i, v^i, \delta q^i, \delta v^i)$ are induced coordinates for TTQ then we define a diffeomorphism $\sigma_Q : TTQ \longrightarrow TTQ$ locally given by

$$\sigma_Q(q^i, v^i, \delta q^i, \delta v^i) = (q^i, \delta q^i, v^i, \delta v^i)$$

(i) Prove that σ_Q is, in fact, a global diffeomorphism such that $\sigma_Q^2 = Id$.
(ii) Prove that $\sigma_Q : (TTQ, \tau_{TQ}, TQ) \longrightarrow (TTQ, T\tau_Q, TQ)$ is a vector bundle isomorphism.

σ_Q is called the **canonical involution** of TTQ.

7.17.4 **The Helmholtz conditions for the non–autonomous case** (Crampin et al. [23]).

The Helmholtz conditions are the conditions that must be satisfied by a regular matrix $g_{ij}(t, q, \dot{q})$ in order that a gives second– order ordinary differential equation $\ddot{q}^i = f^i(t, q, \dot{q})$ when written in the form

$$q_{ij}\ddot{q}^j + h_i = 0 \ \ (h_i = -g_{ij} f^j)$$

becomes the Euler–Lagrange equations for the Lagrangian $L = L(t, q, \dot{q})$.

Putting

$$\alpha^i_j = -(1/2)(\partial f^i/\partial \dot{q}^j); \ \ \beta^i_j = -(\partial f^i/\partial q^j),$$

$$\phi^i_j = \beta^i_j - \alpha^i_k \alpha^k_j - \gamma(\alpha^i_j),$$

where γ is a semispray given by

$$\gamma = \partial/\partial t + \dot{q}^i(\partial/\partial q^i) + f^i(\partial/\partial \dot{q}^i).$$

The necesssary and sufficient conditions for the existence of a Lagrangian for the equations $g_{ij}\ddot{q}^j + h_i = 0$ are that the functions g_{ij} should satisfy the following Helmholtz conditions:

$$g_{ij} = g_{ji}, \ \ \gamma(g_{ij}) = g_{ik}\alpha^k_j + g_{jk}\alpha^k_i,$$

$$g_{ik}\phi^k_j = g_{jk}\phi^k_i; \ \ \partial g_{ij}/\partial \dot{q}^k = \partial g_{ik}/\partial \dot{q}^j.$$

Proceeding along the same lines as the autonomous case show that the geometrical approach of Section 7.5 for the time–dependent case may be applied for the Helmholtz conditions, that is, that there is a 2–form with suitable properties which are exactly equivalently to the Helmholtz conditions.

<u>7.17.5</u> Show that $\wedge^1_{X^v} = \wedge^1_{X^c} = \wedge^1_C = 0$, where X^v, X^c and C are, respectively, the vertical, complete lift of X and the Liouville vector field.

<u>7.17.6</u> Show that the local expression of a tensor field R which preserves $\wedge^1_\xi$ is

$$R = a^i_j[(\partial/\partial q^i) \otimes dq^j + (\partial/\partial v^i) \otimes dv^j)$$

$$+\xi(a^i_j)(\partial/\partial v^i) \otimes dq^j,$$

where $a^i_j = a^i_j(q, v)$.

<u>7.17.7</u> Prove that χ_ξ contains the complete lifts of arbitrary vector fields on M.

<u>7.17.8</u> Let G be a Lie group with Lie algebra g and $\{e_1, \ldots, e_r\}$ a basis for g. Consider normal coordinates (π^i) on G, i.e., if $a = \exp(\sum_{i=1}^n \pi^i e_i)$,

then a has coordinates (π^i). Suppose that $a = \exp a^i e_i$, $b = \exp b^i e_i$, i.e., $\pi^i(a) = a^i$, $\pi^i(b) = b^i$. Then prove that

$$\pi^i(ab) = a^i + b^i + \frac{1}{2} C^i_{jk} a^j b^k + \text{ terms of degree } \geq 3, \quad (*)$$

where C^i_{jk} are the structure constants with respect to the basis $\{e_i\}$ ((*) is called the **Campbell–Hausdorff formula**) (*Hint*: use Exercise 1.22.24).æ

Chapter 8

Presymplectic mechanical systems

8.1 The first–order problem and the Hamiltonian formalism

Supose that (S,ω) is an arbitrary symplectic manifold. Then the linear mapping

$$S_\omega : \chi(S) \longrightarrow \wedge^1 S$$

is an isomorphism. Hence, for every 1–form α on S there is a unique vector field X_α on S such that

$$i_{X_\alpha}\omega = \alpha.$$

For a presymplectic manifold (S,ω) we may also consider the linear mapping $S_\omega : \chi(S) \longrightarrow \wedge^1 S$, and a vector bundle homomorphism, also denoted by S_ω, from TS to T^*S. But now S_ω is not necessarily an isomorphism. So, if α is a 1–form on S, a condition for solving $i_X\omega = \alpha$ is that $\alpha \in Im\ S_\omega$. We call this condition by the **range condition**. As $Ker\ S_\omega$ is not trivial (ω is presymplectic), the range condition says that a solution is obtained modulo $Ker\ S_\omega$. In fact, $Ker\ S_\omega$ is an obstruction to solve uniquely the equation.

In the following we will resume some results of the presymplectic formulation. For further details the reader is invited to see two different viewpoints on the subject. One is due to Gotay et al. [69] where the rank of the presymplectic form ω is constant everywhere. The other is an approach where the rank of ω has a continous variation (Pnevmatikos [104]). The presymplectic situation occurs also when a manifold K is embedded into a symplectic manifold (S,ω). If $\varphi : K \longrightarrow S$ is the embedding, then in general the pair $(K,\bar{\omega})$, where $\bar{\omega} = \varphi^{\star}\omega$, is not a symplectic manifold. Supposing that rank of $\bar{\omega}$ is constant along K then $(K,\bar{\omega})$ is a presymplectic manifold which we will call **constraint manifold.** We may have this situation in the Hamiltonian case where a canonical symplectic form is defined on the cotangent bundle of a given configuration manifold. In the Lagrangian case we may have the presymplectic or symplectic situation if the Lagrangian is degenerate or not. Throughout this chapter **all submanifolds will be considered as embedded submanifolds.**

8.1.1 The presymplectic constraint algorithm

Suppose that (P_1,ω_1) is a presymplectic manifold and α_1 is a given 1–form on P_1. The **first–order problem** is the question of finding a vector field X on P_1 such that $i_X\omega_1 = \alpha_1$. More clearly, if $\alpha \in Im\ S_{\omega_1}$ then for every point $z \in P_1$ there is a tangent vector X at z such that $i_{X_1}\omega = \alpha_1$ at z. But in general there are only some points $z \in P_1$ for which $\alpha_1(z)$ is in $Im\ S_{\omega_1}$ (otherwise the problem is insolvable). So if we assume that the set of such points form a manifold P_2, embedded into P_1 by a mapping $\varphi_2 : P_2 \longrightarrow P_1$ (and identifying P_2 with $\varphi_2(P_2)$, TP_2 with $(T\varphi_2)(TP_2)$) then we may consider the equation on P_2 given by

$$(i_X\omega_1 - \alpha_1)/_{P_2} = 0.$$

As such equation possesses solutions we must demand that the integral curves be constrained to lie in P_2, i.e., the vector field X must be tangent to P_2 at every point $z \in P_2$ (otherwise the system will try to evolve off P_2). This requirement will not necessarily be satisfied and so we consider a third submanifold P_3 of P_2 defined by

$$P_3 = \{z \in P_2/\alpha_1(z) \in S_{\bar{\omega}_1}(TP_2)\}.$$

It is clear that the repetition of such arguments generates a sequence of manifolds

$$\ldots P_{i+1} \longrightarrow P_i \longrightarrow P_{i-1} \longrightarrow \ldots \longrightarrow P_2 \longrightarrow P_1$$

where

$$P_i = \{z \in P_{i-1}/\alpha_1(z) \in S_{\bar{\omega}_1}(TP_{i-1})\},$$

where, for simplicity, we have put $\bar{\omega}_1 = \varphi_{i-1}^\star \circ \ldots \circ \varphi_1^\star(\omega_1)$.

We call this process **geometric algorithm of presymplectic systems**, due to Gotay et al. [69]. We call P_r the **r–ary constraint manifold**.

The process gives the following possibilities:

1. The algorithm produces an integer r such that $P_r = \phi$,
2. The algorithm produces an integer r such that $P_r \neq \phi$ but dim $P_r = 0$.
3. There is an integer r such that $P_r = P_t$ for all $t \geq r$ and dim $P_r \neq 0$.
4. The process is not finite.

The **first case** means that the dynamical equations have no solutions. The **second case** says that the constraint set consists of points and the unique solution is $X = 0$ (there is no dynamics). The **third case** gives a **final** constraint manifold in the sense that if N is any other submanifold along which the dynamical equations are verified then $N \subset P_r$. The **last case** says that the system has an infinite degree of freedom. In such a case we may stipulate that if P_∞ is the intersection of all P_r then we may reproduce one of the other possibilities.

Definition 8.1.1 *Let (W, ω) be a symplectic manifold, $N \subset W$ a submanifold of W. We say that N is a manifold of* **regular consistent constraints** *if for every 1–form α on W the equation $i_X\omega = \alpha$ when restricted to N admits a solution X tangent to N.*

Thus the major interest in the study of the presymplectic systems is to obtain a final submanifold defined by regular consistent constraints. So, at least one solution X to the original equation restricted to the final constraint manifold P_r exists and this solution is tangent to P_r at every point. This solution X is not unique for we can add to it any element of Ker S_{ω_1}.

Remark 8.1.2 We may also re–characterize the definition of the constraint manifold. From the initial equation of motion we have

$$i_Y i_X \omega_1 = i_Y \alpha_1$$

for all $Y \in \chi(P_1)$. Thus $i_Y\alpha_1 = 0$ if and only if $i_Y i_X \omega_1 = 0$. For the above sequence of manifolds if we consider the orthocomplement with respect to ω_1

$$T_z P_{i-1}^{\perp} = \{Y \in T_z P_1 / i_Y i_X \omega_1(z) = 0 \text{ for all } X \in T_z P_{i-1}\}$$

then we define

$$P_i = \{z \in P_{i-1} / i_Y \alpha_1(z) = 0 \text{ for all } Y \in T_z P_{i-1}^{\perp}\}$$

(where we have omitted the use of the embeddings φ_i, φ_{i-1}, etc...).

The process is then constructed as follows: if X, solution of the original equation, is tangent to P_2, then for every vector field Y on P_1

$$0 = (i_Y i_X \omega_1 - i_Y \alpha_1) \circ \varphi_2$$

$$= -\varphi_2^*(i_{T\varphi_2(\hat{X})} i_Y \omega_1) - (i_Y \alpha_1) \circ \varphi_2$$

$$= -i_X(\varphi_2^*(i_Y \omega_1)) - \varphi_2^*(i_Y \alpha_1),$$

where $\hat{X}$ is such that $T\varphi_2(\hat{X}) = X$. Thus, if Y is such that $\varphi_2^*(i_Y\omega_1) = 0$, that is, $Y \in TP_2^{\perp}$, then $\varphi_2^*(i_Y\alpha_1) = 0$. This may not always be the case and so we must restrict our attention to these points of P_2 where $i_Y\alpha_1 = 0$, i.e.,

$$P_3 = \{z \in P_2 / (\varphi_2 \circ \varphi_3)^*(i_Y \alpha_1) = 0, \text{ for all } Y \in T\varphi_2(T_z P_2^{\perp})\},$$

and so on.

To conclude this section let us summarize the approach proposed by Skinner and Rusk [111], [112] for degenerate Lagrangian systems.

We have seen before in Section 7.14, that the Euler–Lagrange equation

$$i_X \omega_L = dE_L \tag{8.1}$$

is equivalent to a first–order equation

$$i_Y \omega_{\oplus} = dD \tag{8.2}$$

on the Whitney sum $W = T^*M \oplus TM$ (Skinner's theorem). This theorem is valid even if the Lagrangian considered $L : TM \longrightarrow R$ is degenerate. In such a case Eq.(8.2) is a presymplectic equation and we cannot say that there is a (unique) solution Y globally defined on W. Therefore we may reproduce the procedure developed before for presymplectic systems.

Let us first recall that the semibasic form α_L on TM defines a C^∞ map $Leg : TM \longrightarrow T^*M$ (which is not a diffeomorphism as L is degenerate) locally given by

$$(q,v) \longrightarrow (q, \partial L/\partial v).$$

If we take $\varphi = Leg$ as it was considered in Section 7.14 and as $Graph\ \varphi \subset W = T^*M \oplus TM$, we have a diffeomorphism $\bar{\varphi} : TM \longrightarrow Graph\ \varphi \subset W$. Now, let V^{π_1} be the vertical bundle with respect to $\pi_1 : W \longrightarrow T^*M$, i.e., $V^{\pi_1} = Ker\ T\pi_1$. Then, if $Z \in V^{\pi_1}$, $i_Z \omega_\oplus = 0$ (locally, $Z = (\delta v)\partial/\partial v$ and $\omega_\oplus = dq \wedge dp$). But $i_Z dD$ does not necessarily vanishes. Therefore we may define a manifold W_1 by

$$W_1 = \{z \in W / i_Z dD = 0 \text{ for all } Z \in (V^{\pi_1})_z\}.$$

For the points $z \in W_1$, the first–order equation can be solved. Supposing that W_1 is embedded into W by a mapping $i_{W_1} : W_1 \longrightarrow W$ and setting $i_{W_1}(W_1) = W_1$, one has $W_1 = Graph\ \varphi \subset W$ and

$$\pi_2 \circ i_{W_1} = (\bar{\varphi})^{-1}, \ \pi_1 \circ i_{W_1} \circ \bar{\varphi} = \varphi.$$

To find a consistent solution of (8.1), i.e., a final constraint submanifold S for which the solution X is a vector field on S $(X : S \longrightarrow TS)$ we apply the above geometric algorithm. For instance, $W_r \subset W$ is the submanifold obtained at the r–th step, i.e., from

$$TW_{r-1}^{\perp} = \{X \in TW / \omega_\oplus(X,Y) = 0 \text{ for all } Y \in TW_{r-1}\}$$

we define

$$W_r = \{z \in W_{r-1} / dD(Y)(z) = 0 \text{ for all } Y \in T_z W_{r-1}^{\perp}\}$$

and we set $W = W_0$. Then, if the algorithm works, we find a final constraint manifold S on which is defined a solution of (8.1). For further details see Skinner and Rusk [111], [112].

8.1.2 Relation to the Dirac–Bergmann theory of constraints

Around 1950 Dirac [47], [48] proposed a Hamiltonian theory for Lagrangian systems which cannot be directly put into this form (see also Bergmann [6]). This means that the condition on the non–singularity of the Lagrangian function with respect to the velocities breaks down. Taking into account the Implicit (or Inverse) function theorem we may say that in order to obtain an unconstrained Hamiltonian formalism in Classical Mechanics we need all momentum variables to be independent of the velocity variables. If we drop the non–singularity condition then there are a certain number of momenta which are not entirely independent of the velocities (of course we may develop a Hamiltonian formalism including these momenta but this is against to the usual procedure, see the Appendix A for further details in local coordinates). Therefore we need to restrict our attention to a certain subset K of the phase space manifold for which the momenta are, as in the standard theory, independent of the velocities (we may think that this corresponds to the regular part of the theory). This subset is taken as a C^∞ manifold and is called the constraint manifold.

Geometrically, Dirac–Bergmann constraint theory may be viewed as follows. Suppose that (W, ω) is an arbitrary symplectic manifold and K is a manifold embedded in W by a mapping $\varphi : K \longrightarrow W$. Set $\bar{\omega} = \varphi^\star \omega$.

Let h be a C^∞ function on K. We want to know: what are the conditions for h to be extended into a C^∞ function H on W (at least locally) such that the corresponding Hamiltonian vector field X_H is tangent to K? We may also ask, if a Hamiltonian H is initially given on W, under what conditions the motion equation $i_X\omega = dH$ restricted to K (i.e., $i_X\omega = d(\varphi^\star H)$) admits a solution X tangent to K at every point? We will examine here only the first case: the extension of h to H.

Recall that if H is an extension of h to all W (or on an open neighborhood) then from the isomorphism S_ω one obtains a unique solution X_H solving $i_X\omega = dH$, but such X_H may not be tangent to K (and so X_H is not an ordinary differential equation along K). The restriction of the dynamical equations to K does not assure us that there is a solution of the problem. We must have the above range condition verified: $d(\varphi^\star H) \in S_{\bar{\omega}}(TK)$. In such a case we cannot assure the unicity of X_H.

Let us suppose that a symplectic manifold (W, ω) is initially given with a submanifold P such that $(P, \bar{\omega})$ is presymplectic. Then after applying the algorithm method we assume that a final constraint manifold K in (W, ω) is obtained (of the above third type) and a C^∞ function h is defined on it. In fact, to simplify things, we suppose that K is a primary constraint manifold

Let (x, y) be a coordinate system defined on a neighborhood $V \subset W$ of some point $z \in K$ such that $y = y^b = (y^{r+1}, \ldots, y^{2n})/_{U=K\cap V} \equiv 0$, with dim $K = r$ and dim $W = 2n$. We extend h to V by

$$H = h + u_b y^b, \tag{8.3}$$

with u_b being arbitrary multipliers to be determined. We have the equation

$$i_X \omega = dH \tag{8.4}$$

extension of

$$i_X \bar{\omega} = dh. \tag{8.5}$$

Equation (8.4) has of course a solution X_H defined for all points in V and in particular for those in $U = K \cap V$. But X_H may not be tangent to U. As we have supposed that dh satisfies the range condition ($dh \in S_{\bar{\omega}}(TU)$) our main problem is the tangency question for this first-order problem (i.e., the consistency of the constraints). We also suppose that the extension H is C^∞ on V.

Let z be a point in K and consider the orthocomplement $T_zK^\perp$ of T_zK in (T_zW, ω_z):

$$T_zK^\perp = \{X \in T_zW / \text{ for all } Y \in T_zK,\ (i_X\omega)(z)(Y) = 0\}.$$

Locally the symplectic complement $T_zK^\perp$ is generated by the family $\{X_{y^b}\}$ of Hamiltonian vector fields corresponding to $\{dy^b\}$ (see Lemma 6.4.1). In the following we represent X_{y^b} by X_b, $r + 1 \leq b \leq 2n$. We consider the following set

$$K(W, \omega) = \{F \in C^\infty(W)/X_F(z) \in T_zK \text{ for all } z \in K\},$$

where X_F is such that $i_{X_F}\omega = dF$.

Proposition 8.1.3 *We have*

$$K(W, \omega) = \{F \in C^\infty(W)/dF(X) = XF = 0, \text{ for all } z \in K \text{ and } X \in T_zK^\perp\}.$$

Proof: It is sufficient to show the proposition for the family $\{X_b\}$. If F is a C^∞ function on W then X_F is tangent to K in a neighborhood of a point z if and only if $X_F(y^b) = 0$, since for any integral curve $f(t)$ of X_F passing through z we have

$$X_F(f(t)) = (d/dt)(f(t))/_{t=0}.$$

The local characterization of K by the functions y^b tell us that all y's are constant along the trajectories of X_F. So

$$X_F(y^b(f(t))) = (d/dt)(y^b(f(t)) = 0.$$

But

$$X_F(y^b) = -X_b(F) = -dF(X_b)$$

which gives the above characterization of $K(W, \omega)$.□

Definition 8.1.4 *The function $F \in K(W,\omega)$ is called of* **first class**. *Functions which are not of first class are called of* **second class**.

Remark 8.1.5 The above definition is due to Lichnerowicz [89]. The Lichnerowicz's method may be summarized as follows: in the place of searching a final constraint manifold K solving the problem we may study the class of functions for which the vector fields are first–order differential equations along K fixed. First class functions on constraint manifolds were characterized by Dirac as functions which are in involution with respect to the Poisson brackets. So, if (x, y) is a local coordinate system in an open neighborhood V of W such that $y = (y^b) \equiv 0$ along $U = V \cap K$, then F is a first class function if and only if for all b one has $\{F, y^b\} = 0$. In fact

$$\{F, y^b\} = \omega(X_F, X_b) = X_b(F) = dF(X_b).$$

We will say that (y^b) are **first class constraint functions** if $y^b \in K(W, \omega)$ for all b. So a first class constraint manifold of (W, ω) is a submanifold K of W locally characterized by the vanishing of functions which are in $K(W, \omega)$ with respect to the Poisson brackets.

Consider again a local coordinate system $(x, y) = (x^i, y^b) = (x^1, \ldots, x^r, y^{r+1}, \ldots, y^{2n})$ on V such that $U = V \cap K$ is defined by $y = 0$. If $\{\partial/\partial x^i, \partial/\partial y^b\}$ is the local basis induced by such coordinates then the restriction to U of $(\partial/\partial y^b)$ generates a subbundle S of $TW/_U$ supplementary of TU in $TW/_U$. The Hamiltonian vector fields X_b of dy^b with respect to the symplectic form ω generates a subbundle $TU^\perp$ of $TW/_U$ and therefore we may put

$$X_b = T_b + S_b$$

for the direct decomposition $TW/_U = TU \oplus S$, where T_b and S_b are sections of TU and S, respectively. A direct computation shows that

$$X_b(H) = T_b(h) + S_b(H)$$

and $X_b(H) = 0$ if and only if $T_b(h) = -S_b(H)$. Taking the differential of (8.3) we see that

$$dH(X_b) = X_b(H) = dh(X_b) + u_a dy^a(X_b)/_U$$

Thus

$$X_b(H) = T_b(h) + u_a S_b(y^a) \tag{8.6}$$

$$X_b(H) = T_b(h) + u_a\{y^b, y^a\} \tag{8.7}$$

$$S_b(H) = u_a S_b(y^a) = u_a\{y^b, y^a\} \tag{8.8}$$

Proposition 8.1.6 *Let (x, y) be a coordinate system as above. A necessary and sufficient condition for a C^∞ function h on $(K, \bar{\omega})$ to be extended along a neighborhood V,*

$$H = h + u_b y^b, \quad r + 1 \leq b \leq 2n,$$

with the property that the vector field X_H is tangent to K at every point in U is that the system

$$u_a\{y^b, y^a\} = -T_b(h) \tag{8.9}$$

admits a solution $u = (u_a)$ for such coordinates.

Proof: If (8.9) is verified then from (8.7) we have that for every vector field $Z \in TU^\perp$, $Z(H) = 0$ and from Proposition 8.1.2, X_H is tangent to U.

If H is in $K(W, \omega)$ then $X_b(H) = 0$ and so $S_b(H) = -T_b(h)$. We develop S_a along the basis $\{\partial/\partial y^b\}/_U$ and we obtain

$$-T_b(h) = \{y^b, y^a\}(\partial H/\partial y^a)/_U.$$

Therefore the functions $\{(\partial H/\partial y^a)/_U\}$ are the desired solutions. □

Thus, from Proposition 8.1.5 we see that the geometrical algorithm says that (8.9) admits a solution at the final constraint manifold. But Proposition 8.1.5 gives us another local method for searching such final constraint manifold (if there is one). The main question is in fact: when are there such coordinate systems (x, y)? If for a given system the matrix $\pi = (\{y^a, y^b\})$ is of maximal rank then of course the problem is solvable. But in general this will not be so. We must discard the case where the system (8.9) is unsolvable for every coordinate system. In such a case Dirac [48] says: "it would mean that the Lagrangian equations of motion are inconsistent and we are excluding such a case".

Suppose that for a given coordinate system the rank of π is not maximal. Since $T_b(h)$ in (8.9) is arbitrary it may occurs that all components of such a vector field are not in the image of the regular part of π. Our system is independent of the u's and $X(H) = 0$ if $T_b(h) = 0$. This equation will not be satisfied except in an open subset U_1 of U. This subset is characterized by a coordinate system $(x_\alpha, \bar{y}^A)$, where now $1 \leq \alpha \leq k$, $;k+1 \leq A \leq r$. Starting again with

$$X_A(H) = T_A(h) + S_A(H)$$

for such situation, we analyze the system

$$-T_A(h) = u_a\{\bar{y}^A, y^a\},$$

where

$$S_A(H) = u_a\{\bar{y}^A, y^a\}$$

is obtained from

$$dH(X_A) = dh(X_A) + u_a y^a(X_A).$$

If the above system is independent of the u's then we continue the process into another subset of U_1 and so on. This algorithm originates a final local constraint submanifold U_r for which at least some of the components of the corresponding $T(h)$ are in the image of the regular part of the respective matrix constructed for such domain.

If we suppose, for the original situation, that for our system (x^i, y^b), s elements of $T_b(h)$ are in the image of the regular part of π, for simplicity (s is the rank of π) then from Linear Algebra we know that the system (8.9) has a solution of type

$$u_a = \bar{u}_a + v_j \bar{u}_{ja},$$

where $\bar{u}_a$ is fixed and $v_j \bar{u}_{ja}$ is a linear combination of all solutions of the homogeneous equations associated to (8.9). If we put

$$f^j = \bar{u}_{ja} y^a$$

then

$$H = h + \bar{u}_a y^a + v_j f^j$$

is the final expression of the extended Hamiltonian. We may verify that H is of first class. Finally, from the above algorithm we see that at the end of the process the Hamiltonian will be of type

$$H = h + \text{first class (primary/secondary) constraints}$$

$$+\text{second class (primary/secondary) constraints.}$$

This finish our geometrical interpretation of Dirac–Bergmann formulation of the constraint theory.

Remark 8.1.7 Let (W, ω) be a symplectic manifold, K a submanifold of W. We recall that K is respectively isotropic, coisotropic, Lagrangian if its tangent space at every point z of K is respectively of the corresponding type as a subspace of $T_z W$. A coisotropic submanifold is a first class constraint manifold. A Lagrangian submanifold is a second class constraint manifold. Isotropic submanifolds are trivial constraint manifolds.

8.2 The second–order problem and the Lagrangian formalism

8.2.1 The constraint algorithm and the Legendre transformation

Let J be the canonical almost tangent structure on the tangent bundle of an m–dimensional manifold M and $\omega = -dd_J L$ the (pre)symplectic form on

W induced by the Lagrangian $L : TM \longrightarrow R$. Then we may also consider the problem of finding a solution X for the problem $i_X\omega_L = \alpha$, where α is a 1–form on TM. Naturally, if L is a regular function (i.e., ω_L is symplectic) then X exists and it is unique. So, let us suppose that L is degenerate (i.e., ω_L is presymplectic). Repeating the above procedure of the preceding section, we wish to know if there exists a submanifold K of TM and a vector field X on TM solving the equation along K, with X being tangent to K at every point of K. As the points of TM for which the dynamical equation is inconsistent are those for which $i_Y\alpha \neq 0$ for all $Y \in T(TM)^\perp$, we define

$$K_1 = \{z \in TM/(i_Y\alpha)(z) = 0, \text{ for all } Y \in T(TM)^\perp\}.$$

We repeat the same procedure to find a final constraint manifold K_r (non–empty) defined by

$$K_r = \{z \in K_{r-1}/ \text{ for all } Y \in TK_{r-1}^\perp,\ (i_Y\alpha)(z) = 0\}.$$

The constrained algorithm for the Lagrangian situation is, however, insufficient since the solution X must be a semispray (second–order differential equation) on the final constraint manifold. Therefore we must incorporate to the original problem the equation $JX = C$, the geometrical condition for a dynamical system to be a semispray. This new problem is sometimes known as the **second–order problem**. Thus this new situation is now characterized by the set of equations

$$i_X\omega_L = \alpha, \ JX = C,$$

on some submanifold of TM (in fact, on some submanifold of the final constraint manifold obtained from the constraint algorithm for the first–order problem).

We will examine this problem in the forthcoming paragraphs but let us first analyse the relation between Lagrangian and Hamiltonian systems for the degenerate Lagrangian case. Thus, we wish to know if there is a Hamiltonian counterpart of degenerate Lagrangian systems. A study in this direction was proposed by Gotay et al. [68], which we will see now.

We first observe that the form ω_L is related to the canonical symplectic form ω_M on T^*M by

$$Leg^*\omega_M = \omega_L,$$

where $Leg : TM \longrightarrow T^*M$ is the Legendre transformation (see Chapter 7). If we suppose $L : TM \longrightarrow R$ regular (i.e., Leg is a diffeomorphism) then $E_L \circ Leg^{-1}$ is the Hamiltonian on T^*M (E_L is the energy for L) and the Hamiltonian counterpart of the Lagrangian L is defined. However, if Leg is not a diffeomorphism of TM onto T^*M we cannot define the Hamiltonian to be $E_L \circ Leg^{-1}$. Defining a function h_1 on the image $Leg\,(TM)$ implicitly by

$$h_1 \circ Leg = E_L$$

then this definition will be well–defined on $Leg\,(TM)$ if and only if for any two points $x, y \in TM$ one has: $Leg\,(x) = Leg\,(y)$ implies $E_L(x) = E_L(y)$. Thus we restrict our attention to a special class of Lagrangian functions verifying the following property:

Definition 8.2.1 *A Lagrangian $L : TM \longrightarrow R$ is said to be* **almost regular** *if the Legendre mapping $Leg : TM \longrightarrow T^*M$ is a submersion onto its image and the fibers $Leg^{-1}(Leg\,(x))$ are connected for all $x \in TM$.*

We now prove that every almost regular Lagrangian system has a special Hamiltonian counterpart. From the above implicit definition it is sufficient to show that $h_1 \circ Leg = E_L$ is a first integral (or a constant of motion) for all $Y \in Ker\,(T\,Leg)$. A reason for this is that for almost regular Lagrangian systems the image $Leg\,(TM)$ can be canonically identified with the leaf space of the foliation of TM generated by the involutive distribution $Ker\,(T\,Leg)$.

The distribution $Ker\,(T\,Leg)$ defines a local foliation on TM in the sense that for each $z \in TM$ there is a (local) submanifold Q of TM passing through z such that $TQ = Ker\,(T\,Leg)/_Q$ (in fact, as $Ker\,(T\,Leg)$ is integrable, from Frobenius theorem one has that $Ker\,(T\,Leg)$ defines globally a foliation). Considering the family of almost regular Lagrangians it is possible to show that $Leg\,(TM)$ may be identified via a diffeomorphism to the leaf space of the foliation generated by the distribution $Ker\,(T\,Leg)$.

Let $Y \in Ker\,(T\,Leg)$. As Y is a vertical vector field there is (locally) a vector field Z such that $JZ = Y$. Thus

$$L_Y E_L = L_{JZ} E_L = i_{JZ} dE_L = i_Z i_J dE_L = -i_Z i_C \omega_L. \tag{8.10}$$

The Liouville vector field C is a vertical vector field and so there is (locally) a vector field X such that $JX = C$. Therefore

$$\omega_L(Z, C) = \omega_L(Z, JX) = -\omega_L(JZ, X) = \omega_L(X, Y)$$

$$= (Leg^\star \omega_M)(X,Y) = \omega_M((T\,Leg)X,\ (T\,Leg,Y)) = 0, \tag{8.11}$$

as $Y \in Ker(T\ Leg)$. But

$$L_Y E_L = i_Y dE_L = dE_L(Y) = Y(E_L)$$

and so $Y(E_L) = 0$, showing that E_L is a first integral for all $Y \in Ker(T\ Leg)$.

Let us study now the relation between the motion equations for both formalisms.

Definition 8.2.2 *Consider a Lagrangian system represented by a triple (TM, ω_L, L), where $L : TM \longrightarrow R$ is the Lagrangian and $\omega_L = -dd_J L$. Let $Leg : TM \longrightarrow T^*M$ be the Legendre transformation and set $M_1 = Leg(TM)$, $\omega_1 = \omega/M_1$ and $h_1 \circ Leg = E_L$, where ω_M is the canonical symplectic form on T^*M and E_L is the energy associated to L. We say that the Hamiltonian system (M_1, ω_1, h_1) is* **first–order equivalent** *to (TM, ω_L, L) if*
(1) for every solution X_L of the equation $i_X\omega_L = dE_L$ the vector field $(T\ Leg)(X_L)$ satisfies the Hamilton equations $i_X\omega_1 = dh_1$.
(2) If X_{h_1} is a solution of the Hamilton equations on M_1 then every $X_L \in (T\ Leg)^{-1}(X_{h_1})$ is a solution of the Euler– Lagrange equations.

The following theorem is due to Gotay and Nester [68].

Theorem 8.2.3 (The equivalence theorem). *Let (TM, ω_L, L) be an almost regular Lagrangian system. Then (TM, ω_L, L) admits a Hamiltonian counterpart (M_1, ω_1, h_1) such that both systems are equivalent.*

Proof: First we apply the constraint algorithm to find a final constraint manifold where a solution X_L for the original Euler–Lagrange equations $i_X\omega_L = dE_L$, tangent to this manifold is obtained. For the sake of simplicity let us suppose that such X_L is obtained at the first step, i.e., at $P_2 \longrightarrow P_1 = TM$, where P_2 is embedded in P_1 via $\varphi_2 : P_2 \longrightarrow P_1$. Let $M_1 = Leg(P_1)$ and $\psi_1 : M_1 \longrightarrow T^*M$ the corresponding embedding of M_1 into T^*M. We set $Leg_1 : P_1 \longrightarrow M_1$, i.e., $Leg_1 \circ \psi_1 = Leg$. Also we define $Leg_2 : P_2 \longrightarrow M_2$ by $Leg_1 \circ \varphi_2 = \psi_2 \circ Leg_2$:

Let us recall that

$$P_2 = \{y \in P_1 = TM / i_Z dE_L(y) = 0, \text{ for all } Z \in Ker\ \omega_L = (TP_1)^\perp\}$$

is a secondary constraint manifold. As $h_1 \circ Leg = E_L$ and $(TP_1)^\perp = (TM_1)^\perp \oplus Ker(T\ Leg)$, using a local basis of vector fields on TP_1 which locally span $(TP_1)^\perp$ such that their prolongations by Leg_1 exist and (locally) span $(TM_1)^\perp$, we see that there is a secondary Hamiltonian constraint manifold M_2 defined by

$$M_2 = \{Z \in M_1 / i_{T\,Leg_1 Z} dh_1(z) = 0\}.$$

As Leg_1 is a submersion, Leg_2 is also a submersion.

Suppose that X_L satisfies the Euler–Lagrange equations. Then, if $(T\ Leg_1)(X_L)$ exists, we have

$$0 = (i_{X_L}\omega_L - dE_L)/_{P_2}$$

$$= (i_{X_L} Leg_1^\star \omega_1 - d\, Leg_1^\star h_1)/_{P_2}$$

$$= Leg_1^\star (i_{T\,Leg_1 X_L}\omega_1 - dh_1)/_{M_2}.$$

As Leg_1 is a submersion, $T\ Leg_1 X_L$ solves the Hamilton equations

$$(i_{T\,Leg_1 X_L}\omega_1 = d\,h_1)/_{M_2}.$$

Conversely, requirement (2) of Definition 8.2.2 is satisfied by running the computations backwards. □

We remark that the above procedure may be adapted directly from the r–th step in the Lagrangian version transforming it to the Hamiltonian description via Leg_r (M_r would be then the primary constraint manifold).

8.2.2 Almost tangent geometry and degenerate Lagrangians

Let us consider again the Euler–Lagrange equations of motion

$$i_X \omega_L = dE_L. \tag{8.12}$$

When L is a regular function there is a unique solution X which is automatically a semispray. Indeed, as we have seen before, from (8.12) one obtains

$$i_{JX}\omega_L = i_C\omega_L$$

and the non–degeneracy of ω_L implies $JX = C$.

If $L : TM \longrightarrow R$ is not a regular function, i.e., ω_L is presymplectic then even if there is a vector field X solving (8.12) we cannot assure the second–order condition $JX = C$, i.e., that X is a semispray.

In this section we develop a geometric formalism in order to study degenerate Lagrangians.

In the following, for simplicity, we will put

$$Ker\ S_{\omega_L} = Ker\ \omega_L.$$

Suppose that (P, Q) is an almost product structure on TM adapted to ω_L, i.e., we have

$$Ker\ P = Ker\ \omega_L$$

Then $Im\,P$ and $Im\,Q$ are ω_L–orthogonal, i.e.,

$$\omega_L(PX, QY) = 0, \text{ for all } X, Y \in \chi(M).$$

Also, if ω_L has rank $2r \leq 2m$, where dim $M = m$, then $Ker\ P = Im\,Q$ is a subbundle of TTM with rank $=$ co–rank $\omega_L = 2m - 2r = 2(m - r)$. As

$$S_{\omega_L} : Im\ P \longrightarrow Im\ P^\star$$

is an isomorphism, given a 1–form α on TM, there exists a unique vector field $X \in Im\,P$ such that

$$i_X\omega_L = P^\star\alpha.$$

Thus, there is a unique vector field $\xi_L \in Im\,P$ such that

$$i_{\xi_L}\omega_L = P^\star(dE_L),\ E_L = CL - L. \tag{8.13}$$

ξ_L is called **P–Euler–Lagrange vector field**. Obviously, the solution of (8.13) is parametrized by the kernel of S_{ω_L}, i.e., ξ_L is a solution on TM modulo $Ker\ \omega_L$.

Suppose that the adapted almost product structure (P, Q) commutes with the canonical almost tangent structure J on TM, i.e.,

$$JP = PJ \text{ (and hence } JQ = QJ).$$

Then it is easy to see that P is locally given by

$$P(\partial/\partial q^i) = P_i^j(\partial/\partial q^j) + \bar{P}_i^j(\partial/\partial v^j),$$

$$P(\partial/\partial v^i) = P_i^j(\partial/\partial v^j),$$

$1 \leq i, j \leq m$.

Thus, P has associated matrix

$$\begin{bmatrix} P_i^j & 0 \\ \bar{P}_i^j & P_i^j \end{bmatrix}$$

Also,

$$P^*(dq^i) = P_j^i dq^j,$$

$$P^*(dv^i) = \bar{P}_j^i dq^j + P_j^i dv^j.$$

Definition 8.2.4 *We say that a vector field $\xi \in Im\, P$ is a* **P–semispray** *(or P–2nd* **order differential equation***) if $J\xi = PC$.*

Proposition 8.2.5 *The solution ξ_L of (8.13) is a P-semispray.*

Proof: First, we compute the action of i_J on both sides of (8.13):

$$i_J i_{\xi_L}\omega_L = i_J P^*(dE_L) = -i_{J\xi_L}\omega_L = (J^* P^*)(dE_L), \qquad (8.14)$$

because

$$[i_J, i_{\xi_L}] = i_J i_{\xi_L} - i_{\xi_L} i_J = -i_{J\xi_L} \text{ and } i_J\omega_L = 0.$$

Now, recalling that

$$i_C d_J + d_J i_C = i_J,$$

we have

$$i_C\omega_L = -i_C dd_J L = i_C d_J dL = (i_J - d_J i_C)dL$$

$$= d_J L - d_J(CL) = d_J(L - CL) = -d_J E_L.$$

Therefore

$$i_C\omega_L = -J^\star(dE_L),$$

and we have

$$i_{PC}\omega_L = -(P^\star J^\star)(dE_L) = -(J^\star P^\star)(dE_L). \tag{8.15}$$

In fact, for every vector field Y

$$(i_{PC}\omega_L)(Y) = \omega_L(PC, Y) = \omega_L(C, PY)$$

$$= -J^\star(dE_L)(PC) = -(P^\star J^\star)(dE_L)(Y).$$

From (8.14) and (8.15) one obtains

$$i_{J\xi_L}\omega_L = i_{PC}\omega_L. \tag{8.16}$$

But $J\xi_L$, $PC \in Im\,P$ and so (8.16) implies $J\xi_L = PC$.□

Let us suppose that the almost product structure (P, Q) is integrable. Then $Im\,P$ is an integrable distribution. Let S be an integrable manifold of $Im\,P$. Then S is a submanifold of TM, $\phi : S \longrightarrow TM$ (for sake of simplicity, we always suppose that all submanifolds are embedded). Also, the couple $(S, \phi^\star\omega_L)$ is symplectic and the following equality holds:

$$\phi^\star P^\star(dE_L) = d(E_L \circ \phi).$$

To show this, let z be a point of S and $X \in T_zS$. Then

$$(\phi^\star P^\star)(dE_L)(X) = P^\star(dE_L(T\phi(X))) = dE_L(P(T\phi(X)))$$

$$dE_L(T\phi(X)) = d(E_L \circ \phi)(X).$$

The vector field ξ_L (as well as PC) is tangent to the submanifold S and so may be restricted to a vector field on S. Thus, for every point $z \in S$

$$J_z(\xi_L(z)) = (J/_{T_zS})(\xi_L(z)) = (PC)(z).$$

However we cannot say that $\xi_{L/S}$ is a semispray. In fact, $\xi_{L/S}$ is only a Hamiltonian vector field associated to the function $E_L \circ \phi$ on the symplectic manifold $(S, \phi^*\omega_L)$, solution of equation

$$i_X(\phi^*\omega_L) = d(E_L \circ \phi).$$

To go on, we must study equation (8.13) in local coordinates and obtain some local relations useful for our main result. Locally one has

$$\omega_L = (\partial^2 L/\partial q^i \partial q^j) dq^i \wedge dq^j + (\partial^2 L/\partial v^i \partial v^j) dq^i \wedge dv^j,$$

$$E_L = v^i(\partial L/\partial v^i) - L.$$

Suppose that $\xi_L \in Im\, P$. Then there is a vector field X on TM such that $\xi_L = PX$, with

$$X = X^i(\partial/\partial q^i) + \bar{X}^i(\partial/\partial v^i).$$

Proposition 8.2.6 *Consider eq.(8.13). Then in a local coordinate system (q^i, v^i) the following relations hold:*

$$X^i P_i^j = v^i P_i^j, \tag{8.17}$$

$$(\partial^2 L/\partial v^i \partial q^j) v^k P_k^i - (\partial^2 L/\partial v^j \partial q^i) v^k P_k^i - (\partial^2 L/\partial v^i \partial v^j) \xi_L(v^i)$$

$$= (\partial^2 L/\partial v^i \partial q^k) v^i P_j^k - (\partial L/\partial q^k) P_j^k + (\partial^2 L/\partial v^i \partial v^k) v^i \bar{P}_j^k, \tag{8.18}$$

$$(\partial^2 L/\partial v^i \partial v^j) v^k P_k^i = (\partial^2 L/\partial v^i \partial v^k) v^i P_j^k. \tag{8.19}$$

Proof: The equalities (8.17), (8.18) and (8.19) are obtained from a straightforward commputation. We show (8.17) and we indicate how (8.18) and (8.19) are obtained. As $\xi_L = PX$, then

$$\xi_L = X^i P(\partial/\partial q^i) + \bar{X}^i P(\partial/\partial v^i)$$

$$= X^i P_i^j (\partial/\partial q^j) + (X^i \bar{P}_i^j + \bar{X}^i P_i^j)(\partial/\partial v^j).$$

But

$$PC = P(v^i(\partial/\partial v^i)) = v^i P_i^j (\partial/\partial v^j),$$

$$J\xi_L = X^i P_i^j (\partial/\partial v^j).$$

As $J\xi_L = PC$, eq.(8.17) is verified. The other equalities (8.18) and (8.19) are obtained in the following manner: we compute $i_{\xi_L}\omega_L$ which gives

$$i_{\xi_L}\omega_L = ((\partial^2 L/\partial v^i \partial q^j) X^k P_k^i - (\partial^2 L/\partial v^j \partial q^i) X^k P_k^i$$

$$-(\partial^2 L/\partial v^i \partial v^j)\xi_L(v^i))dq^j$$

$$+(\partial^2 L/\partial v^i \partial v^j) X^k P_k^i dv^j,$$

where in the development of $i_{\xi_L}\omega_L$ we have used the relations

$$i_{\xi_L}(dq^i) = \xi_L(q^i) = X^k P_k^i,$$

$$i_{\xi_L}(dv^i) = \xi_L(v^i).$$

If we take the differential dE_L and we apply P^* to dE_L we have

$$P^*(dE_L) = \{(v^i(\partial^2 L/\partial v^i \partial q^k) - (\partial L/\partial q^k))P_j^k$$

$$+v^i(\partial^2 L/\partial v^i \partial v^k)\bar{P}_j^k\}dq^j$$

$$+v^i(\partial^2 L/\partial v^i \partial v^k) P_j^k dv^j.$$

As $i_{\xi_L}\omega_L = P^*(dE_L)$ we obtain (8.18) and (8.19) using (8.17). □
Next, we shall prove the main result of this section.

Theorem 8.2.7 *Let $L : TM \longrightarrow R$ be a degenerate Lagrangian and (P, Q) an adapted almost product structure to the presymplectic form ω_L. If (P, Q) is the complete lift of an integrable almost product structure (P_0, Q_0) on M then for every integral manifold S_0 of $Im\, P_0$ the restriction of L to TS_0 is regular. Therefore the equation*

$$(i_X\omega_L - dE_L)/_{TS_0} = 0$$

admits a unique solution vector field $\xi_L : TS_0 \longrightarrow TTS_0$ which is a second-order differential equation.

Proof: Suppose that ω_L has rank $2r$ and let (P_0, Q_0) be an integrable almost product structure on M such that the complete lift (P_0^c, Q_0^c) is adapted to ω_L. Then P_0 has rank r and we have

$$P_0^c J = J P_0^c = P_0^v,$$

$$Q_0^c J = J Q_0^c = Q_0^v,$$

(where P_0^v (resp. Q_0^v) is the vertical lift of P_0 (resp. Q_0) to TM). As (P_0, Q_0) is integrable, we may choose a coordinate system in M, say (q^i), such that P_0 and Q_0 are represented by the matrices

$$\begin{bmatrix} \delta_i^j & 0 \\ 0 & 0 \end{bmatrix}, \quad 1 \leq i, j \leq r,$$

and

$$\begin{bmatrix} 0 & 0 \\ 0 & \delta_i^j \end{bmatrix}, \quad m - r + 1 \leq i, j \leq m,$$

respectively, where δ_i^j = Kronecker's delta. For the induced coordinates (q^i, v^i) on TM, the lift P_0^c is represented by the matrix

$$\begin{bmatrix} \begin{matrix} \delta_i^j & 0 \\ 0 & 0 \end{matrix} & 0 \\ 0 & \begin{matrix} \delta_i^j & 0 \\ 0 & 0 \end{matrix} \end{bmatrix}.$$

In such situation, equalities (8.18) and (8.19) of Proposition 8.2.6 become

$$(\partial^2 L/\partial v^i \partial q^j)v^i - (\partial^2 L/\partial v^j \partial q^i)v^i - \sum_{i=1}^{m}(\partial^2 L/\partial v^i \partial v^j)\xi_L(v^i)$$

$$= (\partial^2 L/\partial v^i \partial q^j)v^i - (\partial L/\partial q^j), \quad 1 \leq j \leq r, \tag{8.20}$$

$$(\partial^2 L/\partial v^i \partial q^j)v^i - (\partial^2 L/\partial v^j \partial q^i)v^i - \sum_{i=1}^{m}(\partial^2 L/\partial v^i \partial v^j)\xi_L(v^i) = 0, \tag{8.21}$$

$$r + 1 \leq j \leq m,$$

$$\sum_{i=1}^{r}(\partial^2 L/\partial v^i \partial v^j)v^i = \sum_{i=1}^{r}(\partial^2 L/\partial v^i \partial v^j)v^i, \quad 1 \leq j \leq r, \tag{8.22}$$

$$\sum_{i=1}^{r}(\partial^2 L/\partial v^i \partial v^j)v^i = 0, \quad r + 1 \leq j \leq m. \tag{8.23}$$

The equations (8.20)–(8.23) may be re-written as follows:

$$\sum_{i=1}^{r}(\partial^2 L/\partial v^j \partial q^i)v^i + \sum_{i=1}^{m}(\partial^2 L/\partial v^i \partial v^j)\xi_L(v^i)$$

$$- (\partial L/\partial q^j) = 0, \quad 1 \leq j \leq r, \tag{8.24}$$

$$\sum_{i=1}^{r}(\partial^2 L/\partial v^i \partial q^j)v^i - \sum_{i=1}^{r}(\partial^2 L/\partial v^j \partial q^i)v^i$$

$$- \sum_{i=1}^{m}(\partial^2 L/\partial v^i \partial v^j)\xi_L(v^i) = 0, \quad r + 1 \leq j \leq m, \tag{8.25}$$

$$\sum_{i=1}^{r}(\partial^2 L/\partial v^i \partial v^j)v^i - \sum_{i=1}^{m}(\partial^2 L/\partial v^i \partial v^j)v^i = 0, \quad 1 \leq j \leq r, \tag{8.26}$$

$$\sum_{i=1}^{r}(\partial^2 L/\partial v^i \partial v^j)v^i = 0, \quad r + 1 \leq j \leq m. \tag{8.27}$$

From (8.24)–(8.27), one deduces

$$\sum_{i=r+1}^{m} (\partial^2 L/\partial v^i \partial v^j) v^i = 0, \quad 1 \le j \le r.$$

Thus, for $1 \le j \le r$, one has

$$\sum_{i=1}^{r} (\partial^2 L/\partial v^j \partial q^i) v^i + \sum_{i=1}^{m} (\partial^2 L/\partial v^i \partial v^j) \xi_L(v^i) - (\partial L/\partial q^j) = 0, \tag{8.28}$$

$$\sum_{i=r+1}^{m} (\partial^2 L/\partial v^i \partial v^j) v^i = 0, \tag{8.29}$$

and, for $r+1 \le j \le m$,

$$\sum_{i=1}^{r} (\partial^2 L/\partial v^i \partial q^j) v^i - \sum_{i=1}^{r} (\partial^2 L/\partial v^j \partial q^i) v^i - \sum_{i=1}^{m} (\partial^2 L/\partial v^i \partial v^j) \xi_L(v^i) = 0, \tag{8.30}$$

$$\sum_{i=1}^{r} (\partial^2 L/\partial v^i \partial v^j) v^i = 0. \tag{8.31}$$

The local expression of ξ_L in such a case is

$$\xi_L = X^i P_i^j (\partial/\partial q^j) + \bar{X}^i P_i^j (\partial/\partial v^j).$$

As $X^i P_i^j = v^i P_i^j$, then, when $1 \le j \le r$, we have

$$X^i = v^i, \text{ for } 1 \le i \le r.$$

As there is no conditions for $r+1 \le j \le m$, we deduce

$$\xi_L = \sum_{i=1}^{r} v^i (\partial/\partial q^i) + \sum_{i=1}^{r} \bar{X}^i (\partial/\partial v^i),$$

where $\bar{X}^i = \bar{X}^i(q^j, v^j)$, $1 \le j \le m$, i.e., $\bar{X}^i$ depends on all coordinates q's and v's.

Let S_0 be an integral manifold of $Im\, P_0$. Then S_0 is locally determined by equations of type

$$q^{r+1} = \text{const.}, \ldots, q^m = \text{const.},$$

since $Im\, P_0$ is spanned by $\{\partial/\partial q^1, \ldots, \partial/\partial q^r\}$. The submanifold of TM locally characterized by

$$q^{r+1} = \text{const.}, \ldots, q^m = \text{const.},\ v^{r+1} = 0, \ldots, v^m = 0$$

is an integral manifold of $Im\, P_0^c$, since $Im\, P_0^c$ is spanned by

$$\{(\partial/\partial q^i)^c,\ (\partial/\partial q^i)^v;\ 1 \leq i \leq r\} = \{\partial/\partial q^i,\ \partial/\partial v^i;\ 1 \leq i \leq r\}$$

and it is clear that such submanifold is TS_0. Therefore $P_0^c C$ restricted to TS_0 is precisely the Liouville vector field on TS_0, say C_0, and now ξ_L/TS_0 is a 2nd order differential equation on TS_0. Thus, the initial Lagrangian L may be replaced by the Lagrangian $L \circ \phi$, where $\phi : TS_0 \longrightarrow TM$ is the induced embedding of $\phi_0 : S_0 \longrightarrow M$, i.e., $\phi = T\phi_0$. Therefore this restricted Lagrangian $L \circ \phi$ **is regular**. We finish the proof by showing that the corresponding dynamical equations are

$$\frac{d}{dt}\left(\frac{\partial L}{\partial \dot{q}^j}\right) - \frac{\partial L}{\partial q^j} = 0,\ 1 \leq j \leq r. \tag{8.32}$$

For this, we consider a path $\sigma : R \longrightarrow S_0$, $\sigma(t) = (q^j(t))$, of ξ_{L/TS_0}. Then the induced curve

$$\dot{\sigma}(t) = (q^j(t), \dot{q}^j(t))$$

in TS_0 is an integral curve of ξ_{L/TS_0}. From (8.28) we have

$$\sum_{i=1}^{r}(\partial^2 L/\partial \dot{q}^j \partial q^i)\dot{q}^i + \sum_{i=1}^{m}(\partial^2 L/\partial \dot{q}^i \partial \dot{q}^j)\xi_L(\dot{q}^i) - (\partial L/\partial q^j) = 0,$$

along $\dot{\sigma}(t)$. But from eq.(8.29) one has

$$\sum_{i=r+1}^{m} (\partial^2 L/\partial \dot{q}^i \partial \dot{q}^j)\dot{q}^i = 0.$$

Since $\xi_L(\dot{q}^i) = 0$ when $r+1 \leq i \leq m$ and $\xi_L(\dot{q}^i) = \ddot{q}^i$ when $1 \leq i \leq r$ (on TS_0) one deduces that

$$\sum_{i=1}^{v}\{(\partial^2 L/\partial \dot{q}^j \partial q^i)\dot{q}^i + (\partial^2 L/\partial \dot{q}^j \partial \dot{q}^i)\ddot{q}^i(\partial L/\partial q^j)\} = 0,$$

or, equivalently

$$\frac{d}{dt}(\partial L/\partial \dot{q}^j) - (\partial L/\partial q^j) = 0,\ 1 \leq j \leq r,$$

which are the Euler–Lagrange equations on TS_0.

This shows that ξ_L/TS_0 is a semispray on TS_0, solution of the corresponding Euler–Lagrange equations.

Let us now show how we obtain the corresponding canonical formalism for degenerate Lagrangians from the present point of view.

Suppose that (P_0, Q_0) is an integrable almost product structure on M. We may choose local coordinates (q^i) on M such that

$$Im\, P_0 =< \partial/\partial q^i/1 \leq i \leq r >,\ Im\, Q_0 =< \partial/\partial q^i,\ r+1 \leq i \leq m > .$$

Thus the complete lift $(P_0^{\bar{c}}, Q_0^{\bar{c}})$ of (P_0, Q_0) to T^*M is an integrable almost product structure on T^*M and we have

$$Im\, P_0^{\bar{c}} =< \partial/\partial q^i,\ \partial/\partial p_i;\ 1 \leq i \leq r >,$$

$$Im\, Q_0^{\bar{c}} =< \partial/\partial q^i,\ \partial/\partial p_i;\ r+1 \leq i \leq m >$$

(see Section 5.5).

(Here we put $F^{\bar{c}}$ to denote the complete lift of a tensor field F on M to T^*M).

As we have suppposing that ω_L has rank $2r$ and (P_0, Q_0) is an integrable almost product on M such that its complete lift (P_0^c, Q_0^c) to TM is adapted to ω_L, we have

$$Ker\, \omega_L = Im\, Q_0^c,$$

$$\omega_L = (\partial^2 L/\partial q^i \partial v^j) dq^i \wedge dq^j + (\partial^2 L/\partial v^i \partial v^j) dq^i \wedge dv^j.$$

Therefore, when $r+1 \leq k \leq m$, we have

$$0 = i_{\partial/\partial q^k}\omega_L = ((\partial^2 L/\partial v^k \partial q^j) - (\partial^2 L/\partial v^j \partial q^k))dq^j + (\partial^2 L/\partial v^j \partial v^k)dv^j,$$

$$0 = i_{\partial/\partial v^k}\omega_L = -(\partial^2 L/\partial v^j \partial v^k)dq^j.$$

Therefore

$$\left.\begin{array}{rcl} \partial^2 L/\partial v^k \partial q^j & = & \partial^2 L/\partial v^j \partial q^k, \\ \partial^2 L/\partial v^j \partial v^k & = & 0, \end{array}\right\} \tag{8.33}$$

$1 \leq j \leq m$, $r+1 \leq k \leq m$.

Thus, the Hessian matrix of L is

$$\begin{bmatrix} \partial^2 L/\partial v^i \partial v^j & 0 \\ 0 & 0 \end{bmatrix},$$

where $\det(\partial^2 L/\partial v^i \partial v^j) \neq 0$, $1 \leq i,j \leq r$.

Let $\alpha_L = d_J L$. Then α_L is semibasic and defines a map

$$Leg : TM \longrightarrow T^*M$$

such that

$$\pi_M \circ Leg = \tau_M \text{ and } Leg^*\omega_M = \omega_L.$$

In local coordinates, we have

$$Leg : (q^i, v^i) \longrightarrow (q^i, p_i), \; p_i = \partial L/\partial v^i.$$

One can easily see that

$$(T\,Leg)(\partial/\partial q^i) = \partial/\partial q^i + (\partial^2 L/\partial q^i \partial v^j)\partial/\partial p_j,$$

$$(T\,Leg)(\partial/\partial v^i) = (\partial^2 L/\partial v^i \partial v^j)\partial/\partial p_j.$$

Therefore, from (8.33), we see that the Jacobian matrix of Leg is

$$\left[\begin{array}{c|cc} I_r & \multicolumn{2}{c}{0} \\ \hline (\partial^2 L/\partial q^j \partial v^i) & (\partial^2 L/\partial v^i \partial v^j) & 0 \\ \cline{2-3} & 0 & 0 \end{array}\right].$$

Now, since $P_0^{\bar{c}}$ is locally given by

$$P_0^{\bar{c}}(\partial/\partial q^i) = \begin{cases} \partial/\partial q^i, & \text{if } 1 \leq i \leq r, \\ 0, & \text{if } r+1 \leq i \leq m \end{cases}$$

$$P_0^{\bar{c}}(\partial/\partial p_i) = \begin{cases} \partial/\partial p_i, & \text{if } 1 \leq i \leq r, \\ 0, & \text{if } r+1 \leq i \leq m \end{cases}$$

then we have

$$(P_0^{\bar{c}})^* \omega_M = \sum_{i=1}^{r} dq^i \wedge dp_i$$

and

$$Leg^*(P_0^{\bar{c}})^* \omega_M = (P_0^c)^* \omega_L.$$

Let us suppose now that there is a C^∞ map $H : T^*M \longrightarrow R$ such that $H \circ Leg = E_L$. Then there is a unique vector field $\bar{\xi}_H \in Im\, P_0^{\bar{c}}$ such that

$$i_{\bar{\xi}_H} \omega_M = (P_0^{\bar{c}})^*(dH).$$

This comes from the fact that $\omega/P_0^{\bar{c}}$ is symplectic, i.e., $(Im\, P_0^{\bar{c}},\ \omega/P_0^{\bar{c}})$ is a symplectic subbundle of $T(T^*M)$ (see Definition 5.1.14).

If S_0 is an integral manifold of $Im\, P_0$, then TS_0 is an integral manifold of $Im\, P_0^c$ and the cotangent bundle T^*S_0 is an integral manifold of $Im\, P_0^{\bar{c}}$. If we set

$$\varphi : TS_0 \longrightarrow TM,\ \psi : T^*S_0 \longrightarrow T^*M$$

for the corresponding embeddings, then $\varphi^* \omega_L$ and $\psi^* \omega_M$ are, respectively, symplectic forms on TS_0 and T^*S_0. Also:

- ξ_L/TS_0 is the Euler–Lagrange vector field associated to the regular Lagrangian $L \circ \varphi$ on TS_0;
- $\bar{\xi}_H/T^*S_0$ is the Hamiltonian vector field associated to the Hamiltonian $H \circ \psi$ on T^*S_0; and
- $Leg : TM \longrightarrow T^*M$ may be restricted to a map, denoted also by Leg from TS_0 to T^*S_0. This map is precisely the Legendre transformation determined by the Lagrangian $L \circ \varphi$.

From these considerations it is now easy to obtain the corresponding Hamiltonian form of the Euler–lagrange equations (8.32).

Example

Let us consider $M = R^2$, $TM = TR^2$ and the Lagrangian

$$L(x, y, u, v) = (1/2)(u^2 - x^2)$$

Then

$$d_J L = u dx, \; \omega_L = -dd_J L = dx \wedge du.$$

Therefore ω_L is a presymplectic form on TR^2 of rank 2. If we set

$$\xi = \xi^1(\partial/\partial x) + \xi^2(\partial/\partial y) + \bar{\xi}^1(\partial/\partial u) + \bar{\xi}^2(\partial/\partial v),$$

then

$$i_\xi \omega_L = \xi^1 du - \bar{\xi}^1 dx.$$

A direct computation shows that

$$E_L = CL - L = (1/2)(u^2 + x^2), \; dE_L = x dx + u du.$$

From the equality $i_\xi \omega_L = dE_L$ we deduce

$$\xi^1 = u, \; \bar{\xi}^1 = -x.$$

Then ξ assumes the general expression

$$\xi = u(\partial/\partial x) + \xi^2(\partial/\partial y) - x(\partial/\partial u) + \bar{\xi}^2(\partial/\partial v).$$

Now,

$$Ker\, \omega_L = \{Y/Y = Y^2(\partial/\partial y) + \bar{Y}^2(\partial/\partial v)\}.$$

So, let (P, Q) be an almost product structure on TR^2 adapted to ω_L, i.e.,

$$Ker\, \omega_L = Im\, Q = Ker\, P.$$

We have

$$Im\, P = \{Y/Y = Y^1(\partial/\partial x) + Y^2(\partial/\partial u)\}$$

Therefore P is given by

$$P(\partial/\partial x) = \partial/\partial x,\ P(\partial/\partial y) = 0,$$

$$P(\partial/\partial u) = \partial/\partial u,\ P(\partial/\partial v) = 0,$$

and then it is represented by the matrix

$$P = \begin{bmatrix} 1 & 0 & 0 & 0 \\ 0 & 0 & 0 & 0 \\ 0 & 0 & 1 & 0 \\ 0 & 0 & 0 & 0 \end{bmatrix}.$$

This shows that $PJ = JP$. In fact, P is the complete lift of the almost product structure (P_0, Q_0) on R^2 represented by the matrix

$$P_0 = \begin{pmatrix} 1 & 0 \\ 0 & 0 \end{pmatrix}.$$

Since $P^*(dE_L) = x\,dx + u\,du$, then the P–Euler–Lagrange vector field $\xi_L \in Im\,P$ such that

$$i_{\xi_L}\omega_L = P^*(dE_L)$$

is given by

$$\xi_L = u(\partial/\partial x) - x(\partial/\partial u).$$

Now, if we consider the straight line $\sigma(t) = (x(t), 0)$, we have on $T\sigma(t) = (x(t), 0, \dot{x}(t), 0)$ that

$$\xi_L = \dot{x}(\partial/\partial x) - x(\partial/\partial u).$$

As the image of the Liouville vector field C under P is precisely the Liouville vector field restricted to $T\sigma(t)$, i.e.,

$$C = \dot{x}(\partial/\partial u),$$

we have the second–order condition $J\xi_L = C$ along $T\sigma(t)$.

We observe that there exist degenerate Lagrangian for which this method don't works. For instance, consider the Lagrangian

$$L = (1/2)(yu - xv - x^2 - y^2).$$

Then

$$\omega_L = dx \wedge dy,$$

$$Ker\, \omega_L = \{Y/Y = \bar{Y}^1(\partial/\partial u) + \bar{Y}^2(\partial/\partial v)\}.$$

Suppose that there exists an adapted almost product structure (P, Q). Then $Im\, Q = Ker\, P = Ker\, \omega_L$. But $JP \neq PJ$. In fact

$$JX = J(X^1(\partial/\partial x) + X^2(\partial/\partial y) + \bar{X}^1(\partial/\partial u) + \bar{X}^2(\partial/\partial v))$$

$$= X^1(\partial/\partial u) + X^2(\partial/\partial v) \in Im\, Q$$

and so $P(JX) = 0$. As

$$Im\, P = \{Y/Y = Y^1(\partial/\partial x) + Y^2(\partial/\partial y)\},$$

we see that

$$J(PX) = X^1(\partial/\partial u) + X^2(\partial/\partial v) \neq P(JX).$$

8.2.3 Other approaches

In this section, we summarize two different approaches for the problem of finding simultaneously solutions of equations

$$i_X\omega_L = dE_L, \quad JX = C$$

for degenerate Lagrangians. The first study is due to Gotay and Nester [68] and consists in finding, under certain hypotheses, a constraint manifold S and a vector field X on S such that

$$(i_X\omega_L - dE_L)/_S = 0, \quad (JX - C)/_S = 0. \tag{8.34}$$

Definition 8.2.8 *A Lagrangian system (TM, ω_L, L) is said to be* **admissible** *if the leaf space $\widehat{TM}$ of the foliation $\mathcal{F}$ on TM defined by the integrable distribution $D = Ker\ \omega_L \cap V(TM)$ admits a manifold structure such that the canonical projection $p : TM \longrightarrow \widehat{TM}$ is a submersion.*

Suppose that P is the final constraint manifold associated to the Lagrangian system (TM, ω_L, L) by the algorithm. If the Lagrangian system is admissible then so is the constrained system in the following sense:

Proposition 8.2.9 *D restricts to an involutive distribution on TP such that $\mathcal{F}_P = \mathcal{F}/P$ foliates P. Furthermore, the leaf space $\hat{P} = P/\mathcal{F}_P$ is a manifold embedded in $\widehat{TM}$ and the induced projection $\pi : P \longrightarrow \hat{P}$ is a submersion.*

Proof: Let $Y \in D$ be a vertical vector field and Z a vector field such that locally $JZ = Y$. From (8.10), we have

$$L_Y(dE_L) = -d(\omega_L(C, Z))$$

and as C is vertical there is X such that locally $JX = C$. Thus

$$L_Y(dE_L) = d(\omega_L(X, JZ)) = d(\omega_L(X, Y)). \tag{8.35}$$

Now, the final constraint manifold P is obtained from a sequence of submanifolds

$$\dots \longrightarrow P_{i+1} \stackrel{\varphi_{i+1}}{\longrightarrow} P_i \longrightarrow \dots$$

and will show that $D/_P \subset TP$ by induction on the constraint manifolds P_i. Of course $D \subset T(TM) = TP_1$. Suppose that $D/_{P_i} \subset TP_i$. The constraint manifold P_{i+1} is characterized by the vector fields $Y \in D$ such that

$$Y(i_W dE_L)/P_{i+1} = 0,$$

for all $W \in (TP_i)^\perp$ (and so Y is tangent to P_{i+1}). Now

$$Y(i_W dE_L) = i_{[Y,W]} dE_L + i_W L_Y dE_L \tag{8.36}$$

and from (8.35), as $Y \in D \subset Ker\, \omega_L$, the second term in (8.36) vanishes.

Letting $Z \in TP_i$ be arbitrary, one has along P_i

$$\omega_L([Y, W], Z) = -L_Y \omega_L(W, Z) + L_Y[\omega_L(W, Z)]$$

$$-\omega_L(W_1[Y, Z]) = 0$$

by the assumptions on Y, Z and W. Consequently $[Y, W]/_{P_i} \in (TP_i)^\perp$ and so $(i_{[Y,W]} dE_L)/_{P_{i+1}} = 0$ which implies $Y(i_W dE_L) = 0$, i.e.,

$$D/_{P_{i+1}} \subset TP_{i+1}.$$

$D/_P$ gives rise to a foliation $\mathcal{F}_P$ of P. The leaf space space $\hat{P}$ of $\mathcal{F}_P$ can be identified with the image of P under the map $\pi = P \circ \varphi$, where $\varphi : P \longrightarrow TM$ is the embedding. Consequently, $\mathcal{F}_P$ inherits a manifold structure from $\mathcal{F}$ such that $\pi : P \longrightarrow \hat{P}$ is a submersion and $\varphi : \mathcal{F}_P \longrightarrow \mathcal{F}$ is an embedding. □

The main theorem of Gotay and Nester [68] is the following.

Theorem 8.2.10 *Let us consider an admisible Lagrangian system with final constraint manifold P embedded in TM. Then there exists at least one submanifold S of P and a unique (for fixed S) vector field X on S which simultaneously satisfies*

$$(i_X\omega_L - dE_L)/S = 0, \ (JX - C)/S = 0.$$

Every such manifold S is diffeomorphic to $\mathcal{F}_P$.

The proof of the theorem will be broken into two steps: the first consists in a local argument, in the sense that, there exists a unique point at each leaf of $\mathcal{F}_P$ at which every solution X satisfies $JX = C$. In the second step it is shown that these points define a submanifold of P diffeomorphic to $\mathcal{F}_P$ and that there exists a unique X solution of the Euler–Lagrange equations tangent to this submanifold

First step: Local existence and uniqueness

We start by supposing that $X \in TP$ is a solution of $(i_X\omega_L - dE_L)/_P = 0$ of the admissible system (TM, ω_L, L).

Definition 8.2.11 *The* **deviation vector field** W_Y *of a vector field Y is the vector field $W_Y = JY - C$.*

It is not hard to see that $i_J(i_Y\omega_L - dE_L) = i_{W_Y}\omega_L$ and if Y solves the Euler–Lagrange equations then $W_Y \in D/_P$.

A vector field X is **prolongable** if it projects to $\widehat{TM}$, the leaf space of the foliation $\mathcal{F}$ on TM defined by the involutive distribution $D = Ker\ \omega_L \cap V(TM)$. This means that $Tp(X) = \hat{X}$ is well–defined. Now, if one replaces "almost regular" by "admissible" and $(Leg(TM), \omega_1, dh_1)$ by $(\widehat{TM}, \widehat{\omega_L}, \widehat{dE_L})$ where $\omega_L = p^*\widehat{\omega_L}$, $dE_L = p^*(\widehat{dE_L})$ then a generalization of the equivalence Theorem suffices to show that prolongable solutions of $(i_X\omega_L - dE_L)/_P = 0$ always exist when (TM, ω_L, L) is an admissible system. More clearly, we have

Lemma 8.2.12 *There exists at least one solution $\hat{X}$ of the Euler–Lagrange equations*

$$(i_{\hat{X}}\hat{\omega}_L - d\hat{E}_L)/_{\hat{P}} = 0.$$

Any vector field $X \in T_p^{-1}\{\hat{X}\}$ will then solve the Euler–Lagrange equations.

Any vector field X is **semi–prolongable** if it is prolongable modulo $V(TM)$.

Proposition 8.2.13 *Let (TM, ω_L, L) be admissible and X a vector field on the final constraint manifold P. Suppose that X is a semi–prolongable solution of $(i_X\omega_L - dE_L)/_P = 0$. Then there exists a unique point in each leaf of $\mathcal{F}_P = \mathcal{F}/_P$ at which X is a semispray.*

Proof: For each point $z \in P$ we denote by $\mathcal{F}_P(z)$ the leaf of $\mathcal{F}_P$ through z. Our assertion is: $n_X = T\tau_M(X(z))$ is the required point. To show this it is necessary to prove that: (i) n_X is independent of the choice of z; (ii) η_X is in $\mathcal{F}_P(z)$; (iii) X is a semispray at η_X. Let us put $X = Y + Z$, as X is semi–prolongable (Y is the prolongable part of X and Z is vertical).

As D is vertical there is a map $\rho : T(\widehat{TM}) \longrightarrow TM$ such that the following diagram

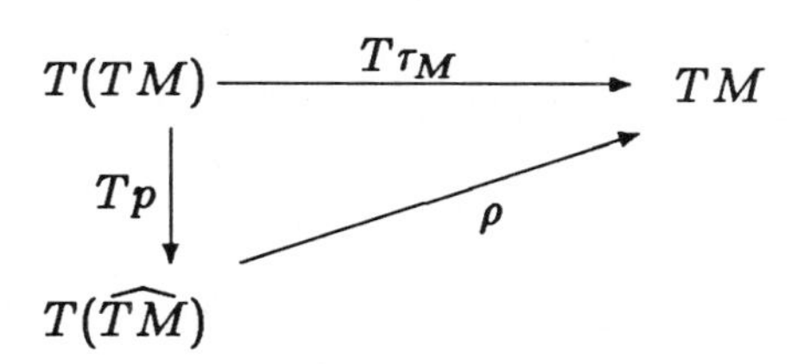

is commutative. Consequently,

$$T\tau_M(X(z)) = T\tau_M(Y(z)) = (\rho \circ Tp)(Y(z)).$$

But $TP(Y(z))$ is independent to the choice of $z \in \mathcal{F}_P(z)$ and so η_X does not depend on z.

To show (ii) we consider the vertical integral curves of $(-1)W_X$ and as $(-1)W_X \in D/_P$ if Y is a solution of $(i_X\omega_L - dE_L)/_P = 0$, these trajectories are contained in the leaves of $\mathcal{F}_P$. Locally

$$C(q,v) = v^i\partial/\partial v^i \text{ and } (JX)(q,v) = \eta_X^i(\partial/\partial v^i) \tag{8.37}$$

and the equation determining the integrals of $(-1)W_X$ is

$$dv^i(t)/dt = v^i(t) - \eta^i_X.$$

But, by (i) the functions η^i_X are constant on $\mathcal{F}_P(z)$ and the integral curve starting at $z = (q_0, v_0)$ is thus

$$v^i(t) = \eta^i_X + e^t(v^i_0 - \eta^i_X).$$

As $t \longrightarrow -\infty$, $v^i(t) \longrightarrow \eta^i_X$ so that η_X is a limit point of $v(t)$. But $v(t) \in \mathcal{F}_P(z)$ for all t and as p is continous one has $\mathcal{F}_P(z)$ closed. Consequently $\eta_X \in \mathcal{F}_P(z)$.

Finally, X is a semispray at η_X because $W_X(\eta_X) = 0$ and from (8.37) we see that X is a semispray only at η_X.□

Second step: Global existence and uniqueness

Two semi–prolongable solutions X, Y of $(i_X\omega_L - dE_L)/_P = 0$ are said to be J–**equivalent** if $J(X - Y) = 0$. We have, of course, an equivalence relation and J–equivalence classes of vector fields X are denoted by $[X]$. If follows that if $X, Y \in [X]$, then $W_X = W_Y$, so that η_X depends only upon $[X]$, denoted by $\eta_{[X]}$.

Now, let $S_{[X]}$ be the union of such $\eta_{[X]}$, for each leaf of $\mathcal{F}_P$. This set depends only upon $[X]$. The map

$$\alpha_{[X]} : \hat{P} \longrightarrow P$$

defined by

$$\alpha_{[X]}(\hat{z}) = T\tau_M(X(z))$$

is C^∞ and injective and it is independent of the choice of $z \in P^{-1}(\hat{z})$. Also, a computation in local coordinates suffices to show that $T\alpha_{[X]}$ is regular and, as $S_{[X]}$ is the image of $\hat{P}$ under the action of $\alpha_{[X]}$ one has that $S_{[X]}$ is a submanifold of P diffeomorphic to $\hat{P}$. On the other hand, if $[X] \neq [Y]$ then $S_{[X]} \neq S_{[Y]}$, i.e., each J–equivalence class $[X]$ of solutions of the Euler–Lagrange equations defines a unique $S_{[X]}$. Now, the solutions of $(i_X\omega_L - dE_L)/_P = 0$ are unique modulo vector fields in $Ker\ \omega_L \cap TP$; thus the set of all submanifolds $S_{[X]}$ is parametrized by $(ker\ \omega_L \cap TP)/V(TM)$.

Let X be a semi–prolongable solution of the Euler–Lagrange equations. Then X verifies

$$(i_X\omega_L - dE_L)/_{S_{[X]}} = 0,\ (JX - C)/_{S_{[X]}} = 0 \tag{8.38}$$

on a submanifold $S[X]$ of P but $X/_{S[X]}$ is not necessarily tangent to $S_{[X]}$. Thus we search a $X \in [X]$ verifying (8.38) and $X \in TS_{[X]}$. If we fix a prolongable solution X and if we set $\hat{X} = TP(X)$ by Lemma 8.2.12, $\hat{X}$ satisfies $(i_{\hat{X}}\hat{\omega}_L - d\hat{E}_L)/_{\hat{P}} = 0$. Now

$$T\alpha_{[X]}(\hat{X}) \in TS_{[X]}$$

and since $p \circ \alpha_{[X]} = Id/_{\hat{P}}$ we have

$$i_{T\alpha_{[X]}(\hat{X})}\omega_L/_{S[X]} = i_{T\alpha_{[X]}(\hat{X})}(p^\star\hat{\omega}_L)/_{S[X]}$$

$$= p^\star(i_{TP(T\alpha_{[X]}(\hat{X}))}\hat{\omega}_L/_{S[X]}$$

$$= p^\star(i_{\hat{X}}\hat{\omega}_L/_{\hat{P}}) = p^\star[dE_L/_{\hat{P}}]$$

$$= dE_L/_{S[X]}$$

by Lemma 8.2.12. As

$$Tp(X) = (Tp \circ T\alpha_{[X]})(\hat{X}),$$

we have

$$X - T\alpha_{[X]}(\hat{X}) \in D/_P$$

and hence it is vertical. Therefore

$$W_{T\alpha_{[X]}(\hat{X})} = W_X = 0$$

showing the following

Proposition 8.2.14 *The vector field $T\alpha_{[X]}(\hat{X})$ is tangent to $S_{[X]}$ and solves (8.38) simultaneously.*

Proposition 8.2.15 *There is a unique vector field $Y \in TS_{[X]}$ satisfying (8.35) simultaneously.*

Proof: If Y and Z are two solutions then $W_Y = W_Z = 0$ and so $J(Y - Z) = 0$. From the first equation of (8.38) we have $Y - Z \in Ker\ \omega_L$. Thus $Y - Z \in D/_P$ which gives a contradiction since $TS_{[X]} \cap D = 0$ and $\pi : P \longrightarrow \hat{P}$ is a diffeomorphism. □

Now, let us examine a second approach due to Cantrijn et al. [11].

Let $L : TM \longrightarrow R$ be a Lagrangian function. Then it is easy to see that L is regular if and only if $Ker\ \omega_L \cap V(TM) = 0$. Thus, if L is degenerate $Ker\ \omega_L \cap V(TM)$ is non–trivial. Furthermore, $Ker\ \omega_L \cap V(TM)$ defines an integrable distribution on TM and if we consider the restriction $J/_{Ker\ \omega_L}$ then $Ker\ (J/_{Ker\ \omega_L}) = Ker\ \omega_L \cap V(TM)$ and $J(Ker\ \omega_L) \subset Ker\ \omega_L \cap V(TM)$. Let E_L be the energy of L and take $Y \in S_\omega^{-1}(dE_L)$, i.e.,

$$i_Y \omega_L = dE_L.$$

From the equality

$$i_{JY} \omega_L = i_C \omega_L$$

we see that $X = JY - C \in Ker\ \omega_L$ and as $X \in V(TM)$ we have $X \in Ker\ \omega_L \cap V(TM)$. The following result is due to Cariñena et al. (1985).

Proposition 8.2.16 *Suppose $dim\ Ker\ \omega_L = 2\ dim\ (Ker\ \omega_L \cap V(TM))$. Then there is a vector field $Z \in Ker\ \omega_L$ such that $JZ = X$.*

Proof: The following diagram is commutative:

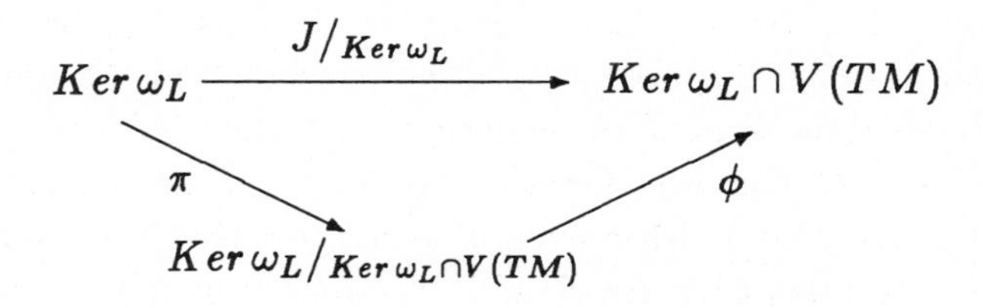

where ϕ is an injective homomorphism and π is the canonical projection. Now, in general

$$dim\ (Ker\ \omega_L) \leq 2dim\ (Ker\ \omega_L \cap V(TM))$$

and as we are supposing the equality, ϕ is also surjective. Thus the assertion is proved. □

The vector field $Y - Z$ verifies

$$J(Y - Z) = JY - JZ = JY - X = C$$

Thus, if we assume that the degenerate Lagrangian L admits a **global dynamics**, i.e., a globally defined vector field Y such that $i_Y \omega_L = dE_L$, then there is a semispray $\xi = Y - Z$ on TM which verifies $i_\xi \omega_L = dE_L$. In such a case, a general solution is any particular solution plus a vector field belonging to $Ker\ \omega_L$. Now, from the proof of Proposition 8.2.16, we have the following types of Lagrangians:

Type I: if $dim\ (Ker\ \omega_L) = dim\ (Ker\ \omega_L \cap V(TM)) = 0$,

Type II: if $dim\ (Ker\ \omega_L) = 2dim\ (Ker\ \omega_L \cap V(TM)) \neq 0$.

Type III: if $dim\ (Ker\ \omega_L) < 2dim\ (Ker\ \omega_L \cap V(TM))$.

Clearly, type I Lagrangians are regular Lagrangians. We consider in the rest of this section only type II Lagrangians. In such a case, if the Lagrangian L admits a global dynamics - which will be of type $A + B$, A being any particular solution and $B \in Ker\ \omega_L$ - then there exists a semispray ξ which is a global solution. We also assume that all degenerate Lagrangians of type II admits a global dynamics. The third assumption will concern $Ker\ \omega_L$. Consider the quotient space

$$\widehat{TM} = TM/_{Ker\ \omega_L}.$$

If we assume that the foliation defined by $Ker\ \omega_L$ is a fibration with projection $\hat{\pi} : TM \longrightarrow \widehat{TM}$ then there is a unique symplectic form $\hat{\omega}_L$ on $\widehat{TM}$ such that $\omega_L = \hat{\pi}^* \hat{\omega}_L$ (for a proof see Guillemin and Sternberg [74], Theorem 2.5.2). Suppose that $Ker\ \omega_L$ verifies such condition. Then it is possible to show that J projects onto a tensor field of type $(1, 1)$ $\hat{J}$ on $\hat{T}M$ if $Im(L_Z J) \subset Ker\ \omega_L$, for all $Z \in Ker\ \omega_L$ (we call this condition by the **kernel condition**). Moreover, $\hat{J}$ is an integrable almost tangent structure on $\widehat{TM}$, the subspaces $Im\ \hat{J}_x$ are Lagrangian with respect to $\hat{\omega}_L(x)$, for each $x \in \widehat{TM}$ and if ξ is the above semispray then the projected vector field $\hat{\xi}$ verifies

$$L_{\hat{\xi}} \hat{\omega}_L = 0,\ L_{\hat{\xi}} \hat{J} \circ \hat{J} = \hat{J},\ \hat{J} \circ L_{\hat{\xi}} \hat{J} = -\hat{J}.$$

Therefore (see Remark 7.4.4) we have

Theorem 8.2.17 *(Cantrijn et al. [11]). Let $L : TM \longrightarrow R$ be a degenerate Lagrangian of type II, with a global dynamics, such that the foliation* *defined by $Ker\ \omega_L$ is a fibration. Suppose that $Ker\ \omega_L$ verifies the kernel condition. Then $(\widehat{TM}, \hat{J}, \hat{\xi})$ is a regular Lagrangian system (called "**reduced Lagrangian system**").*

Now, let $\hat{L}$ be a (local Lagrangian) for $\hat{\xi}$. Putting $L_1 = \hat{L} \circ \hat{\pi}$, we can easily prove that the Lagrangian L_1 is **gauge (locally) equivalent** to the original L in the sense that they determine the same dynamics.

Remark 8.2.18 Actually, we can relate this approach with the one given in Section 8.2.2. From the assumptions in Section 8.2.2, one has a global solution for the Euler–Lagrange equations. Moreover, the assumption that $JP = PJ$ (and hence $JQ = QJ$) is equivalent to say that L is of type II. In fact

$$Im\, Q = Ker\ \omega_L$$

implies that

$$Im\ (J/_{Ker\ \omega_L}) = J(Ker\ \omega_L) = J(Im\ Q)$$

$$= Im\, Q \cap V(TM) = Ker\ \omega_L \cap V(TM).$$

Furthermore, the assumption that (P, Q) is integrable is equivalent to the assumption that $Ker\ \omega_L$ defines a fibration.

8.3 Exercises

In Section 7.15 we have examined the set of 1–forms $\wedge^1_\xi$, where ξ is a semispray. If we associate to each ξ an appropriate 1–form $\alpha \in \wedge^1_\xi$ and if $\alpha = dL$, for some regular function L then we obtain the Euler–Lagrange equations. Now, let (P, Q) be an almost product structure on TM which commutes with J and ξ a P–semispray (see Definition 8.2.4). We set

$$E_\xi = I_d - L_\xi \circ J^\star; \quad \tilde{E} = P^\star \circ E_\xi.$$

$$\wedge^1_\xi = Ker\ E_\xi; \quad \tilde{\wedge}^1_\xi = Ker\ \tilde{E}_\xi.$$

A form $\alpha \in \tilde{\wedge}^1_\xi$ is called P–regular if $\omega_\alpha = -d(J^\star\alpha)$ is presymplectic and P is adapted to ω_α and ξ is said to be P–regular Euler–Lagrange vector field if there is a P–regular $\alpha \in \tilde{\wedge}^1_\xi$ such that $\alpha = dL$, (for what follows we refer sections 7.15 and 8.2.2).

8.3.1 Show that such L is degenerate and that the motion equation associated to L is $P^\star(L_\xi(J^\star(dL))) = P^\star(dL)$. Show that this last equation is $i_\xi \omega_L = P^\star(dE_L)$. Thus, if P is as in Theorem 8.2.7 one obtains the regular equations of motion for L.

8.3.2 Compute the local expressions of $E_\xi \alpha$ and $\tilde{E}_\xi \alpha$ (*Hint*: as $J\xi = PC$, take $\xi = PX$, where X is a vector field such that $JX = C$. Use the local expression of P given in p. 415 to show that $X^i P^j_i = v^i P^j_i$, where X^i is the $(\partial/\partial q^i)$–component of X and v^i the $(\partial/\partial v^i)$–component of C.

8.3.3 Suppose that R is a tensor field of type (1,1) on TM which commutes with J, P and satisfies $J(L_\xi R) = 0$. Show that $R^\star \circ E_\xi = E_\xi \circ R^\star$, $R^\star \circ \tilde{E}_\xi = \tilde{E}_\xi \circ R^\star$ and that R preserves $\wedge^1_\xi$ and $\tilde{\wedge}^1_\xi$.

8.3.4 Let P be the complete lift to TM of an integrable almost product structure P_0 on M. Let A_0 be a tensor field of type (1,1) on M which commutes with P_0. Show that A_0^c commutes with J and P_0^c and satisfies $J(L_\xi A_0^c) = 0$. Hence $(A_0^c)^\star$ preserves $\wedge^1_\xi$ and $\tilde{\wedge}^1_\xi$.

8.3.5 Let α be a P–regular element of $\tilde{\wedge}^1_\xi$. Prove that

$$i_\xi P^\star(\alpha - d < PC, \alpha >) = 0.$$

Show that this equation represents a generalization of the usual law of conservation of energy (*Hint* : $E_L = i_{PC} dL - L \Longrightarrow -dE_L = dL - d < PC, dL >$, if ξ is a P–regular Euler–Lagrange vector field and $\alpha = dL$. Show that E_L is constant along the integrals of ξ).

8.3.6 Let P as in Exercise 8.3.4. Let $f : M \longrightarrow R$ be a smooth function. Prove that, if $\alpha \in \tilde{\wedge}^1_\xi$ then $\alpha' = \alpha + d(\xi f) \in \tilde{\wedge}^1_\xi$.

8.3.7 Suppose P is as in Exercise 8.3.4. Let Y be a Lie symmetry of ξ such that $L_Y P_0 = 0$, P_0 being such that $P_0^c = P$. Show that L_{Y^c} preserves $\tilde{\wedge}^1_\xi$. Thus, if $P_0 X = Y$ for some vector field X on M show that

$$L_{(P_0 X)^c} \tilde{\wedge}^1_\xi \subset \tilde{\wedge}^1_\xi$$

8.3.8 Let α be a P–regular element of $\tilde{\wedge}^1_\xi$ and Y a vector field on TM such that

$$(1) \quad P^\star L_Y(J^\star \alpha) = P^\star(df), \; f : TM \longrightarrow R$$

$$(2) \quad i_Y(P^\star(\alpha - d < PC, \alpha >)) = 0.$$

Show that $F = f - < Y, J^\star\alpha >$ is a first integral of ξ. Conversely, prove that to each first integral F of ξ there corresponds a vector field satisfying (1) and (2).

<u>8.3.9</u> Show that, if L is degenerate then $Ker\ \omega_L \cap V(TM)$ defines an integrable distribution on TM. Show that dim $(Ker\ \omega_L) \leq 2$ dim $(Ker\ \omega_L \cap V(TM))$.

<u>8.3.10</u> Suppose that the foliation defined by Ker ω_L is a fibration with projection $\hat{\pi} : TM \longrightarrow TM/\text{Ker } \omega_L$. Show that J projects onto a tensor field of type (1,1) $\hat{J}$ on $TM/\text{Ker } \omega_L$ if $Im(L_Z J) \subset Ker\ \omega_L$ for all $Z \in Ker\ \omega_L$. Prove that if $\hat{J}$ is integrable almost tangent structure then $Im\ \hat{J}_z$ are Lagrangian with respect to $\hat{\omega}_L(z)$, $z \in TM/\text{Ker } \omega_L$.

<u>8.3.11</u> Let g be a degenerate metric on a differentiable m–dimensional manifold M. Let N be the distribution defined by

$$N_x = \{X \in T_xM;\ g_x(X,Y) = 0, \text{ for all } Y \in T_xM\}$$

and L defined by $L(v) = (1/2)g_x(v,v)$, $v \in T_xM$, $x \in M$. Suppose that the elements of N are Killing vector fields for g. Show that

$$\text{Ker}\, \omega_L = N^c.$$

Appendix A

A brief summary of particle mechanics in local coordinates

A.1 Newtonian Mechanics

A.1.1 Elementary principles

Let us consider a single particle and let us denote by $\vec{r}$ the position with respect to the canonical referencial $(\vec{i}, \vec{j}, \vec{k})$, from the origin to the particle. Suposing that $\vec{r}$ is a function depending on the time t, **Newton's second law of motion** says that

$$\vec{F} = m(d^2/dt^2)\vec{r} = (d/dt)\vec{p}, \tag{A.1}$$

where $\vec{F}$ is the **total force** on the particle, m is a constant (the mass of the particle) and $p = m(d/dt)\vec{r}$ is the (linear) momentum. The forces are supposed dependent only on the time, position and velocity. Eq.(A.1) may be extended to a mechanical system involving a great number of particles, say N. If $\{P_1, \ldots, P_N\}$ is a family of particles, then for each α, $1 \leq \alpha \leq N$, Eq.(A.1) takes the following vectorial form:

$$\vec{F}^{\alpha} = m^{\alpha}(d^2/dt^2)\vec{r}^{\alpha} = (d/dt)\vec{p}^{\alpha}, \tag{A.2}$$

where $\vec{r}$ is the position vector of P_{α}, with mass m^{α} and $\vec{p}^{\alpha} = m^{\alpha}(d/dt)\vec{r}$.

We can express the above equations in its Cartesian form. If

$$(x_1^\alpha(t), x_2^\alpha(t), x_3^\alpha(t)) = (x_1^\alpha, x_2^\alpha, x_3^\alpha) \tag{A.3}$$

denotes the position of P_α at the instant t, then the motion equations take the form

$$F_\ell^\alpha = m_\ell^\alpha(d^2/dt^2)x_\ell^\alpha = m_\ell^\alpha \ddot{x}_\ell^\alpha$$

$$= (d/dt)p_\ell^\alpha = \dot{p}_\ell^\alpha;\ 1 \leq \ell \leq 3,\ 1 \leq \alpha \leq N \tag{A.4}$$

where the dots mean the time derivatives and $m_1^\alpha, m_2^\alpha, m_3^\alpha$ represents the mass of P_α under the condition $m_1^\alpha = m_2^\alpha = m_3^\alpha$.

The mechanical system $\{P_1, \ldots, P_N\}$ of N–particles in the Euclidean space R^3 can be considered as being a single particle P in the Euclidean space R^{3N} (in fact, this is not correct: we must take into account that two or more particles cannot occupy the same position at the same time; thus the position of the system is given by $R^{3N} - W$, where W is defined as the set of coordinates for which the particles have same position. Also some constraint relations like the system is restricted to a sphere may be considered. Having in mind these restrictions we now assume that P is in all R^{3N}). For this, we represent the coordinates (A.3) of P_α by

$$(x_{3\alpha-2}, x_{3\alpha-1}, x_{3\alpha}) \tag{A.5}$$

that is, the initial position of the system in R^3 is represented in R^{3N} by

$$(x_1, x_2, x_3, \ldots, x_{3N-2}, x_{3N-1}, x_{3N}).$$

Let us put $k = 3N$. Then the fundamental equations (A.4) are expressed in R^k by

$$F_i = m_i \ddot{x}_i,\ \ 1 \leq i \leq k, \tag{A.6}$$

where $F_1, \ldots, F_k$ are forces acting on the particle. More clearly, the force F_i is acting on some particle P_α in the sense that for each i there is $\alpha \in \{1, \ldots, N\}$ such that $i \in \{3\alpha - 2, 3\alpha - 1, 3\alpha\}$.

As the particle P represents the initial system of N particles, the fundamental equation in R^k, equivalent to eq.(A.1), is

$$\sum_{i=1}^{k}(F_i - m_i(d^2/dt^2)x_i) = 0,$$

that is,

$$\sum_{i=1}^{k}(F_i - m_i\ddot{x}_i) = 0 \tag{A.7}$$

Theorem A.1.1 (Momentum conservation law) *If the sum of all forces acting on the mechanical system $\{P_1, \ldots, P_N\}$ (or P) vanishes, then the total momentum of the system is conserved.*

In fact, $p_i = m_i\dot{x}_i = m_i(d/dt)x_i$ and as we are assuming that m_i is invariant with respect to the time t, $p_i = (d/dt)(m_i x_i)$. So, as $\sum F_i = 0$, one has $\sum p_i =$ constant.

A.1.2 Energies

k–forces $F_1, \ldots, F_k$ acting on a particle $P \in R^k$ are said to be **conservatives** if there is a diferentiable mapping $f : R^k \longrightarrow R$, at least of class C^2, such that $F_i = (\partial f/\partial x_i)$, $1 \leq i \leq k$. The **potential energy** V of P is defined by $V = -f+$ constant. Therefore, a force is **conservative** if $F_i = -(\partial V/x_i)$. The potential energy $V : R^k \longrightarrow R$ is a function depending explicitly on the position coordinates and implicity on the time. So

$$dV = \sum_{i=1}^{k} (\partial V/\partial x_i) dx_i,$$

and

$$- dV = \sum_{i=1}^{k} F_i dx_i. \tag{A.8}$$

The integral

$$\oint F_i dx_i \tag{A.9}$$

vanishes. Reciprocally, the existence of a function V (at least of class C^2) verifying (A.8) is determined by condition (A.9). This is a consequence of Green's Theorem and the value of V at a point x is the value of the integral

$$-\int_{x_0}^{x} \sum_{i=1}^{k} F_i d\xi_i = V(x)$$

along an arbitrary path from x_0 to x.

The kinetic energy of $P \in R^k$, at time t, is defined by the expression

$$T = (1/2)\sum_{i=1}^{k} m_i(\dot{x}_i)^2,$$

From this definition we have

$$\frac{dT}{dt} = \sum_{i=1}^{k} m_i \dot{x}_i \ddot{x}_i = \sum_{i=1}^{k} F_i \dot{x}_i, \tag{A.10}$$

called the **1st form of the energy equation.**

Theorem A.1.2 (Energy Conservation Law) *Let us suppose that on P, act only conservative forces. Then there is a constant function E, called the* **"total energy"** *so that*

$$T(t) + V(t) = E, \text{ for all } t \in R \tag{A.11}$$

In fact, the time derivative of T and V gives

$$\frac{dT}{dt} = \sum m_i \dot{x}_i \ddot{x}_i; \quad \frac{dV}{dt} = \sum \frac{\partial V}{\partial x_i} \dot{x}_i = -\sum F_i \dot{x}_i,$$

and so

$$0 = \sum (m_i \ddot{x}_i - F_i)\dot{x}_i = \frac{d}{dt}(T + V) = \frac{d}{dt} E.$$

In general, we will say that a mechanical system is **conservative** if equality (A.11) holds. This expression is called the **2nd form of the energy equation**. If the system is not conservative, then we may have the 1st form if all forces aren't conservatives. However, if there are some conservative forces, then we put

$$F_i = F_i^c + F_i^{\bar{c}}$$

where F_i^c = conservative force and $F_i^{\bar{c}}$ = non–conservative force. The 1st form takes the expression

$$\frac{dT}{dt} = \sum_i F_i \dot{x}_i = \sum_i F_i^c \dot{x}_i + \sum_i F_i^{\bar{c}} \dot{x}_i = -\frac{dV}{dt} + \sum_i F_i^{\bar{c}} \dot{x}_i,$$

and so,

$$\frac{d}{dt}(T+V) = \sum_i F_i^{\bar{c}} \dot{x}_i,$$

called the **3rd form of the energy equation.**

Up to now we have supposed no restrictions on the motion of the particle. However, in general, there are some restrictions, like, for example, a particle moving on a plane, sphere, etc. We call these restrictions by **constraints**. We will be concerned with only a type of constraints, called **holonomic**, which are characterized by equations of type

$$g_j(x_i) = 0, \quad 1 \leq j \leq r,$$

depending only on the position.

A.2 Classical Mechanics: Lagrangian and Hamiltonian formalisms

A.2.1 Generalized coordinates

Suppose that P is a mechanical system on R^{3N} and the number of holonomic constraints is r. Then the family of functions

$$f_j(x_1, \ldots, x_{3N}) = 0, \quad 1 \leq j \leq r$$

which characterize the constraint relations define a subspace C of R^{3N} of dimension $m = 3N - r$. We call C the **configuration space** (it is assumed to be an embedded manifold in R^{3N}). Locally C is characterized by a family of functions

$$q^A : V \longrightarrow R, \quad 1 \leq A \leq m$$

where V is an open neighborhood of R^{3N} such that the Jacobian of $q^A(x_1, \ldots, x_{3N})$ with respect to the x's is non–vanishing at every point of $V \cap C$. In the following we will identify C with the Euclidean space R^m. We call the functions q^A, $1 \leq A \leq m$, by **generalized coordinates** of the system P. Thus the introduction of the generalized coordinates allows us consider a new set of independent coordinates in terms of which the old coordinates are expressed by the relations

$$x_i = x_i(q^1, \ldots, q^m), \ 1 \leq i \leq 3N$$

as we are supposing the regularity of the Jacobian of q^A with respect to the x's. Thus, we have

$$\frac{dx_i}{dt} = \dot{x}_i = \sum_{A=1}^{m} \frac{\partial x_i}{\partial q^A} \dot{q}^A + \frac{\partial x_i}{\partial t}, \ 1 \leq i \leq 3N$$

and so

$$\frac{\partial \dot{x}_i}{\partial \dot{q}^B} = \frac{\partial x_i}{\partial q^B}, \ 1 \leq B \leq m \tag{A.12}$$

The derivation of the kinetic energy T with respect to $\dot{q}^B$ (resp. q^B) gives

$$\frac{\partial T}{\partial \dot{q}^B} = \sum_{i=1}^{3N} m_i \dot{x}_i \frac{\partial \dot{x}_i}{\partial \dot{q}^B} = \sum_{i=1}^{3N} m_i \dot{x}_i \frac{\partial x_i}{\partial q^B} \tag{A.13}$$

$$\frac{\partial T}{\partial q^B} = \sum_{i=1}^{3N} m_i \dot{x}_i \frac{\partial \dot{x}_i}{\partial q^B} = \sum_{i=1}^{3N} m_i \dot{x}_i \frac{d}{dt} \left(\frac{\partial x_i}{\partial q^B} \right) \tag{A.14}$$

where the last expression in (A.14) is obtained from

$$\frac{d}{dt} \left(\frac{\partial x_i}{\partial q^B} \right) = \sum_{C=1}^{m} \frac{\partial}{\partial q^C} \left(\frac{\partial x_i}{\partial q^B} \right) \dot{q}^C + \frac{\partial}{\partial t} \left(\frac{\partial x_i}{\partial q^B} \right)$$

$$= \frac{\partial}{\partial q^B} \left(\sum_{C=1}^{m} \frac{\partial x_i}{\partial q^C} \dot{q}C + \frac{\partial x_i}{\partial t} \right) = \frac{\partial \dot{x}_i}{\partial q^B}.$$

Now, if we derive (A.14) with respect to the time t one has

$$\frac{d}{dt} \left(\frac{\partial T}{\partial \dot{q}^B} \right) = \sum_{i=1}^{3N} m_i \ddot{x}_i \frac{\partial x_i}{\partial q^B} + \sum_{i=1}^{3N} m_i \dot{x}_i \frac{d}{dt} \left(\frac{\partial x_i}{\partial q^B} \right)$$

Taking into account (A.14) one obtains

$$\frac{d}{dt} \left(\frac{\partial T}{\partial \dot{q}^B} \right) - \frac{\partial T}{\partial q^B} = \sum_{i=1}^{3N} m_i \ddot{x}_i \frac{\partial x_i}{\partial q^B} \stackrel{def}{=} Q_B \tag{A.15}$$

since

$$Q_B = \sum_{i=1}^{3N} F_i \frac{\partial x_i}{\partial q^B}.$$

If the mechanical system is conservative we have

$$Q_B = -\sum_{i=1}^{3N} \frac{\partial V}{\partial x_i}\frac{\partial x_i}{\partial q^B} = -\frac{\partial V}{\partial q^B}$$

and so (A.15) takes the form

$$\frac{d}{dt}\left(\frac{\partial T}{\partial \dot{q}^B}\right) - \frac{\partial}{\partial q^B}(T - V) = 0.$$

A.2.2 Euler–Lagrange and Hamilton equations

As the potential energy is supposed to be a function depending only on position (and time) we have $(\partial V/\partial \dot{q}^B) = 0$ and the above expression is

$$\frac{d}{dt}\left(\frac{\partial}{\partial \dot{q}^B}(T - V)\right) - \frac{\partial}{\partial q^B}(T - V) = 0.$$

Setting $L = T - V$, we have

$$\frac{d}{dt}\left(\frac{\partial L}{\partial \dot{q}^B}\right) - \frac{\partial L}{\partial q^B} = 0 \qquad (A.16)$$

known by the name of **Euler–Lagrange equations** for a conservative mechanical system. The function $L = L(t, q^A, \dot{q}^A)$, $1 \leq A \leq m$, is called the **Lagrangian** of the system. In the non–conservative situation (A.16) assumes the form

$$\frac{d}{dt}\left(\frac{\partial L}{\partial \dot{q}B}\right) - \frac{\partial L}{\partial q^B} = K_B$$

with $K_B = Q_B - (\partial V/\partial q^B)$.

In the following we suppose that L is a function depending implicitly in t, i.e., $L : R^m \times R^m \longrightarrow R$. Let us set

$$v^A = \dot{q}^A; \quad L_A = (\partial L/\partial v^A)(q^B, v^B)$$

and assume that

$$\left(\frac{\partial L_A}{\partial v^B}\right), \quad 1 \leq B \leq m, \quad 1 \leq A \leq m$$

is non–singular at a point $(\bar{q}^B, \bar{v}^B)$. Consider the Euclidean space $R^m \times R^m \times R^m$ with coordinates (q^A, v^A, p^A). For $(\bar{q}^B, \bar{v}^B)$, let

$$L_A(\bar{q}^B, \bar{v}^B) = \bar{p}^A, \quad 1 \leq B \leq m, \quad 1 \leq A \leq m \tag{A.17}$$

and define the mappings

$$F_A(q^B, v^B, p^A) = L_A(q^B, v^B) - p^A, \quad 1 \leq A \leq m$$

Setting

$$G(q^B, v^B, p^B) = (F_1(q^B, v^B, p^1), \ldots, F_m(q^B, v^B, p^m))$$

we have from (A.17)

$$G(\bar{q}^B, \bar{v}^B, \bar{p}^B) = (0, \ldots, 0).$$

On the other hand the matrix $(\partial G/\partial v^A)(q^B, v^B, p^B)$ is

$$\begin{pmatrix} \frac{\partial F_1}{\partial v^1} & \cdots & \frac{\partial F_1}{\partial v^m} \\ \vdots & & \vdots \\ \frac{\partial F_m}{\partial v^1} & \cdots & \frac{\partial F_m}{\partial v^m} \end{pmatrix}_{(q,v,p)} = \begin{pmatrix} \frac{\partial L_1}{\partial v^1} & \cdots & \frac{\partial L_1}{\partial v^m} \\ \vdots & & \vdots \\ \frac{\partial L_m}{\partial v^1} & \cdots & \frac{\partial L_m}{\partial v^m} \end{pmatrix}_{(q,v)}$$

which is non–singular at $(\bar{q}^B, \bar{v}^B, \bar{p}^B)$. Thus, from the Implicit function theorem there is locally a unique function $(\psi_1, \ldots, \psi_M)$ such that

$$\psi_A(q^B, p^B) = v^A$$

So, there are m–cordinates $(p^1, \ldots, p^m)$ such that the velocities $v^1, \ldots, v^m$ are locally dependent functions of $(q^1, \ldots, q^m, p^1, \ldots, p^m)$. In the following we suppose that such result is valid at least on an open subset U of $R^m \times R^m$. Furthermore, we will set (see(A.17))

$$p^A = \frac{\partial L}{\partial v^A}, \quad 1 \leq A \leq m. \tag{A.18}$$

These coordinates are called **generalized momenta**. Hence, the introduction of such coordinates is a consequence of the non–degeneracy of the Hessian matrix of L with respect to the velocities, (at least on an open neighborhood of $R^m \times R^m$). In such a case we say that L is **regular** (if the non–degeneracy property is valid on all $R^m \times R^m$ then we say that L is **hyperregular**).

Let us consider again the (conservative) Euler–Lagrange equations:

$$\frac{d}{dt}\left(\frac{\partial L}{\partial \dot{q}^A}\right) - \frac{\partial L}{\partial q^A} = 0, \quad 1 \leq A \leq m.$$

If we develop this expression we obtain

$$\sum_{A=1}^{m} \frac{\partial^2 L}{\partial \dot{q}^A \partial \dot{q}^B}\ddot{q}^B + \frac{\partial^2 L}{\partial \dot{q}^A \partial q^B}\dot{q}^B + \frac{\partial^2 L}{\partial \dot{q}^A \partial t} - \frac{\partial L}{\partial q^A} = 0$$

i.e., the above equations are of type

$$\sum_{B=1}^{m} a_{AB}\ddot{q}^B = f_A(q^B, \dot{q}^B) \tag{A.19}$$

where

$$(a_{AB}) = \left(\frac{\partial^2 L}{\partial \dot{q}^A \partial \dot{q}^B}\right).$$

Now, let us set $\dot{q}^A = r_A$. Then we may replace in an equivalent way the m Euler–Lagrange equations (A.19) by the $2m$ differential equations of first–order

$$\dot{q}^A = r_A, \; \dot{r}_A = f_A(q^B, r^B) \tag{A.20}$$

This is a well–known method to search solutions of (A.19). However we will develop a different view point which consists in the utilization of new coordinates due to Hamilton. For this, consider the transformation

$$(q^A, \dot{q}^A) \longrightarrow (q^A, p^A)$$

which is at least a local diffeomorphism and (A.18) permits to express the velocities $\dot{q}$'s in terms of the momenta p's. Conversely, the inverse of the above transformation gives the momenta in terms of the positions and velocities.

To give such inverse consider the function

$$H = \sum_{A=1}^{m} p^A \dot{q}^A - L \tag{A.21}$$

This function depends on $(q^A, \dot{q}^A, p^A)$ but as we are supposing L regular

$$H = H(q^A, \varphi^A(q^B, p^B),\ p^A) = H(q^A, p^A)$$

Thus

$$\frac{\partial H}{\partial p^B} = \dot{q}^B,\ \ 1 \leq B \leq m$$

gives the regularity condition for the p's be functions of the q's and $\dot{q}$'s.

The function $H : R^m \times R^m \longrightarrow R$ defined by (A.21) is called **Hamiltonian function** of the mechanical system. The space of the variables (q^A, p^A) is called **phase space** of the mechanical system. Now, consider the differential of (A.21):

$$dH = \sum_{A=1}^{m} (p^A d\dot{q}^A + \dot{q}^A dp^A) - \frac{\partial L}{\partial q^A} dq^A - \frac{\partial L}{\partial \dot{q}^A} d\dot{q}^A$$

$$= \sum_{A=1}^{m} \dot{q}^A dp^A - \frac{\partial L}{\partial q^A} dq^A \tag{A.22}$$

On the other hand

$$dH = \sum_{A=1}^{m} \frac{\partial H}{\partial q^A} dq^A + \frac{\partial H}{\partial p^A} dp^A. \tag{A.23}$$

Thus, (A.22) and (A.23) give

$$\frac{\partial H}{\partial p^A} = \dot{q}^A;\ \ \frac{\partial H}{\partial q^A} = -\frac{\partial L}{\partial q^A} \tag{A.24}$$

but from the Euler–Lagrange equations we have

$$\frac{\partial L}{\partial q^A} = \frac{d}{dt}\left(\frac{\partial L}{\partial \dot{q}^A}\right) = \frac{d}{dt}(p^A) = \dot{p}^A$$

and so (A.24) is

$$\frac{\partial H}{\partial p^A} = \dot{q}^A;\ \ \frac{\partial H}{\partial q^A} = -\dot{p}^A. \tag{A.25}$$

These equations are called **Hamilton equations, canonical form of Euler–Lagrange equations** or simply **canonical equations**.

We have used above an important mapping called **Legendre transformation**. It can be considered as a transformation which associates a regular function L defined on $R^m \times R^m$ a function H from $R^m \times R^m$ to R defined by (A.21). We represent by $Leg_L : R^m \times R^m \longrightarrow R^m \times R^m$ the map $(q^A, \dot{q}^A) \longrightarrow (q^A, p^A)$, where p^A is given by (A.18) and the inverse $(Leg_L)^{-1} : (q^A, p^A) \longrightarrow (q^A, \dot{q}^A)$ where

$$\dot{q}^A = \frac{\partial H}{\partial p^A}$$

with $L = \sum p^A \dot{q}^A - H$, i.e., $(Leg_L)^{-1}$ is Leg_H.

Let us consider a holonomic conservative mechanical system and suppose $H = H(t, q^A(t), p^A(t))$. Then

$$\frac{d}{dt}H = \frac{\partial H}{\partial t} + \sum_{A=1}^{m} \frac{\partial H}{\partial q^A} d\dot{q}^A + \sum_{A=1}^{m} \frac{\partial H}{\partial p^A} d\dot{p}^A$$

$$= \frac{\partial H}{\partial t} + \sum_{A=1}^{m} (-\dot{p}^A)\dot{q}^A + \sum_{A=1}^{m} \dot{q}^A \dot{p}^A \qquad \text{(from (A.25)}$$

and so

$$\frac{d}{dt}H = \frac{\partial H}{\partial t} = -\frac{\partial L}{\partial t}$$

Thus, if H does not depend explicitly on t, H assumes the same value along the motion of the system, i.e., $H =$ constant. If we consider the kinetic energy

$$T = (1/2) \sum_A m_A (\dot{q}^A)^2$$

as

$$p^A = \frac{\partial L}{\partial \dot{q}^A} = \frac{\partial T}{\partial \dot{q}^A} = \sum_A m_A \dot{q}^A$$

$$2T = \sum_A m_A \dot{q}^A \dot{q}^A = \sum_A p^A \dot{q}^A$$

we have

$$H = 2T - L = 2T - T + V = T + V = \text{constant} = E,$$

the total energy of the system. Thus a sufficient condition for the Hamiltonian be the total energy of the mechanical system is: (a) the system is conservative; (b) the Lagrangian (resp. Hamiltonian) is only implicitly dependent on the time. The converse is not generally true (see Goldstein for example).

Finally a **first integral** of a system of ordinary differential equations is a function which is constant along every integral curve (solution) of the system. Let $\varphi = \varphi(q^A, p^A)$ be a function. For every integral curve $(q^A(t), p^A(t))$ we have

$$\frac{d\varphi}{dt} = \frac{\partial \varphi}{\partial q^A}\frac{dq^A}{dt} + \frac{\partial \varphi}{\partial p^A}\frac{\partial p^A}{dt} = \frac{\partial \varphi}{\partial q^A}\frac{\partial H}{\partial p^A} - \frac{\partial \varphi}{\partial p^A}\frac{\partial H}{\partial q^A}$$

($+\frac{\partial \varphi}{\partial t}$ if φ depend explicitly on t).

Thus, φ is a first integral if

$$\frac{\partial \varphi}{\partial q^A}\frac{\partial H}{\partial p^A} - \frac{\partial \varphi}{\partial p^A}\frac{\partial H}{\partial q^A} = 0 \qquad \text{(A.26)}$$

The expression (A.26) is represented by $\{\varphi, H\}$, called **Poisson brackets**. Thus, for autonomous mechanical systems we have $\{H, H\} = 0$, i.e., the Hamiltonian is a first integral.

Appendix B

Higher order tangent bundles. Generalities

The theory of Jet manifolds introduced by Ch. Ehresmann around 1950 is an important topic in modern differential geometry. We shall give a brief outline of this theory and we shall examine a particular case concerning higher order tangent bundles the place where the geometric formulation of higher order Lagrangian particle mechanics is developed. For a more general description involving Lagrangians depending in many independent variables with higher order derivatives see our book in this series (de León and Rodrigues [38]). Our article [41] gives also another approach for this subject.

B.1 Jets of mappings (in one independent variable)

Let M be a manifold of dimension m and R the Euclidean space with coordinate t. In what follows we shall assume that all the mappings are C^∞–class. Let $f : R \longrightarrow M$ and $g : R \longrightarrow M$ two mapping such that $f(t) = g(t)$, for a fixed $t \in R$. We say that f and g are $\sim_k$**–related** (or that they are **tangent to the** k–th **order**) at $t \in R$ if for all function $h : M \longrightarrow R$ the function

$$h \circ f - h \circ g : R \longrightarrow R$$

is "flat" of order k at t, i.e., this function and all their derivatives up to order k, included, vanishes at t. The equivalence classes determinedd by $\sim_k$ are called **jets of order** k, or simply, k**–jets**, with source t and same target.

The k–jet of a mapping $f : R \longrightarrow M$ at $t \in R$ is denoted by $j_t^k f$ or $\tilde{f}^k(t)$. The set of all k– jets at t is denoted by $J_t^k(R,M)$ and we set $J^k(R,M)$ for the union

$$J^k(R,M) = \bigcup_{t \in R} J_t^k(R,M)$$

(we remark that we can define also in a similar way k–jets of local mappings). It can be shown that $J^k(R,M)$ has a $(k+1)m$ dimensional manifold structure. $J^k(R,M)$ can be fibered in different ways:

- **a source projection** $\alpha^k : J^k(R,M) \longrightarrow R$; $\alpha^k(\tilde{f}^k(t)) = t$
- **a target projection** $\beta^k : J^k(R,M) \longrightarrow M$; $\beta^k(\tilde{f}^k(t)) = f(t)$

and the projection $\rho_r^k : J^k(R,M) \longrightarrow J^r(R,M)$; $\rho_r^k(\tilde{f}^k(t)) = \tilde{f}^r(t)$, where $r \leq k$. The mapping which associates to each point $t \in R$ the k–jet of $f : R \longrightarrow M$ at t is called the **k–jet prolongation** of f and is represented by $j^k f$ or $\tilde{f}^k$. hence $\tilde{f}^k : R \longrightarrow J^k(R,M)$ is defined by $t \longrightarrow \tilde{f}^k(t)$ and $\tilde{f}^k$ is a section of the fibred manifold $(J^k M, \alpha^k, R)$.

B.2 Higher order tangent bundles

Let us now consider the particular situation $t = 0 \in R$. Then the submanifold $J_0^k(R,M)$ is denoted by $T^k M$ and called the **tangent bundle of order k** of M. One has, of course, the same type of fibrations as above. In fact, if $r \leq k$, we have the canonical projection $\rho_r^k : T^k M \longrightarrow T^r M$ given by $\rho_r^k(\tilde{\sigma}^k(0)) = \tilde{\sigma}^r(0)$ and the target projection $\beta^k : T^k M \longrightarrow M$ given by $\beta^k(\tilde{\sigma}^k(0)) = \sigma(0)$, where $\sigma : R \longrightarrow M$ is a function. Obviously $\rho_0^k = \beta^k$ and $T^0 M$ is identified with M if $k = 0$ and $T^1 M$ with TM if $k = 1$.

We shall now describe the local coordinates for $T^k M$. Let U be a chart of M with local coordinates y^A, $1 \leq A \leq m$, $\sigma : R \longrightarrow M$ a curve in M such that $\sigma(0) \in U$ and set $\sigma^A = y^A \circ \sigma$, $1 \leq A \leq m$. Then the k–jet $\tilde{\sigma}^k(0)$ is uniquely represented in $(\beta^k)^{-1}(U) = T^k U$ by

$$(y^A, z_1^A, \ldots, z_k^A), \ 1 \leq A \leq m$$

where

$$y^A = \sigma^A(0), \ z_i^A = (1/i!)(d^i \sigma^i / dt^i)(0), \ 1 \leq i \leq k$$

(we will set $y^A = z_0^A$). Then we have a chart $(\beta^k)^{-1}(U)$ in T^kM with local coordinates $(z_0^A, z_1^A, \ldots, z_k^A)$. The factor $(1/i!)$ appears only for technical reasons. We may consider the following coordinate system in $(\beta^k)^{-1}(U)$:

$$(q^A, q_1^A, \ldots, q_k^A), \ 1 \leq A \leq m,$$

where $q^A = \sigma^A(0)$ and $q_i^A = i!\ z_i^A$, $0 \leq i \leq k$, $1 \leq A \leq m$.

Now, let σ be a curve in M. We shall denote by $\tilde{\sigma}^k$ the **canonical prolongation** of σ to T^kM defined as follows:

$$\tilde{\sigma}^k(t) = \tilde{\sigma}_t^k(0),$$

where $\sigma_t(s) = \sigma(s+t)$. If $k = 1$ we put $\tilde{\sigma}^1 = \dot{\sigma}$. Along the prolongation $\tilde{\sigma}^k$ to T^kM of a curve σ in M we have $q_i^A = (d^i\sigma^A/dt^i)$, $0 \leq i \leq k$.

Let $h : M \longrightarrow R$ be a function. The **lift** of h to T^kM is defined as follows: for each $i \in \{1, 2, \ldots, k\}$ the $< i >$–lift $h^{<i>}$ of h is

$$h^{<i>}(\tilde{\sigma}^k(0)) = (1/i!)(d^i/dt^i)(h \circ \sigma)(0),$$

for every curve σ in M. Clearly $z_i^A = (z^A)^{<i>}$. Let X be a vector field on M. Then the $< i >$**–lift of** X to T^kM, denoted by $X^{<i>}$ is the unique vector field on T^kM such that

$$X^{<i>} f^{<j>} = (Xf)^{<j-i>}$$

for every function $f : M \longrightarrow R$ and $j \in \{1, 2, \ldots, k\}$. Locally, if $X = X^A(\partial/\partial z^A)$ then

$$X^{<i>} = \sum_{j=1}^{k} (X^A)^{<j-i>} (\partial/\partial z_j^A)$$

(we set $f^{<\ell>} = 0$ if $\ell > k$). In particular $(\partial/\partial z^A)^{<j>} = (\partial/\partial z_j^A)$. We are now able to define the $< j >$**–lift of a tensor field** F of type (1,1) on M to a tensor field $F^{<j>}$ of type (1,1) on T^kM. The $< j >$**–lift** of F to T^kM is the unique tensor field of type (1,1) $F^{<j>}$ on T^kM such that

$$F^{<j>} X^{<i>} = (FX)^{<j+i>}$$

for every vector field X on M and $i \in \{1, 2, \ldots, k\}$. We call $F^{<0>}$ the **complete lift** of F to T^kM.

B.3 The canonical almost tangent structure of order k

The use of the above lifting procedure on the identity operator I_M of M permits us to define a unique tensor field J_1 which endows T^kM with an almost tangent structure of order k. It is the tensor field of type (1,1) giving by $J_1 = I_M^{<1>}$ with rank $J_1 = km$ and $J_1^k = J_1 \circ \ldots \circ J_1$ (k–times) $\neq 0$ and $J_1^{k+1} = 0$. The composition of J_1 r–times gives more $(k-1)$ tensor fields which are locally expressed by

$$J_r = J_1^r = \sum_{i=0}^{k-r} (\partial/\partial z_{r+i}^A) \otimes dz_i^A, \quad 2 \leq r \leq k$$

(in particular

$$J_1 = \sum_{i=0}^{k-1} (\partial/\partial z_{i+1}^A) \otimes dz_i^A).$$

An exterior calculus generated by these tensor fields may be constructed: an **interior product** i_{J_r} defined by

$$(i_{J_r}\omega)(X_1, \ldots, X_p) = \sum_{i=1}^{p} \omega(X_1, \ldots, J_r X_i, \ldots, X_p) \tag{B.1}$$

where ω is a p–form and X_i are vector fields on T^kM, and an **exterior differentiation** d_{J_r} defined by

$$d_{J_r} = i_{J_r} d - d i_{J_r}, \tag{B.2}$$

where d is the usual exterior differentiation.

B.4 The higher–order Poincaré–Cartan form

Let $\varphi_k : T^kM \longrightarrow T(T^{k-1}M)$ be the mapping given by $j_0^k\sigma \longrightarrow j_0^1\tau$, where $\tau : R \longrightarrow T^{k-1}M$ is defined by $t \longrightarrow \tau(t) = j_0^{k-1}\sigma_t$ with $\sigma_t(s) = \sigma(s+t)$. Then locally

$$\varphi_k(z^A, z_i^A) = \sum_{i=0}^{k-1} (i+1) z_{i+1}^A (\partial/\partial z_i^A).$$

We use this map to construct the **Tulczyjew differential operator,** represented by d_T, which maps each function f on T^kM into a function d_Tf on $T^{k+1}M$ defined by

$$d_Tf(\dot{j}_0^{k+1}\sigma) = df(\dot{j}_0^k\sigma)(\varphi_{k+1}(\dot{j}_0^{k+1}\sigma)).$$

Locally, we have

$$d_Tf(z^A, z_i^A) = \sum_{i=0}^{k}(i+1)z_{i+1}^A(\partial f/\partial z_i^A).$$

In particular,

$$d_T(z_i^A) = (i+1)z_{i+1}^A \text{ and } d_T^r(z^A) = d_T \circ \ldots \circ d_T(z^A) = r!z_r^A.$$

The operator d_T extends in a very natural way to an operator which maps p–forms on T^kM into p–forms on $T^{k+1}M$. Also, we have $dd_T = d_Td$. We may now construct the canonical vector fields on T^kM, generalizing the Liouville vector field on the tangent bundle TM. In order to do this let us recall that the vertical lift of a vector field Y on $T^{k-1}M$ to T^kM with respect to the projecction $\rho^k : T^kM \longrightarrow M$ is the unique vector field Y^{v_k} on T^kM given by

$$Y^{v_k}((\rho_s^k)^*d_T^sf) = \begin{cases} 0 \text{ if } s=0 \\ sY((\rho_{s-1}^k)^*(d_T^{s-1}f)), \text{ if } 1 \leq s \leq k, \end{cases}$$

for every function f on M. Locally, if $Y = \sum_{i=0}^{k-1} Y_i^A(\partial/\partial z_i^A)$ then $Y^{v_k} = \sum_{i=0}^{k-1} Y_i^A(\partial/\partial z_{i+1}^A)$. Now we construct the canonical vector field C_1 on T^kM as follows:

$$C_1(\dot{j}_0^k\sigma) = (\varphi_k(\dot{j}_0^k\sigma))^{v_k}$$

and locally we have

$$C_1 = \sum_{i=0}^{k-1}(i+1)z_{i+1}^A(\partial/\partial z_{i+1}^A).$$

One obtains a family of vector fields C_r on T^kM, $2 \leq r \leq k$, defined by $C_r = J_1C_{r-1}$ and locally

$$C_r = \sum_{i=0}^{k-r}(i+1)z_{i+1}^A(\partial/\partial z_{r+i}^A).$$

We may transport the geometric structures defined on T^kM to $R \times T^kM$ (which may identified with $J^k(R, M)$): we set

$$\bar{J}_r = J_r - C_r \otimes dt, \quad 1 \leq r \leq k$$

and we may define in a similar way to (B.1) and (B.2) the operators $i_{\bar{J}_r}$ and $d_{\bar{J}_r}$. One has, for instance

$$d_{\bar{J}_r} f = i_{\bar{J}_r} df = (\bar{J}_r)^*(df) = (J_r)^*(df) - (C_r f)dt.$$

With such structures we define the **Poincaré–Cartan form for a higher–order** time–dependent Lagrangian $L : R \times T^kM \longrightarrow R$ by

$$\Omega_L = \sum_{i=0}^{k-1} (-1)^i (1/(i+1)!) d_T^i (d_{\bar{J}_{i+1}} L) + Ldt.$$

If we develop this definition in local coordinates $q_i^A = i! z_i^A$, $0 \leq i \leq k$ and if we set $\theta_i^A = dq_i^A - q_{i+1}^A dt$, then one obtains the expression

$$\Omega_L = \sum_{i=0}^{k-1} p_A^{i+1} \theta_i^A + Ldt$$

where

$$p_A^{i+1} = \sum_{j=0}^{k-i-1} (-1)^j d^j/dt^j (\partial L/\partial q_{i+j+1}^A), \quad 0 \leq i \leq k-1.$$

It is possible to show that this approach maintains the main ideas of tangent bundle geometry and we suggest the reader the references quoted in the introduction of this Appendix.

Bibliography

[1] ABRAHAM, R. and MARSDEN, J.
Foundations of Mechanics, 2nd ed., Benjamin, New York, 1978.

[2] AMBROSE, W. PALAIS, R. and SINGER, W.
Sprays, *An. Acad. Bras. Cienc., 32 (1960), 163–178.*

[3] ANDERSON, J. and BERGMANN, P.
Constraints in Covariant Field Theories, *Phys. Rev., 83 (1951), 1018–1025.*

[4] ARNOLD, V.
Méthodes Mathématiques de la Mécanique Classique, Ed. Mir, Moscou, 1974.

[5] BENENTI, S.
Symplectic relations in Analytical Mechanics, in "Modern Developments in Analytical Mechanics", *Proceedings of the IUTAM–ISIMM Symposium, Ed. S. Benenti, M. Francariglia and A. Lichnenowicz, Torino (1983), 39–91.*

[6] BERGMANN, P. and GOLDBERG, I.
Dirac Bracket Transformations in Phase Space, *Phys. Rev., 98, 2 (1955), 531–538.*

[7] BERNARD, D.
Sur la géométrie différentielle des G–structures, *Ann. Inst. Fourier, 10 (1960), 151–270.*

[8] BLAIR, D.
Contact manifolds in Riemannian geometry, Lect. Notes in Math., 509, Springer, Berlin, 1976.

[9] BOOTHBY, W.M.
An Introduction to differentiable manifolds and Riemannian Geometry, Academic Press, New York, 1975.

[10] BRUCKHEIMER, M.R.
Thesis, University of Southampton, 1960.

[11] CANTRIJN, F., CARIÑENA, J., CRAMPIN, M. and IBORT, L.
Reduction of Degenerate Lagrangian Systems, *J. Geom. Phys., 3 (1986), 353–400.*

[12] CARIÑENA, J. and IBORT, L.
Geometric theory of the equivalence of Lagrangians for constrained systems, *J. Phys. A: Math. Gen., 18 (1985), 3335–3341.*

[13] CARTAN, E.
Leçons sur les invariants intégraux, Hermann, Paris, 1921.

[14] CHERN, S.S.
The geometry of G–structures, *Bull. Amer. Math. Soc., 72 (1966), 167–219.*

[15] CHOQUET, Y.
Géométrie Différentielle et Systèmes Extérieurs, Dunod, Paris, 1968.

[16] CLARK, R.S. and BRUCKHEIMER, M.
Sur les structures presque tangentes, *C.R. Acad. Sc. Paris, 251 (1960), 627–629.*

[17] CLARK, R.S. and GOEL, D.S.
On the geometry of an almost tangent manifold, *Tensor, N.S., 24 (1972), 243–252.*

[18] CLARK, R.S. and GOEL, D.S.
Almost cotangent manifolds, *J. Differential Geom. 9 (1974), 109–122.*

[19] CLARK, R.S. amd GOEL, D.S.
Almost tangent manifolds of 2nd order, *Tohoku Math. J., 24 (1972), 79–92.*

[20] CRAMPIN, M.
On the differential geometry of the Euler–Lagrange equations and the

inverse problem of Lagrangian dynamics, *J. Phys. A: Math. Gen., 14 (1981), 2567–2575.*

[21] CRAMPIN, M.
Tangent bundle geometry for Lagrangian dynamics, *J. Phys. A: Math. Gen., 16 (1983), 3755–3772.*

[22] CRAMPIN, M.
Defining Euler–Lagrange fields in terms of almost tangent structures, *Phys. Lett., 95A (1983), 466–468.*

[23] CRAMPIN, M., PRINCE, G. and THOMPSON, G.
A geometrical version of the Helmoltz conditions in time–dependent Lagrangian dynamics, *J. Phys. A: Math. Gen. 17 (1984), 1437–1447.*

[24] CRAMPIN, M. and THOMPSON, G.
Affine bundles and integrable almost–tangent structures, *Math. Proc. Camb. Phil. Soc., 98 (1985), 61–71.*

[25] CRUYMEYROLLE, A. and GRIFONE, J.
Symplectic Geometry, Research Notes in Math., 80, Pitmann, London, 1983.

[26] DE ANDRES, L.C., DE LEON, M. and RODRIGUES, P.R.
Connections on tangent bundles of higher order, *to appear in Demonstratio Mathematica.*

[27] DE ANDRES, L.C., DE LEON, M. and RODRIGUES, P.R.
Canonical connections associated to regular Lagrangians of higher–order, *to appear in Anais Acad. Bras. Ciências, 1989.*

[28] DE BARROS, C.
Sur la géométrie différentielle des formes différentielles extérieures quadratiques, *Atti Congr. Int. Geometria Differenziale, Bologna (1967), 1–26.*

[29] DE LEÓN, M.
Connections and f–structures on T^2M, *Kôdai Math. J., 4 (1981), 189–216.*

[30] DE LEÓN, M. and LACOMBA, E.
Les sous–variétés lagrangiennes dans la dynamique lagrangienne

d'ordre supériéur, *C.R. Acad. Sci. Paris, 307, ser. II (1988), 1137–1139.*

[31] DE LEÓN, M. and LACOMBA, E.
Lagrangian submanifolds and higher–order mechanical systems, *Preprint.*

[32] DE LEÓN, M., MENDEZ, I. and SALGADO, M.
Connections of order k and associated polynomial structures on T^kM, *An. Stiint. Univ. Al. I. Cuza Iasi, 33, 3 (1987), 267–276.*

[33] DE LEÓN, M., MENDEZ, I. and SALGADO, M.
p–Almost tangent structures, *Rend. Circ. Mat. Palermo, Serie II, 37 (1988), 282–294.*

[34] DE LEÓN, M., MENDEZ, I. and SALGADO, M.
Integrable p–almost tangent manifolds and tangent bundles of p'–velocities, *Preprint.*

[35] DE LEÓN, M., MENDEZ, I. and SALGADO, M.
p–Almost cotangent structures, *Preprint.*

[36] DE LEÓN, M., MENDEZ, I. and SALGADO, M.
Regular p–almost cotangent structures, *J. Korean Math. Soc., 25, No. 2 (1988), 273–287.*

[37] DE LEÓN, M. and RODRIGUES, P.R.
Formalisme hamiltonien symplectique sur les fibres tangents d'ordre supérieur, *C.R. Acad. Sci. Paris, 301, ser. II (1985), 455–458.*

[38] DE LEÓN, M. and RODRIGUES, P.R.
Generalized Classical Mechanics and Field Theory, North–Holland Mathematical Studies, Notas de Matematica, No. 112, Amsterdam, 1985.

[39] DE LEÓN, M. and RODRIGUES, P.R.
Higher order almost tangent geometry and non–autonomous Lagrangian dynamics. *Supp. Rend. Circolo Mat. Palermo, ser. II, 16 (1987), 157–171.*

[40] DE LEÓN, M. and RODRIGUES, P.R.
Second order differential equations and non–conservative Lagrangian mechanics, *J. Phys. A. Math. Gen., 20 (1987), 5393–5396.*

[41] DE LEÓN, M. and RODRIGUES, P.R.
A contribution to the global formulation of the higher–order Poincaré–Cartan form, *Lett. Math. Phys., 14, 4 (1987), 353–362.*

[42] DE LEÓN, M. and RODRIGUES, P.R.
Dynamical connections and non–autonomous Lagrangian systems, *Ann. Sc. l'Univ. Toulouse, IX (1988), to appear.*

[43] DE LEÓN, M. and RODRIGUES, P.R.
Almost contact structures and time–dependent Lagrangian systems, *Portugaliae Mathematica, to appear.*

[44] DE LEÓN, M. and RODRIGUES, P.R.
Degenerate Lagrangian systems and their associated dynamics, *Rendiconti di Matematica, ser. VII, vol. 8 (1988), 105–130.*

[45] DE LEÓN, M. and RODRIGUES, P.R.
Second Order Differential Equations and Degenerate Lagrangians, *Preprint.*

[46] DE LEÓN, M. and RODRIGUES, P.R.
On mechanical systems of higher–order with constraints, *Preprint.*

[47] DIRAC, P.
Generalized Hamiltonian Dynamics, *Can. J. Math., 2 (1950), 129–148.*

[48] DIRAC, P.
Lectures on Quantum Mechanics, Belfer Graduate School of Science Monog. Ser., No. 2, 1964.

[49] DOMBROWSKI, P.
On the geometry of the tangent bundle, *J. Reine Ang. Math. 210 (1962), 73–88.*

[50] DUISTERMAAT, J.
Fourier Integral Operators, Courant. Inst. Math. Sc., N. York University, New York, 1973.

[51] EHRESMANN, Ch.
Les prolongements d'une variété diférentielle: I.-Calcul des Jets, prolongement principal, *C.R. Acad. Sc. Paris, 233 (1951), 598–600.*

[52] ELIOPOULOS, H.
Structures presque–tangentes sur les variétés différentielles, *C.R. Acad. Sc. Paris, 255 (1962), 1563–1565.*

[53] ELIOPOULOS, H.
On the general theory of differentiable manifolds with almost tangent structure, *Canad. Math. Bull., 8 (1965), 721–748.*

[54] ELIOPOULOS, H.
Structures r–tangentes sur les variétés différentielles, *C.R. Acad. Sc. Paris, 263 (1966), 413–416.*

[55] FLANDERS, H.
Differential Forms with applications to the Physical Sciences, Acad. Pres, New York, 1963.

[56] FRÖLICHER, A. and NIJENHUIS, A.
Theory of vector–valued differential forms, *Ind. Math., 18 (1956), 338–385.*

[57] FUJIMOTO, A.
Theory of G–structures, Publ. Study Group of Geometry, 1, Tokyo Univ., Tokyo, 1972.

[58] GALLISSOT, T.
Les formes extérieures en Mécanique, *An. Inst. Fourier, Grenoble, 4 (1952), 145–297.*

[59] GALLISSOT, T.
Les formes extérieures et la Mécanique des milieux continus, *C.R. Acad. Sc. Paris, 244A (1957), 2347–2349.*

[60] GARCIA, P.L.
Geometria simplética en la teoría clásica de campos, *Coll. Math., 19 (1968), 1–66.*

[61] GARCIA, P.L.
The Poincaré–Cartan invariant in the calculus of variations, *Symp. Math., 14 (1974), 219–246.*

[62] GELFAND, I. and FOMIN, J.
Calculus of Variations, Prentice–Hall, Englewood Cliff, 1963.

[63] GODBILLON, C.
Géométrie Différentielle et Mécanique Analytique, Hermann, Paris, 1969.

[64] GOLDSCHMIDT, H. and STERNBERG, Sh.
The Hamilton–Jacobi formalism in the calculus of variations, *Ann. Inst. Fourier, Grenoble, 23 (1973), 203–267.*

[65] GOLDSTEIN, H.
Classical Mechanics, Addison–Wesley, Mass., 1950.

[66] GOTAY, M.
Presymplectic manifolds, geometric constraint theory and the Dirac–Bergmann theory of constraints, *Ph. D. thesis, Univ. of Maryland, 1–198, 1979.*

[67] GOTAY, M. and NESTER, J.
Presymplectic Lagrangian systems I: the constraint algorithm and the equivalence theorem, *Ann. Inst. Henri Poncaré, 30, 2 (1978), 129–142.*

[68] GOTAY, M. and NESTER, J.
Presymplectic Lagrangian systems II: the second–order equation problem, *Ann. Inst. Poincaré, 30, 1 (1980), 1–13.*

[69] GOTAY, M., NESTER, J. and HINDS, G.
Presymplectic manifolds and the Dirac–Bergmann theory of constraints, *J. Math. Phys., 19, 11 (1978), 2388–2399.*

[70] GRAY, A. and HERVELLA, L.M.
The Sixteen Classes of Almost Hermitian Manifolds and their Linear Invariants, *Ann. Mat. Pura Appl. (IV) 123 (1980), 35–58.*

[71] GREUB, W.
Linear Algebra 3rd Ed., Springer, Heidelberg, 1967.

[72] GRIFFITHS, Ph.
Exterior Differential systems and the Calculus of Variations, Prog. in Math., 25, Birkhäuser, 1983.

[73] GRIFONE, J.
Estructure presque tangente et connexions I, II, *Ann. Inst. Fourier, Grenoble, 22, 3 (1972), 287–334 and 22, 4 (1972), 291–338.*

[74] GUILLEMIN, V. and STERNBERG, Sh.
Symplectic techniques in Physics, Cambridge Univ. Press, Cambridge, 1984.

[75] HOFFMAN, K. and KUNZE, R.
Linear Algebra, Prentice Hall, N. Delhi, 1967.

[76] HUSEMOLLER, D.
Fibre bundles, 2nd Ed., Springer, New York, 1975.

[77] KLEIN, J.
Espaces variationnels et Mécanique, *Ann. Inst. Fourier, Grenoble, 12 (1962), 1–124.*

[78] KLEIN, J.
Opérateurs différentiels sur les variétés presque–tangentes, *C.R. Acad. Sc. Paris, 257A (1963), 2392–2394.*

[79] KLEIN, J.
Les systèmes dynamiques abstraits, *Ann. Inst. Fourier, Grenoble, 13, 2 (1963), 191–202.*

[80] KLEIN, J. and VOUTIER, A.
Formes extérieures génératrices de sprays, *Ann. Inst. Fourier, Grenoble, 18 (1968), 241–250.*

[81] KOBAYASHI, S. and NOMIZU, K.
The Foundations of Differential Geometry, I, II, Willey Intersc., New York, 1963 and 1969.

[82] KOILLER, J.
Reduction of some classical non–holonomic systems with symmetry, *Preprint.*

[83] KOSZUL, J.
Lectures on Fibre Bundles and Differential Geometry, Tata Inst. Bombay, 1960.

[84] KOWALSKI, O.
Curvature of the induced Riemannian metric on the tangent bundle of a Riemannian manifold, *J. reine angew. Math. 250 (1971), 124–129.*

[85] LEECH, J.
Classical Mechanics, Methuen's Physical Monographs, Methuen, John Wiley, New York, 1965.

[86] LEHMANN-LEJEUNE, J.
Integrabilité des G-structures définies par une 1-forme 0-deformable à valeurs dans le fibré tangent, *Ann. Inst. Fourier, 16 (1966), 229-287.*

[87] LIBERMANN, P.
Sur le probleme d'equivalence de certaines structures infinitésimales régulières, *Thèse de doctorat d'État, Strasbourg, 1953.*

[88] LIBERMANN, P. and MARLE, Ch.
Symplectic geometry and Analytical Mechanics, Reidel Publ., Dordrecht, 1987.

[89] LICHNEROWICZ, A.
Variété symplectique et dynamique attachée à une sous-variété, *C.R. Acad. Sc. Paris, 280A (1975), 523-527.*

[90] LICHNEROWICZ, A.
Les variétés de Poisson et leurs algébres de Lie associées, *J. Diff. Goem., 12 (1977), 253-300.*

[91] LOOMIS, L. and STERNBERG, Sh.
Advanced Calculus, Addison-Wesley, Mass., 1968.

[92] MAC LANE, S.
Hamiltonian Mechanics and Geometry, *Amer. Math. Monthy, 77 (1970), 570-586.*

[93] MARLE, Ch.
Contact manifolds, canonical manifolds and the Hamilton-Jacobi method in Analytical Mechanics, in *"Proceedings of the IUTAM-ISSIM Symposium", Torino, 1983, 255-272.*

[94] MARMO, G. et al.
Liouville dynamics and Poisson brackets, *J. Math. Phys., 22, 4 (1981), 835-842.*

[95] MARMO, G., SALETAN, E., SIMONI, A. and VITALE, B.
Dynamical Systems, John Wiley & Sons, New York, 1985.

[96] MILNOR, J. and STASHEFF, J.D.
Characteristic Classes, Princeton Univ. Press, Princeton, New Jersey, 1974.

[97] MOLINO, P.
Riemannian Foliations, Progress in Math., 73, Birkhäuser, Boston, 1988.

[98] MOSER, J.
On the volume elements on a manifold, *Trans. Am. Math. Soc., 120 (1965), 286–294.*

[99] NEIMARK, Y. and FUFAEV, N.A.
Dynamics on nonholonomic systems, AMS Translations, 33, 1972.

[100] NEWLANDER, A. and NIRENBERG, L.
Complex analytic coordinates in almost complex manifolds, *Ann. of Math., 65 (1957), 391–404.*

[101] NIJENHUIS, A.
X_{n-1}-forming sets of eigenvectors, *Indag. Math., 13 (1951), 200–212.*

[102] OUBIÑA, J.A.
New classes of almost contact metric structures, *Publ. Math., 32 (1985), 187–193.*

[103] PLANCHART, E.
Geometría Sympléctica, VII Escuela Latinoamericana de Matemáticas, Univ. Simón Bolívar, 1984.

[104] PNEVMATIKOS, S.
Singularités in Géométrie Symplectique, in **Symplectic Geometry**, *Res. N. in Math., 80, Ed. A. Crumeyrolle and J. Grifone, Pitman Books, London, 1983, 184–216.*

[105] PRINCE, G.
Toward a classification of dynamical symmetries in classical mechanics, *Bull. Austral. Math. Soc., 27 (1983), 53–71.*

[106] ROBINSON, R.C.
Lectures on Hamiltonian systems, Monografías de Matematica, No. 7, Inst. Mat. Pura Appl., Rio de Janeiro, 1971.

[107] SANTILLI, R.M.
Foundations of Theoretical Mechanics, Springer, Berlin, 1978.

[108] SASAKI, S.
On the differential geometry of tangent bundles of Riemannian manifolds, *Tohôku Math. J., 10 (1958), 338–354.*

[109] SARLET, W., CANTRIJN, F. and CRAMPIN, M.
A new look at second–order equations and Lagrangian mechanics, *J. Phys. A: Math. Gen., 17 (1984), 1999–2009.*

[110] SKINNER, R.
First–Order Equations of Motion for Classical Mechanics, *J. Math. Phys., 24 (11), (1983), 2581–2588.*

[111] SKINNER, R. and RUSK, R.
Generalized Hamiltonian Dynamics I.- Formulation on $T^*Q \oplus TQ$, *J. Math. Phys., 24 (11), (1983), 2589–2594.*

[112] SKINNER, R. and RUSK, R.
Generalized Hamiltonian Dynamics. II.- Gauge Transformations, *J. Math. Phys. 24 (11), (1983), 2595–2601.*

[113] SPIVAK, M.
A Comprehensive Introduction to Differential Geometry, I–V, Publish or Perish, Berkeley, 1979.

[114] STERNBERG, Sh.
Lectures on Differential Geometry, Prentice–Hall, Englewood Cliffs, New Jersey, 1964.

[115] THOMPSON, G.
Integrable almost cotangent structures and Legendrian bundles, *Math. Proc. Camb. Phil. Soc., 101 (1987), 61–78.*

[116] TULCZYJEW, W.M.
Lagrangian submanifolds and Hamiltonian dynamics, *C.R. Acad. Sci. Paris, 283 (1976), 15–18.*

[117] TULCZYJEW, W.M.
Lagrangian submanifolds and Lagrangian dynamics, *C.R. Acad. Sci. Paris, 283 (1976), 675–678.*

[118] VAISMAN, I.
Symplectic Geometry and Secondary Characteristic Classes, Progress in Math., 72, Birkhäuser, Boston, 1987.

[119] VILMS, J.
Connections on tangent bundles, *J. Diff. Geometry, 1 (1967), 235–243.*

[120] VILMS, J.
Curvature of nonlinear connections, *Proc. Amer. Math. Soc., 19 (1968), 1125–1129.*

[121] WALKER, A.
Almost product structures, *Proc. Symp. Pure Math., III (1961), 94–100.*

[122] WALKER, A.G.
Connections for parallel distributions in the large, II, *Quart. J. Math. Oxford (2), vol. 9 (1958), 221–231.*

[123] WARNER, F.
Differentiable Manifolds and Lie Groups, Scott, Foresman and Co., Glenview, 1971.

[124] WEBER, R.W.
Hamiltonian Systems with Constraints and their Meaning in Mechanics, *Arch. Rat. Mech. Anal., 91 (1985), 309–335.*

[125] WEINSTEIN, A.
Lectures on Symplectic Manifolds, CBMS Conf. Series No. 29, American Mathematical Society, 1977.

[126] WELLS, R.O.
Differential Analysis on Complex Manifolds, Springer–Verlag, New York, 1980.

[127] WILLARD, S.
General Topology, Addison–Wesley, Reading, Massachussetts, 1970.

[128] WILMORE, T.J.
Parallel distributions on manifolds, *Proc. London Math. Soc. B (6) (1956), 191–204.*

[129] YANO, K.
On a structure defined by a tensor field f of type (1,1) satisfying $f^3 + f = 0$. *Tensor N.S., 14 (1963), 99–109.*

[130] YANO, K. and ISHIHARA, S.
Tangent and Cotangent Bundles: Differential Geometry, Marcel Dekker Inc., New York, 1973.

[131] YANO, K. and KON, M.
Structures on manifolds, Series in Pure Math., vol. 3, World Scientific, Singapore, 1984.

Index